Collins

OCR Gateway GCSE (9–1)

Physics

Teacher Pack

Tony Forsythe
Jonny Friend
Stuart Lloyd
Jen Randall
Series editor: Ed Walsh

William Collins' dream of knowledge for all began with the publication of his first book in 1819. A self-educated mill worker, he not only enriched millions of lives, but also founded a flourishing publishing house. Today, staying true to this spirit, Collins books are packed with inspiration, innovation and practical expertise. They place you at the centre of a world of possibility and give you exactly what you need to explore it.

Collins. Freedom to teach

HarperCollins Publishers
1 London Bridge Street
London SE1 9GF

Browse the complete Collins catalogue at
www.collins.co.uk

First edition 2016

10 9 8 7 6 5 4 3 2 1

© HarperCollins*Publishers* 2016

ISBN 978-0-00-815104-1

Collins® is a registered trademark of HarperCollins Publishers Limited

www.collins.co.uk

A catalogue record for this book is available from the British Library

Commissioned by Lucy Rowland and Lizzie Catford
Edited by Hamish Baxter
Project managed by Elektra Media Ltd
Development and copy edited by Peter Batty
Proofread by Jane Klima
Illustrated by Jerry Fowler
Typeset by Jouve India Private Limited
Cover design by We are Laura
Printed in Great Britain by Martins the Printers
Cover images © Shutterstock/watchara, Shutterstock/Everett Historical

All rights reserved. No part of this book may be reproduced, stored in a retrieval system, or transmitted in any form or by any means, electronic, mechanical, photocopying, recording or otherwise, without the prior permission in writing of the Publisher. This book is sold subject to the conditions that it shall not, by way of trade or otherwise, be lent, re-sold, hired out or otherwise circulated without the Publisher's prior consent in any form of binding or cover other than that in which it is published and without a similar condition including this condition being imposed on the subsequent purchaser.
HarperCollins does not warrant that www.collins.co.uk or any other website mentioned in this title will be provided uninterrupted, that any website will be error free, that defects will be corrected, or that the website or the server that makes it available are free of viruses or bugs. For full terms and conditions please refer to the site terms provided on the website.

Contents

 © HarperCollins*Publishers* Limited 2016

4 Magnetism and magnetic fields

5 Waves in matter

6 Radioactivity

7 Energy

 © HarperCollins*Publishers* Limited 2016

8 Global challenges

 © HarperCollins*Publishers* Limited 2016

Introduction to the course

Welcome to the Collins GCSE Science course. These materials have been developed by a team of experts drawing on many years' experience in teaching, curriculum development and educational publishing. They have been produced to support you, the professional in the classroom, teach more effectively and your students to make good progress.

The aim is to both develop students' interest in science and enable them to attain grades that reflect a mastery in the subject. The course is not intended to be prescriptive or restrictive, shoe horning teachers into a very particular way of teaching, but to be used in a range of ways. On the other hand, it doesn't offer a bewildering array of options which require a lot of filtering and selection of activities.

We are aware that specifications sometimes seem to change at an alarming rate. We have tried to reflect both the key features of the new specifications and the classic elements of high quality teaching.

Key changes for the courses introduced in 2016

Non-modular

The new specifications have a non-unitised structure. Content can be covered in any order deemed appropriate. This doesn't mean that the order doesn't matter. There are a number of principles that can be used when selecting a 'running order', including:

- deciding which topics are more accessible for students at an earlier stage in their GCSE studies
- identifying which topics need to be covered to give access to which other concepts
- using topics that have possibly greater appeal, due to the role of practical work or engaging contexts, to sustain interest over the whole course.

However, what is also important is to build in assessment, tracking and intervention. You need to know if students are developing an understanding that at least corresponds to their ability and to be able to respond.

The Collins course offers a particular route. The exams are designed to assess one part of the content in Paper 1 and the other part in Paper 2. One approach is to teach the 'Paper 1 content' first and then the 'Paper 2 content'. This means that a past Paper 1 could be used as a 'half way assessment point'.

However, the materials don't have to be used in this order. It may be decided that the needs of students are not best served by concentrating all the material on one topic in one place, but rather by revisiting it later on.

 © HarperCollins*Publishers* Limited 2016

Maths

Science has always had a strong relationship with maths and this continues. What has been strengthened in this set of courses is the specificity of the location of the skills and the role of the skills in exams. It has been made very clear in the specifications which skills relate to which topics and the exams will disadvantage candidates who can't demonstrate mastery of the relevant skills.

In the course we have taken a 'two pronged approach'. Mathematical skills are used in context as key concepts are covered. They are reflected in questions that are set and students are shown how they are drawn upon at particular points. However, we know that for some students this won't be sufficient. Therefore, we have developed other spreads that will focus on a particular set of mathematical skills and explore how they develop. These always place the skills into a scientific context, so it is clear why they are relevant, but the text is led by the developing mathematical skill.

Practical skills

Practical work has a strong role in supporting students' developing understanding of science, and many people involved in the British system are passionate about ensuring this role is maintained. In recent years much of the debate is about how it should be assessed, and successive specifications have used different approaches.

In this series there is no direct assessment of practical skills. Instead the approach is being taken that enquiry will be put in a better place if students are assessed on their mastery of skills and processes in the final exams. A minimum of 15% of marks will be allocated to questions relating to the stipulated practical investigations. However, these won't necessarily be AO1 questions; as well as understanding and recalling the procedures, candidates will also be expected to apply the skills to other contexts and to interpret and evaluate evidence.

As well as covering the effective running of the investigations themselves, the Collins course also offers guidance on placing these in a wider context, so that students can see how their skills and understanding can be used and assessed.

Synoptic questions

Another of the key changes is the introduction of synoptic questions, drawing upon more than one set of ideas. The justification behind this is that scientists often need to draw upon ideas and processes from different areas of science. Again, this is reflected in the assessment approach developed.

Extended written responses

Unlike the previous set of specifications, QWC (Quality of Written Communication) is no longer being assessed in science. However, there is a requirement for candidates to be able to display an ability to develop and sustain a longer response. This is reflected in the

© HarperCollins*Publishers* Limited 2016

questions in the end of topic tests as well as in the Collins Connect digital assessment materials.

Assessment

Grades

The grading system is being changed to a numbered system, in which 1 is the lowest grade awarded and 9 the highest. There is no one-to-one correlation between grades in the incoming and outgoing systems, but the following features are being deployed:

- The new grade 4 is equivalent to the old grade C.
- The new grade 7 is equivalent to the old grade A.
- The threshold for performance indicators will be the new grade 5.

Endorsed publications are no longer permitted to carry specific reference to how certain grades may be attained. Nevertheless, although the calibration has altered, what has not changed is what constitutes a better quality of response. Each of the spreads in the Student Book is designed to present students with an increasing challenge and provide access to higher grades. However, we have also been mindful when developing the resources that a wide range of students will be using them, and so there is a focus upon accessibility as well.

Attribution grids

There has long been an important role attached to the use of end of topic tests. When students have been taught something, we (and they) want to know how well they've understood it and so we've included tests. However, we have developed these further, both in terms of structure and use.

In order to be useful, we think that as far as possible tests should be authentic to the form and function of assessment in GCSE. In other words, if students do well in the tests, this should correspond to doing well in the final exams. This is important, both to improve the validity of any data generated and also in terms of communicating to students what it feels like to answer the kind of questions that they will face in the exam. These aren't mock exams, but they should have a similar balance of components.

Each of the end of topic tests has been written to an attribution grid, so that the basis is the same. The tests are marked out of 40 and have ten marks at each of four different levels of demand. Broadly speaking, Foundation Tier students should do the first two bands and Higher Tier students the final three bands. If they do so, each group will be answering questions which have a 2:2:1 ratio of assessment objectives 1, 2 and 3, which mirror the make-up of the final exams. This is important. AO1 questions focus on knowledge and understanding and are sometimes over-represented in internal tests. As many marks go on AO2 (application). Another 20% go on AO3 (interpretation and evaluation); students need to be aware of this. The constant proportions mean that it is possible to analyse student performance against AO and modify subsequent teaching accordingly.

 © HarperCollins*Publishers* Limited 2016

Similarly, the tests also reflect a mix of response types (objective test questions, short written responses and longer written responses), and the presence of questions relating to mathematical skills and to the stipulated practicals.

Intervention

The composition of the tests is designed to inform the process of intervention. They will obviously provide a global score which can be used to draw conclusions about attainment in that topic. However, intervention is then somewhat problematical; if the group has moved on to another topic, the only option might be to set up additional sessions elsewhere in the school day. These may be successful but it is useful to be able to move analysis beyond the perspective of looking for which areas of content cause problems. Some do, but this isn't the only source of underperformance.

With the end of chapter tests (and the Collins Connect tests), the attribution grids mean that student performance can be analysed in other ways (see above), including against:

- assessment objectives
- style of response
- command of mathematical skills
- mastery of enquiry skills.

As these are all generic components of science education, whichever topic is being addressed next, there will be opportunities to focus upon improving performance.

It will then be possible in the next module test to see if the analysis of marks indicates that this has been addressed.

 © HarperCollins*Publishers* Limited 2016

How to use these resources

Flexibility

The first thing to say is that the resources are designed to be used flexibly. There is no assumption that all students will do all of the activities suggested in the order indicated or that no other resources or learning activities will be used. The materials have an open and transparent structure, which clearly indicates key components and supports the teacher to select activities that are right for their students.

Features

Student Book

Topic introduction: every chapter starts with the page on the left showing students ideas they have met previously that they will be making use of, and a page on the right indicates why the ideas in the forthcoming chapter are both interesting and important. The purpose of this is to engage students, indicate that they already know something about it but persuade them that it is worth engaging with.

Main content spread: each spread includes material that is ramped in challenge. There is no set grade equivalence for the start and end but the spread always starts with more accessible material. Each of the sections includes questions to embed understanding of the ideas, learning objectives, key words and a 'did you know?' feature to engage students further.

Key concept spread: in every topic we have identified a key concept and revisited this in a slightly different way. These concepts have a kind of 'gatekeeper' function: if students grasp these it will give them access to several other areas of content and so they are worth focusing upon. The highlighting of the spread means they can easily be referred back to, either from further on in the same chapter or subsequent ones.

Practical spread: the specification indicates particular investigations that students should have undertaken. Questions relating to these will feature in terminal examinations; however, these will be not only AO1 (knowledge and understanding) questions but also AO2 (application) and AO3 (interpretation and evaluation) questions. These spreads have been written to support students to take ideas explored in these experiments and place them in a wider context, thus reflecting the range of skills that examiners will be assessing.

Maths skill spread: mathematical skills have long been a key part of GCSE Science courses and exams. However, the proportion and challenge of these questions is being increased. We have developed spreads to reflect this. Maths is still embedded in the main content, with skills being drawn upon as necessary. However, some students will need more than this, and so these additional features have been written. Each takes a particular set of maths skills and shows how they are used in a scientific context, taken from that topic. As the spread becomes more challenging, it shows how the mathematics can be used to answer more complex questions.

 © HarperCollins*Publishers* Limited 2016

Check your progress spread: at the end of each chapter the main ideas are set out as progressive outcomes. In essence they show what it looks like to be better in the use of a particular idea. They can be used in various ways but students self-assessing is one of the more usual ones.

Worked examples: these provide a commentary on typical student responses to GCSE-style questions, showing what has been done well and how further improvement might be achieved.

End of topic questions: these are written to an attribution grid as described above. The questions provide a broad coverage of ideas within the topic, items with different levels of demand (both in terms of cognitive and conceptual complexity) and needing different types of response. This means that they provide varying degrees of challenge and in various ways; they should indicate both what students can do and also what the next steps in their learning need to be.

Teacher Pack

The support for individual lessons includes features such as learning objectives, outcomes, working scientifically focus, maths skills focus, key words and resources needed. The outline for the lesson structure is based around a five-stage learning cycle; this proposes that in every lesson there should be these components:

- Engage: in every lesson students need to be engaged, so that they become involved in the lesson and see it as being of interest and relevance.
- Challenge and develop: students will already have a certain level of understanding of a topic. The idea of this phase is to challenge this and move their thinking on. This is what will drive progress; students should see that their current mode of thinking can be developed so that they can make sense of more complex ideas or a wider range of contexts. This might be a demonstration, question to discuss or video clip.
- Explain: this is where the teacher uses some means of offering an explanation. It could be formal explanation, a demonstration or commentary on a diagram; it might use targeted questioning. It should offer a way of making sense of a problem or phenomenon. The Student Book often plays a crucial role in this phase.
- Consolidate and apply: this is a crucial phase in which students embed the ideas. It should consist of opportunities to 'take on board' new ideas and own them. Written questions and discussion may support this. The worksheets often have a significant role here.
- Extend: students may be ready to extend their learning to something broader or more challenging.

In the ideas for learning activities there is a choice and the options relate to planning provision for students working at different levels.

 © HarperCollins*Publishers* Limited 2016

Teacher Pack Resources

Each lesson is supported by worksheets, which provide activities for students to apply their understanding. They have a ramped structure, offering different levels of challenge.

Practical worksheets support students with planning, carrying out and writing up practical and investigation work.

Technicians' notes provide equipment lists and set-up instructions for practicals.

Collins Connect

Collins Connect makes Collins GCSE Science available at home and at school online and can be used as a front-of-class teaching tool, as well as a way to set homework and tests.

OCR GCSE (9–1) Science on Collins Connect contains:

- Bookview – an interactive digital version of the Student Books. It is ideal for whiteboard use and gives total fluidity between digital and print, with a page-for-page match.
- Quick starters – activities that pose an interesting question to hook students in at the start of a lesson. You can click to show a hint and then click to show the answer.
- Videos and animations – to support you to make real world links and model abstract or challenging concepts.
- Slideshow presentations – available in PowerPoint so you can edit and adapt them to suit the needs of your class.
- This Teacher Pack – so you can access the lesson plans and supporting resources online.
- Homework activities for every lesson – you can assign automarked homework quizzes to your class. Students can log in from home to complete the activities and get immediate feedback. For every lesson there are also 'creative' homework sheets with more open-ended activities.

You can also assess and track students' progress in GCSE Science using the full suite of digital resources on Collins Connect. Collins Connect provides regular and timely assessments to help you analyse student performance.

- The digital tests come in two versions: 'fully automarked' and 'automarked plus teacher response'. The fully automarked tests enable you to get a very quick gauge of students' knowledge and understanding. The 'automarked plus teacher response' tests include longer written response questions, giving students practice of the full range of question styles they will encounter in the final exams. All tests come with a mark scheme.
- End of chapter tests help you to review students' progress on a topic-by-topic basis.
- End of teaching block tests (at the end of every two chapters) provide additional common review checkpoints to help you review progress through the linear course and check students' retention of earlier learning.

© HarperCollins*Publishers* Limited 2016

- End of year and end of course tests help students to prepare and practice for the final exams.

Differentiation

Differentiation is a key feature of the student books, the lesson plans and the worksheets.

- The student books support differentiation with their ramped structure in each spread. The early section offers an accessible 'entry point' and later text is more challenging. This is reflected in the questions, both in the spreads and at the end of each topic. The progress checking spreads also support an understanding of progressing to higher levels of understanding.
- The teacher packs support differentiation by offering learning activities at different levels of challenge for different phases of the lesson, thus making it easier for teachers to tailor provision to a particular class.
- The worksheets also have a ramped structure.

Four stage plan

One of the ways of running the GCSE course is to use a four-stage plan. The idea behind this is that:

- some topics need to be taught before others and so should be placed earlier in the course
- with some topics there is no preferred running order and departments might want to avoid all groups in the same year doing exactly the same topic at the same time
- having what might be termed 'common assessment points' enables periodic assessment and reporting of progress to take place at points at which all students have covered the same ground – the same assessment can be used and students switched between groups if necessary.
- To provide additional common assessment points, Collins OCR Physics can be divided into four teaching blocks:
    - Teaching block 1: chapters 1 and 2 (Paper 1)
    - Teaching block 2: chapters 3 and 4 (Paper 1)
    - Teaching block 3: chapters 5 and 6 (Paper 2)
    - Teaching block 4: chapters 7 and 8 (Paper 2)

End of teaching block tests are provided on Collins Connect and can be used as common assessment points.

© HarperCollins*Publishers* Limited 2016

The order of the teaching blocks can be altered, if you prefer to teach content from Paper 1 and 2 over the duration of the course. The four-stage plan can be represented diagrammatically as follows:

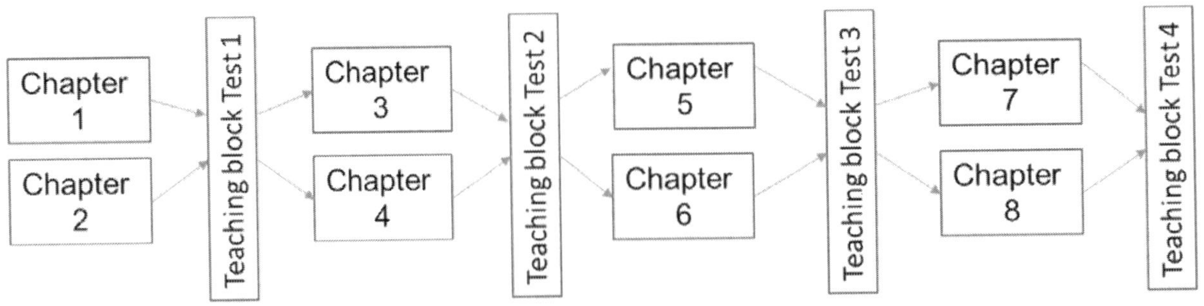

 © HarperCollins*Publishers* Limited 2016

Matter: Introduction

When and how to use these pages

The Introduction in the Student Book notes some of the ideas and skills in this topic area that students will already have met from Key Stage 2 or Key Stage 3, and also provides an indication of what they will be studying in this chapter. *Ideas you have met before* is not intended to be a comprehensive summary of all the prior ideas, but rather points out a few of the key ideas and supports the view that scientific understanding is progressive. Although we might be meeting new contexts, we can often use existing ideas to start to make sense of them.

In this chapter you will find out indicates some of the new ideas that the chapter will introduce. Again, it isn't a detailed summary of content. Its purpose is more to act as a 'trailer' and generate some interest.

The outcomes, then, will be: recognition of prior learning that can be built on, and interest in finding out more.

There is a number of ways of using these. You might, for example:

- use *Ideas you have met before* as the basis for a revision lesson as you start the first new topic
- use *Ideas you have met before* as the centres of spider diagrams, to which students can add examples, experiments they might have done previously, or ideas they found interesting
- make a note of any unfamiliar or difficult terms and return to these in the relevant lessons
- use ideas from *In this chapter you will find out* to ask students such questions as: Why is this important? How could it be used? What might we be doing in this topic?

Overview of the unit

In this chapter, students will learn about the role of particles when thinking about density, changes between states of matter, pressure and volume. They will also learn how to apply their understanding of particle behaviour to the energy in closed systems of solids, liquids and gases, with a particular focus on internal energy, specific heat capacity and latent heat. Students will develop their own models of microscopic particle motion to understand some macroscopic behaviours.

Students will have the opportunity to measure specific heat capacity and consider the effects of latent heat in exploring, both qualitatively and quantitatively, how matter changes state from one phase to the other.

The concept of internal energy is developed from earlier thinking regarding particle behaviour to consider kinetic and potential energy distribution. Students will learn some of the fundamental theory behind the gas laws. They will analyse basic particle theory as well as thinking about and investigating how pressure affects volume and how temperature affects volume, while solving problems for both relationships.

There are important skill development methods to promote creating and reading graphs, including practising turning data into information and interpreting non-linear relationships.

This chapter offers a number of opportunities for students to relate hands-on experience to slightly more abstract ideas. They will use a range of thinking and personal skills to help their learning and support their peers.

Obstacles to learning

Students may need extra guidance with the following terms and misconceptions:

- **Density** Students might think that a suitable liquid particle model is one that shows particles more spread out than those in solids but less densely arranged than those in gases. They will learn that more accurate modelling will show no large gaps between particles in liquids (some liquids are more dense than solids) and will also consider bonds between the particles.
- **Particle motion** Students might think that particles are perfectly still when cold or solid and may need reminding that particles always exhibit some form of vibration.
- **Conservation of mass** Many students believe that a liquid loses mass when evaporating or boiling and do not always consider that mass is conserved with the vapour produced. Students may need some help to realise that we consider the system as a whole.
- **Latent heat** Often the energy used to alter the state of matter is not recognised and students may assume that while a solid melts into a liquid, the temperature will continue to rise unabated.

 © HarperCollins*Publishers* Limited 2016

Practicals in this unit

In this unit, students will do the following practical work:

- create models of three states of matter: solid, liquid and gas
- Practical: measure the volume of an irregularly shaped object
- measure the mass of a liquid while it solidifies
- investigate internal energy in a range of systems
- calculate specific heat capacity
- measure the temperature of stearic acid as it is heated and cooled
- Practical: determine the specific heat capacities of different substances
- make a coin jump by transferring body heat
- investigate the effect on volume of a gas when the pressure on it is increased
- investigate the process of heating ice at a steady rate to past boiling point
- plot the temperature curve when wax is heated

	Lesson title	**Overarching objectives**
1	Key concept: Developing ideas for the structure of the atom	• Understand how ideas about the structure of the atom have changed. • Understand how evidence is used to test and improve models.
2	Atomic structure	• Describe the structure of the atom. • Use symbols to represent particles. • Describe ionisation.
3	Density	• Use the particle model to explain the different states of matter and differences in density. • Calculate density.
4	Key concept: Particle model and changes of state	• Use the particle model to explain states of matter. • Use ideas about energy and bonds to explain changes in state. • Explain the relationship between temperature and energy.
5	Practical: To investigate the densities of regular and irregular solid objects and liquids	• Interpret observations and data. • Use spatial models to solve problems. • Plan experiments and devise procedures. • Use an appropriate number of significant figures in measurements and calculations.
6	Changes of state	• Describe how, when substances change state, mass is conserved. • Describe energy transfer in changes of state. • Explain changes of state in terms of particles.

 © HarperCollins*Publishers* Limited 2016

7	Internal energy	• Describe the particle model of matter. • Understand what is meant by the internal energy of a system. • Describe the effect of heating on the energy stored within a system.
8	Specific heat capacity	• Describe the effect of increasing the temperature of a system in terms of particles. • State the factors that are affected by an increase in temperature of a substance. • Explain specific heat capacity.
9	Specific latent heat	• Explain what is meant by latent heat. • Describe that when a change of state occurs it changes the energy stored but not the temperature. • Perform calculations involving specific latent heat.
10	Maths skills: Drawing and interpreting graphs	• Plot a graph of temperature against time, choosing a suitable scale. • Draw a line of best fit (which may be a curve). • Interpret a graph of temperature against time. • Comment on the specific heat capacity and specific latent heat of a substance.
11	Practical: Investigating specific heat capacity	• Use theories to develop a hypothesis. • Evaluate a method and suggest improvements. • Perform calculations to support conclusions.
12	Particle motion in gases	• Relate the temperature of a gas to the average kinetic energy of the particle. • Explain how a gas has a pressure. • Explain that changing the temperature of a gas held at constant volume changes its pressure.
13	Increasing the pressure of a gas	• Describe the relationship between the pressure and volume of a gas at constant temperature. • Calculate the change in the pressure or volume of a gas held at constant temperature when either the pressure or volume is increased or decreased. • Explain how doing work on a gas can increase its temperature.
14	Pressure in a fluid	• Explain how pressure acts in a fluid • Calculate pressure at different depths in a liquid. • Explain what causes upthrust.
15	Atmospheric pressure	• Show that the atmosphere exerts a pressure. • Explain variations in atmospheric pressure with height. • Describe a simple model of the Earth's atmosphere and atmospheric pressure.

© HarperCollins*Publishers* Limited 2016

16	Maths skills: Handling data	• Recognise the difference between mean, mode and median. • Explain the use of tables and frequency tables. • Explain when to use scatter diagrams, bar charts and histograms.

 © HarperCollinsPublishers Limited 2016

Lesson 1.1: Key concept: Developing ideas for the structure of the atom

Lesson overview

OCR Specification reference

P1.1a

Learning objectives

- Understand how ideas about the structure of the atom have changed.
- Understand how evidence is used to test and improve models.

Learning outcomes

- [O1] Recall the current model of an atom.
- [O2] Research the contribution of two scientists in the development of ideas for the structure of the atom.
- [O3] Contribute research to a group and summarise the group research.

Skills development

- Appreciate the advancement of scientific ideas.
- Research efficiently.
- Summarise research carried out and inform others of research findings.

Maths focus

- Order dates in a timeline.

Resources needed modelling clay (three colours if available), access to a variety of research sources (both textbooks and internet sources would be ideal); Worksheets 1.1.1, 1.1.2, 1.1.3 and 1.1.4

Digital resources access to a variety of research sources

Key vocabulary atom, atomic structure, Democritus, Dalton, Thomson, Rutherford, Bohr, Chadwick, electron, neutron, proton, alpha particle, nuclear model, nucleus, plum pudding model

Teaching and learning

Engage

- Issue a block of modelling clay to each pair of students; ask the students to cut it in half, in half again, and so on, until they cannot cut it any smaller. Could they theoretically keep going forever? [O1]
- **Pair talk:** Discuss whether it would be possible, in theory, to continue to keep cutting the modelling clay into smaller and smaller pieces. Take feedback from a number of different groups, developing ideas as appropriate. [O1]

Challenge and develop

- Ask the students to use the modelling clay from the first activity to build the best model they can of a carbon-6 atom. [O1]
- Inform the students that, until 1932, we didn't know about neutrons, so they should remove these from their model. [O1]
- Next, inform them that, until 1911, we didn't know about protons, so they should remove those too. [O1]
- Finally, state that, until 1897, we didn't know about the electron, so those need to be removed too – oh dear!
- Then ask: "So what did we think matter was made of before we knew about the protons, neutrons and electrons?" Have students discuss this in pairs before introducing the next task.

 © HarperCollins*Publishers* Limited 2016

Explain

- Sort students into groups of three and give each group a research task at an appropriate level:
  - Low demand: Worksheet 1.1.1 [O2]
  - Standard demand: Worksheet 1.1.2 [O2]
  - High demand: Worksheet 1.1.3. [O2]
- After they have completed their research, students should regroup and create a group timeline using their research findings. [O3]

Consolidate and apply

- Following the creation of the timeline in groups, issue Worksheet 1.1.4, which students can use to summarise the research findings from each group. [O2, O3]
- Alternatively, students could be asked to create a storyboard showing the development of ideas about the structure of the atoms. The storyboard should have at least six separate 'episodes'. [O2, O3]

Extend

Ask students able to progress further to do the following:

- Summarise our changing ideas about the structure of the atom in exactly 50 words. [O3]

Plenary suggestions

Ideas hothouse: Ask students to work in pairs to list points about what they know about a particular idea. Then ask the pairs to join together into groups of four and then groups of six to eight to discuss this further and to come up with an agreed list of points. Ask one person from each group to report back to the class.

Ask me a question: Ask students to write a question about something from the last topic and then a mark scheme for the answer. Encourage them to come up with questions worth more than one or two marks and to try their questions out on each other.

Answers to questions

Worksheet 1.1.4

Scientists correctly matched to their development:

Scientist and date(s)	Development
Democritus (470–380 BC)	All matter is made of very small indivisible 'atomos'
Dalton (1808)	All atoms of the same element are the same
J J Thomson (1897)	Discovered the electron and proposed the 'plum pudding' model of the atom
Rutherford (1911)	Discovered protons and the nucleus
Bohr (1913)	Suggested that electrons move around the nucleus in specific layers
Chadwick (1932)	Discovered neutrons

 © HarperCollins*Publishers* Limited 2016

Lesson 1.2: Atomic structure

Lesson overview

OCR Specification reference

P1.1b, c

Learning objectives

- Describe the structure of the atom.
- Use symbols to represent particles.
- Describe ionisation.

Learning outcomes

- [O1] State that the number of protons in an element is the atomic number and that the total number of protons and neutrons is the mass number.
- [O2] Understand that atoms of an element all have the same number of protons but can have different numbers of neutrons, giving different isotopes.
- [O3] Use nuclear notation to show subatomic particles in an isotope.

Skills development

- Represent atoms using accepted notation.
- Use a model to represent ionisation.
- Discuss the limitations of models.

Maths focus

- Subtraction of one number from another to give the number of neutrons.

Resources needed modelling clay (ideally three different colours), copies of the periodic table; Worksheets 1.2.1, 1.2.2 and 1.2.3

Digital resources access to online simulation and means of projection

Key vocabulary atomic number, energy level, mass number, nucleon, relative atomic mass, isotope, ionise, nuclear radiation

Teaching and learning

Engage

- Show students an image of an illuminated dial (search 'illuminated instrument dial') and tell the students that radium has been used in the paint. Now show students an image of a nuclear plant powered by uranium and ask whether radium or uranium is more active.
- Tell the students that radium emits one million times more radiation than uranium. Now ask the students to find radium and uranium on the periodic table. [O1]
- Ask how the elements can be so different when their atomic numbers are so close. What is different about the elements? [O1]

Challenge and develop

- Issue students with a copy of the periodic table and ask students to find the element with atomic number 6. Ask them what information this gives about the element. Ask the students what else the periodic table can tell them about carbon. [O1]
- Issue pairs of students with some modelling clay (ideally three colours) and ask them to build a model of a single carbon atom using the data from the periodic table. Ask some pairs of students to show their model to the rest of the group, explaining how they decided what their model should look like and what each part of the model shows. [O1, O3]
- Alternatively, use a suitable simulation (search 'PhET model of atom') to model atoms, or use students themselves as subatomic particles and model different atoms by arranging the students in the centre of the room (if they are modelling protons or neutrons) or around the edges of the room (in 'layers') if they are modelling electrons. [O3]
- Show students an image of excavated bones and ask them to suggest how the age of the bones could be estimated. They should come up with the idea of 'carbon dating'. Tell students that the ratio of carbon-14 to carbon-12 in each living organism at any given time can be assumed to be constant. Ask the students what we mean by carbon-14 and 'heavy water'. Introduce the idea of atoms of the same element existing with different numbers of neutrons, and define these as isotopes. [O2]

Explain

- Students should now be issued with the atomic structure worksheet at an appropriate level:

 © HarperCollins*Publishers* Limited 2016

- Low demand: Worksheet 1.2.1 [O1, O2, O3]
- Standard demand: Worksheet 1.2.2 [O1, O2, O3]
- High demand: Worksheet 1.2.3 [O1, O2, O3]

Consolidate and apply

- Discuss with students what would happen to an atom if one of its electrons were removed. If necessary, students could return to their model of a carbon atom and count the subatomic particles once an electron had been removed. Define the term 'ionisation' as the gain or loss of one or more electrons from an atom, resulting in an ion. [O1]
- Introduce the term 'nuclear radiation' and model this using ball bearings and a marble: model ionising radiation using small ball bearings as nucleons and electrons and a marble as incoming radiation from the nucleus of a radioactive atom – roll incoming radiation towards the electrons – the structure of the atom is changed – this models ionising radiation in that the structure of the atom becomes altered in some way. [O3]

Extend

Ask students able to progress further to do the following:

- Research how nuclear radiation from an isotope of americium is used in smoke alarms. [O2]

Plenary suggestions

Learning triangle: At the end of the lesson, ask students to draw a large triangle with a smaller inverted triangle that just fits inside it (so they have four triangles). Ask students to think back over the lesson and identify and write in the outer triangles:

- something they saw
- something they did
- something they discussed.

Ask students to write in the central triangle something they learned.

Ideas hothouse: Ask students to work in pairs to list points about what they know about a particular idea. Then ask the pairs to join together into groups of four and then groups of six to eight to discuss this further and to come up with an agreed list of points. Ask one person from each group to report back to the class.

Freeze frame: Ask students to create a 'freeze frame' of an idea in the topic. This involves three or four students arranging themselves as a static image and other students suggesting what it represents.

Answers to questions

Worksheets 1.2.1 and 1.2.2

1. Words in the following order: symbol; number; electrons; relative; protons; equal; isotopes

Worksheets 1.2.1 to 1.2.3

2.

Name of element	Chemical symbol	Mass number (A)	Atomic number (Z)	Number of protons	Number of neutrons	Number of electrons
Carbon	C	12	6	6	6	6
Lithium	Li	7	3	3	4	3
Magnesium	Mg	24	12	12	12	12
Uranium	U	235	92	92	143	92
Radium	Ra	226	88	88	138	88
Iron	Fe	56	26	26	30	26

Worksheets 1.2.2 and 1.2.3

3.

Formula of ion	Number of protons	Number of neutrons	Number of electrons
Na^{+}	11	12	10
Ca^{2+}	20	20	18
F^{-}	9	10	10
O^{2-}	8	8	10

 © HarperCollins*Publishers* Limited 2016

Lesson 1.3: Density

Lesson overview

OCR Specification reference

P1.1d, f

Learning objectives

- Use the particle model to explain the different states of matter.
- Describe differences in density for different states of matter.
- Calculate density for the different states of matter.

Learning outcomes

- [O1] Describe each state of matter by arranging particles in a limited space.
- [O2] Calculate densities for some states of matter.
- [O3] Explain why each state of matter has the properties it exhibits.

Skills development

- Think scientifically: use models in explanations.
- Work scientifically: make predictions.
- Learner development: ask questions.

Maths focus

- Change the subject of an equation.
- Substitute numerical values into algebraic equations using appropriate units for physical quantities.
- Visualise and represent 2D and 3D forms, including 2D representations of 3D objects.

Resources needed large marbles, modelling clay, solid (for example, house brick), liquid (for example, coloured water), gas (for example, steaming kettle), large bag, table tennis balls (some glued or tacked together to form a solid 3D matrix), digital cameras (if available); Worksheet 1.3, Practical sheet 1.3, Technician's notes 1.3

Digital resources camera linked to projector or interactive screen; online animation of states of matter (optional)

Common misconceptions

- Liquids are always less dense than solids. *(Some liquids, such as liquid mercury, have similar or higher densities than solids.)*
- Particles stay perfectly still. *(No particles will ever practically be completely still; even at a fraction above absolute zero all particles have some form of vibration or kinetic energy.)*

Key vocabulary particle model, density, bonds, gas, liquid, solid

Teaching and learning

Engage

- Ask students to **draw** their version of how to model particles in the three states of matter using Worksheet 1.3. Put these to one side for discussion later. [O1]
- Share the three demonstrative examples of a solid, liquid and a gas. Gauge prior knowledge with questions relating to how each example is different from another, drawing towards molecular arrangement for the properties of each, as opposed to basic descriptions. Ask students to **predict**, and justify, which has most mass, which has (or has not) fixed shape and size and, possibly, which can flow. [O1]
- Revisit these justifications and diagrams later in the lesson.

Challenge and develop

- Pass around Practical sheet 1.3, marbles, modelling clay (and digital cameras, if available).

- Divide students into small groups and ask them to **model** a gas, using some rolled-out modelling clay to make a box on the desk and adding as many marbles as necessary. If possible, take photos to share with the class. Ask students whether their groups of marbles always occupy the same size and shape. [O1, O3]
  - Higher demand: Prompt a discussion on where there are weaknesses in these static models or photos: ask how you could give the 'gas' in the model a little more energy; predict differences with another photo and by asking how numbers of collisions may change.
- Now ask students to **model** a liquid with as many marbles as they feel necessary. Take photos to share, if possible. Ask how many marbles students can add and still look for the properties of a liquid. Students should **predict** the size and shape again. They can alter the shape of the modelling clay box to demonstrate how a liquid behaves. Students should **compare** their models with neighbouring groups. [O1]
- Students should now **model** a solid. For *standard* and *lower demand* students, arrange hints on Post-it notes around the classroom, advising students to use a small amount of modelling clay in between marbles. Take photos to share, if possible, emphasising this use of modelling clay to model bonds. Ask once more for **predictions** of size and shape; try removing the modelling clay box to help any students who are struggling. This can be emphasised by comparing the model of table tennis balls glued or tacked together with a bag of loose table tennis balls. [O1]
- **Explore** the worksheet questions. Students should **substitute** values to gain a numerical answer. [O2]

Explain

- Show a suitable website animation or your photos of each modelled state. [O1, O3]
- Students should **discuss** the effect of bonding for solids. Ask them to **explain** how this has an effect on the shape and effect of flow for solids. [O1]
- Ask students to **compare** models of liquids and solids to **describe** one similarity and two differences, examining size, shape, mass, etc. Students should **contrast** any two of the three models against each other. [O1, O3]

Consolidate and apply

- **Compare** students' models with the earlier three demonstrative examples. Students should **review** their original models and draw new models, aiming to **improve** at least one model. [O1, O3]
  - High demand: Estimate a mass and volume to allow some students to determine a density.
- Students can **share** their worksheets with others in the class for others to suggest any other improvement or information on size or shape information.
- Students **complete** the review triangle in the worksheet individually. [O1, O2, O3]
- **Explore** the worked example in the Student Book, practising rearranging equations and substituting values. **Answer** questions 2 and 3 in the Student Book, referring to the models created in the practical. [O2]

Extend

Ask students able to progress further to do the following:

- **Create** questions for their peers that mix units: kilograms to grams, cubic centimetres to cubic metres. [O2]
- **Sketch** and **explain** how cold air sinks below hot air. [O1, O3]

Plenary suggestions

Students use their review triangles to ask their peers questions that they would like answered, based on the lesson.

Arrange students in groups. Call out a state of matter; students arrange themselves into a model of the state of matter (linking arms for bonding where necessary). Ask students to justify their model.

Answers to questions

Worksheet 1.3, Part 3, Calculations

1. 1.5 kg/m^3
2. 148 kg/m$^{3;m}$
3. 10^{-6} m^3; 1 cm

 © HarperCollins*Publishers* Limited 2016

Lesson 1.4: Key concept: Particle model and changes of state

Lesson overview

OCR Specification reference

P1.1e

Learning objectives

- Use the particle model to explain states of matter.
- Use ideas about energy and bonds to explain changes of state.
- Explain the relationship between temperature and energy.

Learning outcomes

- [O1] Describe the particle assembly for each of the three states of matter.
- [O2] Describe how particles bonded to each other change with a transfer of energy to alter the state of matter.
- [O3] Explain how a model of particles that move quickly and have high energy might describe the high temperature of a substance.

Skills development

- Think scientifically: analyse data.
- Work scientifically: interpret evidence.
- Learner development: ask questions.

Maths focus

- Draw graphs.

Resources needed tripod, gauze, heatproof mat, Bunsen burner, 250 ml beaker, thermometer, ice, stopwatch, table tennis balls (or Molymod kits); Worksheet 1.4, Practical sheet 1.4, Technician's notes 1.4

Digital resources projector, access to PhET website

Key vocabulary temperature, internal energy, quantity, states of matter, particle model, bonds, latent heat, fusion, vaporisation

Teaching and learning

Engage

- You might like to begin this lesson by reading out some comments from the circle activity at the end of the last lesson for the class to **debate** as a whole.
- Review students' prior understanding by using the solid, liquid and gas props and ask them to stand for a gas, sit for a liquid and squat for a solid. Recall the nature of bonds for each state of matter with each prop. [O1]
- Students each **explain** to a partner which of these has a higher temperature: a burning match, a fish tank or a cold bath, and also which has, collectively, more energy. [O3]
- Project (students could use their own computers or tablets) one of the previously used PhET website simulations as a stimulus to help lower-attaining students. [O2]

Challenge and develop

- Demonstrate the investigation, promoting care of the hot object, as the Bunsen burners will be in use for several minutes (Practical sheet 1.4 and Technician's notes 1.4). [O2]

 © HarperCollins*Publishers* Limited 2016

- Students should **collect** data every minute, so that they have sufficient data to plot a graph that shows the three states of matter for water, referring to Worksheet 1.4. [O1, O2]
- Students will **present** their data in the form of a graph with space left to write a label for the curve. [O2]

Explain

- Each group will **comment** on and **explain** the four sections of another group's graph. [O2]
- Each group will **predict** the last stage of the plot and extrapolate the gaseous temperature increase phase (use the term 'gradient' when labelling this section – it should be less steep). [O1, O2, O3]
  - High demand: Students should **quantify** the gradients of each part of the curve of their graph. [O1]

Consolidate and apply

- Using the speech bubbles on Worksheet 1.4, students assess each other's plots, use one speech bubble to make a reference to any misconception or add information. Points to look out for would be diagrams of a state change, relating particle energy to temperature or vibration increase. [O1, O2, O3].
  - High demand: Students should relate each section of the plot to their knowledge of specific heat capacity and latent heats of fusion and vaporisation.
- Students re-sketch a plot to describe how the temperature might change for a closed system of hot vapour at 150 °C placed in a freezer at −40 °C. [O1, O2, O3]

Extend

Ask students able to progress further to do the following:

- **Explain** why the slopes of the warming phases are generally steeper for solids and less steep for gases. [O2, O3]

Plenary suggestions

Supply a one-word short phrase 'answer' to each table for students to think of a question that can be asked for other tables.

 © HarperCollins*Publishers* Limited 2016

Lesson 1.5: Practical: To investigate the densities of regular and irregular solid objects and liquids

Lesson overview

OCR Specification reference

Prac P1 (materials)

Learning objectives

- Interpret observations and data.
- Use spatial models to solve problems.
- Plan experiments and devise procedures.
- Use an appropriate number of significant figures in measurements and calculations.

Learning outcomes

- [O1] Plan and carry out an investigation into the density of irregularly shaped objects, gathering data for mass and volume.
- [O2] Use generated data to determine the density of a material, such as modelling clay.
- [O3] Obtain a precise and reliable set of data by investigation, using appropriate numbers of significant figures.

Skills development

- Think scientifically: analyse data.
- Work scientifically: present evidence.
- Learner development: collaborate effectively.

Maths focus

- Calculate means (averages).
- Keep consistent units.

Resources needed real house brick and foam house brick (or example of real and fake concrete if preferred), eureka can, mini whiteboards, pens, 100 ml measuring cylinder, modelling clay, 250 ml beaker, water, balance accurate to 1 g or 0.1 g; Worksheet 1.5, Practical sheet 1.5, Technician's notes 1.5

Digital resources video of Archimedes and Eureka, such as this

Key vocabulary displaced liquid, density, mass, volume, significant figures, resolution

Teaching and learning

Engage

- Create some intrigue towards density by using the props suggested in Technician's notes 1.5, ideally the fake foam brick. Drop one of each prop into some volunteers' hands after asking them, in pairs, to **predict** what they would feel if you increased the height from which the props were dropped. [O1]
- Use mini whiteboards to **answer** some quick questions, starting with, "What weighs most, a tonne of feathers or a tonne of bricks?" Then ask a few more (no calculators), for example, "What is the unit of mass? The unit of volume? The formula for density? The density of a liquid that has a mass of 400 kg and a volume of 0.4 m^3 (higher demand)? A mass of 2000 kg and a volume of 2 m^3 (standard demand)?" [O2]
- Show a video of Archimedes and Eureka, such as this. Ask students to think, discuss in pairs and share: how could we determine the density of an irregularly shaped object in the lab using basic equipment? **What observation will be made?** [O2, O3]

 © HarperCollins*Publishers* Limited 2016

Challenge and develop

- Share thoughts as a class in response to the above episode; challenge any misconceptions. [O1]
- Two students can be led to demonstrate the best procedure by **applying** the points made by their peers earlier, guiding and correcting any poor practice, for example, not reading the measuring cylinder volume at eye level. Develop the idea of a eureka can and explain why it isn't necessarily the most accurate way of determining displaced volume (water left in pipe, reliable maximum volume in the can). [O1]
- Distribute Practical sheet 1.5, drawing attention to the high-demand extension task, as necessary. [O3]
- Students **record** their observations on the sheet. [O1, O3]
- Data is **presented** in the form of a table with averages, taking care to maintain suitable significant figures. [O2]

Explain

- Students **report** their findings orally, **explaining** the value of density that they obtained and how confident they are in this value, based on the tolerance in the repeated dataset. [O1]
- Students **examine** the reliability of their investigation and search for any anomalous values; they suggest improvements (for example, ensuring that no drops of water are left on the sides of the measuring cylinder; making sure that eye-line observations have been made). [O3]
  - High demand: Students should attempt to **quantify** the percentage of variation in the class results.

Consolidate and apply

- Refer to the video and apply students' findings from their investigation. Students should **outline** a brief plan to help the King of Sicily determine whether he has been cheated or not. [O1]
- Students can complete Worksheet 1.5 to reinforce **calculating** densities of irregularly shaped objects and complete the questions from Section 1.5 of the Student Book. [O2]
- On the mini whiteboards, students write two things that went well with their investigation and one thing to improve, explaining why. [O1, O2, O3]

Extend

- Ask students able to progress further to **describe** ways to find the volume of irregularly shaped objects that are less dense than water.
  - High demand: Students can consider and explain why the methods used wouldn't work in micro-gravity.
- Students should practise **designing** their own questions and answers for densities using different units for mass and volume.

Plenary suggestions

Students use Post-it notes to leave questions to pose for the next lesson or questions they would like answered before the next lesson.

Answers to questions

Worksheet 1.5, Section 1

1. 14.33 g/cm^3 (1433 kg/m^3)
2. 8.1 g/cm^3 (8100 kg/m^3)
3. 11.5 g/cm^3 (11.500 kg/m^3)
4. 71.83 kg/m^3
5. 3630 kg/m^3

Worksheet 1.5, Section 2

1. 48 g (0.048 kg)
2. 13 530 kg
3. 78 g (0.078 kg)
4. 660 kg
5. 35.34 g (0.03534 kg)

Worksheet 1.5, Section 3

1. 0.636 cm^3 (6.36×10^{-7} m^3)
2. 7.47 m^3
3. 41 cm^3 (4.1×10^{-5} m^3)
4. 38.45 m^3
5. 7.77 cm^3 (7.77×10^{-6} m^3)

 © HarperCollins*Publishers* Limited 2016

Lesson 1.6: Changes of state

Lesson overview

OCR Specification reference

P1.2a, b

Learning objectives

- Describe how, when substances change state, mass is conserved.
- Describe energy transfer in changes of state.
- Explain changes of state in terms of particles.

Learning outcomes

- [O1] Use the terms 'melting', 'freezing', 'boiling', 'condensing', 'evaporating' and 'sublimating' correctly.
- [O2] Accurately describe the change in particle arrangement for any change of state. Predict the direction of energy transferred in a change of state.
- [O3] Compare differences in bonds and explain the role of heat energy transfer between states of matter.

Skills development

- Think scientifically: analyse data.
- Work scientifically: interpret evidence.
- Learner development: ask questions.

Maths focus

- Consider the highest number of significant figures to use with the balance.

Resources needed flakes of wax, stirring rod, thermometer, balance measuring to 1 g or 0.1 g, Bunsen burner, heatproof mat, tripod, gauze, tongs, vocabulary cards, examples of each state of matter (for example, ice to melt and boil); Worksheet 1.6, Practical sheet 1.6, Technician's notes 1.6

Digital resources projector, access to PhET website video clips of the sublimation of iodine crystals and dry ice

Key vocabulary bond, energy, boil, changes of state, condense, conservation of mass, evaporate, freeze, melt, sublimate

Teaching and learning

Engage

- Before starting, you might like to revisit some of the Post-it notes from the end of Lesson 1.5.
- Issue students with states of matter vocabulary cards (see Technician's notes 1.6); students **explain** to their partners what each term means. [O1]
- Project some images of different states of matter and of matter changing state (for example, chocolate melting, water steaming in a sauna). Ask students to **choose** and hold up the most suitable card linked to each image. [O1]
- Demonstrate the three states of matter (see Technician's notes 1.6). Ask students to choose and hold up the states of matter cards related to your demonstrations. [O1]
- For speed, rather than demonstrating, show some video clips of the sublimation of iodine crystals and of dry ice. Ask students to **choose** and hold up the card for the 'missing' state of matter. [O1]

Challenge and develop

- Ask students to **role-play** melting a liquid, with each student acting as a particle. Repeat the role-play but now ask students to introduce transferring energy in any creative way they can. [O1, O2]

 © HarperCollins*Publishers* Limited 2016

- Project a change of state from the PhET interactive website and ask students to **discuss** in pairs how numbers of particles relate to the mass of the system.
- Ask students to **predict** how energy might affect the solid bonds. **Role-play** once more, including observing the solid bonding. Invite explanations of what happened once the solid bonds had been 'broken'. Challenge the notion of 'no bonds' for a liquid with a what-happens-next scenario once the liquid particles have been given energy with 'no bonds'. [O3]
- Refer to Practical sheet 1.6. Students **heat** some flakes of wax (for example, salol) over a Bunsen burner, using a tripod and gauze on which to place the beaker. Stirring as the wax melts, heat the wax to around 60 °C (not more, to avoid overheating), wearing goggles at all times. Carefully transfer the beaker and contents to a nearby balance using the tongs (you may want to do this, depending on your students). Let the contents cool for a few minutes; then record the mass again. [O1, O2]

Explain

- While the wax cools, students **apply** the model of energy transfer, from the role-play before, to explain what is happening to the cooling contents of their beakers. [O2]
- Students complete the storyboard on Worksheet 1.6, drawing particle diagrams to **demonstrate** their understanding. Students explain the purpose of the card lid suggested in the practical evaluation (Practical sheet 1.6). [O1, O2, O3]
    - High demand: **Explain** kinetic and potential energy in the storyboards.

Consolidate and apply

- Project a picture of a glacier and **ask** students how much of the land's surface is covered by glacial ice (around 10%). [O1]
- Ask students to **explain** with reference to mass conservation what happens as expected global temperatures rise through the decades. [O1]
- Introduce the 'man-made versus natural' nano-debate (fewer than 20 words each) for global warming. [O1]
- Answer questions 1 to 8 from pages 24 and 25 of the Student Book. [O1, O2, O3]

Extend

Ask students able to progress further to do the following:

- **Explain** how evaporation happens before the liquid's boiling point has been reached. There is no need to introduce the Boltzmann factor; just consider the many particles in the liquid.

Plenary suggestions

Play a game of 'Keyword splat' on the board, using such words as 'liquid bonds', 'solid bonds', 'energy', etc., basing questions on your assessment of the students' progression through the lesson.

 © HarperCollins*Publishers* Limited 2016

Lesson 1.7: Internal energy

Lesson overview

OCR Specification reference

P1.2c

Learning objectives

- Describe the particle model of matter.
- Understand what is meant by the internal energy of a system.
- Describe the effect of heating on the energy stored within a system.

Learning outcomes

- [O1] Relate kinetic (and potential) energy to the internal energy of a system.
- [O2] Arrange particles with velocity vector arrows to explain the internal energy of a system.
- [O3] Use macroscopic evidence to justify increases and decreases in the internal energy of a system.

Skills development

- Think scientifically: use models.
- Work scientifically: record observations.
- Learner development: present observations.

Maths focus

- Calculate average velocity.

Resources needed ball and ring expansion kit, clamp, boss and clamp stand, Bunsen burner, heatproof mat, ice pack, thin wooden stick with lead on end, small hammer, block of wood, bicycle pump, kettle, thermometer, 250 ml beaker, ice cubes, dessert spoon, strips of rubber, thin pieces of steel, small springs, large ball-bearing; Worksheet 1.7, Practical sheets 1.7.1, 1.7.2, 1.7.3, 1.7.4, 1.7.5 and 1.7.6, Technician's notes 1.7

Digital resources access to the interactive PhET website, pictures of: a bridge, railway lines, the Eiffel Tower

Key vocabulary internal energy, particle model

Teaching and learning

Engage

- Revisit a Key Stage 3 demonstration of how heat can be thought of as being transferred along a metal. Have students line up at the front of the class and **role-play** being atoms in a metal. They might not model vibrating while waiting in line and concentrate only on their stationary positions (applaud any who exhibit correct modelling). [O3]
- Demonstrate the spring arrangement. Ask students to **suggest** improvements and draw these towards the effect of gravity being negligible between particles. Ask for improvements to this demonstration. Highlight energy transformations between potential and kinetic energy. [O1]
  - High demand: Students **estimate** the kinetic energy of the ball using their mental mathematical skills, based on the approximate mass and velocity you use. [O1]

Challenge and develop

- Introduce the circus (Practical sheets 1.7.1–1.7.6). Explain the safety instructions as you demonstrate each circus activity. [O3]
- Set a timer for each of the activities based on your knowledge of your class, probably about three to five minutes at each station. At the end of each station time, give students two further minutes to update their worksheets.

 © HarperCollins*Publishers* Limited 2016

- (The extension questions are aimed at higher-attaining students.) For Practical sheet 1.7.1, the ball will need to be adequately cooled by an adult between each group. [O3]

Explain

- Ask students to complete the 'double bubble map' in the first section of Worksheet 1.7 to **show** differences that can be observed macroscopically between high internal energy and low internal energy systems. [O1, O2, O3]
- Outline how students can **represent** particle models with 'to-scale' velocity arrows and students complete the second section in Worksheet 1.7. [O2]
  - High demand: Students can attempt to include calculations of some kinetic energy values to deduce the velocity of a particle (average v = 440 m/s, mass = 5 × 10^{-26} kg).

Consolidate and apply

- Refer to the differences between summer and winter and project a picture of a bridge, some railway lines and the Eiffel Tower. Invite predictions about changes across a year and ask pairs to invent some designs to counteract foreseen problems. [O2, O3]
- Form a Socratic dialogue, with students taking turns to sit in the hot seat. Begin by asking if the statement 'all internal energy changes are described with macroscopic features' is true or false. [O1, O2, O3]

Extend

Ask students able to progress further to do the following:

- Consider what could happen to particles with extremely high kinetic energy once they collide (to encounter concepts of plasma and ionisation).

Plenary suggestions

Roll a sentence: Each student must supply a word to complete a viable sentence relating to describing the effect of heating a substance in relation to the particle theory. Stop at every fifth word to allow some thinking time to form the next part of the sentence. Give the class 'phone a friend', 50:50 and 'ask the audience' options if they get stuck.

Answers to questions

Worksheet 1.7, Extension questions

1. electrons
2. elastic (ideally) – otherwise energy would be dissipated
3. fluids

 © HarperCollins*Publishers* Limited 2016

Lesson 1.8: Specific heat capacity

Lesson overview

OCR Specification reference

P1.2d (part), e

Learning objectives

- Describe the effect of increasing the temperature of a system in terms of particles.
- State the factors that are affected by an increase in temperature of a substance.
- Explain specific heat capacity.

Learning outcomes

- [O1] Show which direction heat is being transferred in a system with two bodies.
- [O2] Calculate the specific heat capacity of a substance mathematically.
- [O3] Explain why it is generally harder to increase the temperature of a solid than that of a liquid or gas.

Skills development

- Think scientifically: analyse data.
- Work scientifically: present evidence.
- Learner development: collaborate effectively.

Maths focus

- Rearrange formulae.
- Calculate percentages.

Resources needed 250 ml beakers, density cubes, a rod, tripod, gauze, Bunsen burner, heatproof mat, thermometer, balance, balloons, ice, boiling tube; Worksheet 1.8, Practical sheet 1.8, Technician's notes 1.8

Digital resources internet and projector, images of different materials

Key vocabulary energy, transfer, specific heat capacity

Teaching and learning

Engage

- Show students two balloons, one filled with water, the other air. Ask students, in groups of three, to **predict** what will happen when the air-filled balloon is held over a candle flame. The three students each have a word that they must use (heat, rubber and air) and should reply sharing their prediction. [O1, O3]
- Demonstrate both balloons, asking what **effect** the water has on the other side of the rubber. [O1, O3]

Challenge and develop

- Carry out Practical 1.8. Ensure that the strings are tied tightly to the density cubes and rods beforehand. Students should set a water bath, heating the water to boiling, and submerge the density cube for around a minute. After one minute, they carefully transfer the hot density cube to the other beaker and suspend the block midway; after a minute, the block can be left at the bottom of the beaker. While continuing to stir the water, through the card lid, the maximum temperature should be recorded. Using the masses they have **measured**, the specific heat capacity of water and the recorded temperature changes, students calculate the specific heat capacity of the density cube material (for example, brass). [O2]
- Using the standard measured value of around 380 J/(kg °C) for the specific heat capacity of brass (or the value for whichever material of density cube is used), students can **calculate** a percentage difference. [O2]
- Repeat, if time, but this time use a different density cube, for example, aluminium, to compare values. [O2]

 © HarperCollins*Publishers* Limited 2016

- High demand: Students should **discuss** why and how the specific heat capacity for glass should also be taken into account.

Explain

- Ask students to **draw** (perhaps on mini whiteboards) an explanation for the difference in values of the two materials, based on 'particles' and stressing what they have previously learnt about density. [O3]
- Students should **sketch,** using Worksheet 1.8, how they can represent particle models with arrows to show the direction of heat transfer for one of the systems investigated in the practical. [O2]

Consolidate and apply

- Project images of different materials. Students can explore the materials and **justify** why materials are chosen for different purposes based on high or low specific heat capacity, for example, saucepan handles, saucepans, water jackets. [O1]
- Students should answer questions 1–8 on pages 28 and 29 of the Student Book. [O2]

Extend

Ask students able to progress further to **clarify** why gases have very high specific heat capacities and why the value for each gas is determined for constant volume and constant pressure.

Plenary suggestions

Melt a small amount of ice over a Bunsen burner and ask students to predict, after two minutes of thinking and discussing in pairs, which of melting, boiling and evaporating happens at the quickest rate.

 © HarperCollins*Publishers* Limited 2016

Lesson 1.9: Specific latent heat

Lesson overview

OCR Specification reference

P1.2d (part), f

Learning objectives

- Explain what is meant by latent heat.
- Describe that when a change of state occurs it changes the energy stored but not the temperature.
- Perform calculations involving specific latent heat.

Learning outcomes

- [O1] Describe the effect on temperature when energy is supplied to change state.
- [O2] Explain what happens to the energy supplied while changing state.
- [O3] Justify the different changes of state numerically.

Skills development

- Think scientifically: consider the quality of evidence.
- Work scientifically: interpret evidence.
- Learner development: ask questions.

Maths focus

- Select correct units.
- Calculate energy required to change state.

Resources needed different types of chocolate, grater (or grated chocolate), water bath, goggles, 250 ml beaker, boiling tube, thermometer, stopwatch, Bunsen tripod, gauze, heatproof mat, clamp, clamp stand, boss head, immersion heater, ammeter, voltmeter, power pack, connecting leads; Worksheet 1.9, Practical sheet 1.9, Technician's notes 1.9

Digital resources access to the interactive PhET website, temperature probe and data logger

Key vocabulary fusion, vaporisation, latent heat, specific latent heat, specific latent heat of fusion, specific latent heat of vaporisation

Teaching and learning

Engage

- Show some chocolate and ask what ways students personally **think** chocolate has been 'improved' by scientists over the years. Ask which chocolate students think melts the easiest and at what temperature. [O1]
- Melt some grated chocolate in a water bath and project the temperature using a data logger. (Use a camera if you do not have a data logger.) [O1]
- Announce the melting stage to the class and ask for **observations** of the temperature rate of increase. [O2]
  - Low demand: Students can also **simulate** this from the PhET website. [O2]

Challenge and develop

- Students **explore** how the rate of temperature-increase plateaus at a change of state, following the instructions from Practical sheet 1.9. [O2]
- Students **investigate** how a consistent transfer of heat from the Bunsen burner over time affects the temperature of the stearic acid and take readings of the temperature every minute. Carrying out this practical without a data logger will improve communication in student groups, as well as decision-making skills.

 © HarperCollins*Publishers* Limited 2016

By nominating students as reporter, graph plotter, practical controller and tabulator scribe, this practical can be differentiated by task. [O1, O2]

- High demand: Students can conduct this experiment using an immersion heater connected with an ammeter and voltmeter to **calculate** energy transfer to the water bath. Carefully clamp the heater beforehand, to minimise the risk of burning. [O3]

Explain

- Have each group's reporter **articulate** the group's ideas as to why there is a notable plateau of the rate of temperature increase around melting. An acronym of the words 'solid, liquid and bond' can be used as a hint for any *lower-attaining* students. [O1, O2]
- Students explain what latent heat means using the terms 'energy' and 'bonds'. [O2]
- Graphs are **plotted** by all students. The two relationships should be **described** and written down. [O3]

Consolidate and apply

Students talk in pairs and then in fours.

- First, they discuss and **report** how latent heat is useful in sweating. [O1, O2]
- Second, they **predict** an observation regarding energy when the process is reversed. Pass around some reversible hand warmers to demonstrate. [O2]
- Students can attempt the worksheet **calculations** before attempting the questions from the Student Book, page 31. [O3]

Extend

Ask students able to progress further to do the following:

- Discuss why steam can **cause** more painful scalding than boiling water.

Plenary suggestions

Each student should write either a question they would like answered about their learning or an answer to a question they know the answer to on a piece of paper. Students screw their pieces of paper into balls and throw them across the lab (or onto a desk). They then pick up the nearest ball and read out what has been written and select a peer to respond to it.

Answers to questions

Worksheet 1.9, Section 1

1. 2286 J/g (2.23×10^6 J/kg)
2. 0.359 kg (359 g)

Worksheet 1.9, Section 2

1. 1131 s (18 min and 51 s)

© HarperCollins*Publishers* Limited 2016

Lesson 1.10: Maths skills: Drawing and interpreting graphs

Lesson overview

OCR Specification reference

PM1.2 I, ii

Learning objectives

- Plot a graph of temperature against time, choosing a suitable scale.
- Draw a line of best fit (which may be a curve).
- Interpret a graph of temperature against time.
- Comment on the specific heat capacity and specific latent heat of a substance.

Learning outcomes

- [O1] Choose the most suitable scale based on the data given.
- [O2] Select the correct axis to plot data against.
- [O3] Plot line of best fit for data of temperature and time while describing relationships.

Skills development

- Think scientifically: analyse data.
- Work scientifically: present evidence.
- Learner development: collaborate effectively.

Maths focus

- Draw graphs.

Resources needed tripod, gauze, heatproof mat, Bunsen burner, 250 ml beaker, thermometer, paraffin wax, stopwatch, stirring rod, graph paper; Worksheet 1.10, Practical sheet 1.10, Technician's notes 1.10

Digital resources projector

Key vocabulary variable, scale, data, range, line or curve of best fit

Teaching and learning

Engage

- Pass around a few copies of Worksheet 1.10. Each pair of students is to **correct** one thing wrong with the graph before passing the sheet on to a neighbouring pair. Students should spot several deliberate mistakes, for example, irregularly spaced graduations on the *y*-axis, incorrect values on axes, mislabelled *y*-axis, no units, gaps, anomaly. [O1, O2]
- Explore students' current understanding of independent and dependent variables using Worksheet 1.10.
  - Low demand: Students could try the DRY MIX method (dependent, responding, *y*-axis; manipulated, independent, *x*-axis). [O2]

Challenge and develop

- Students should **melt** paraffin wax and **collect** temperature-increase data (Practical sheet 1.10 and Technician's notes 1.10).
- Students **choose** which is the dependent and which is the independent variable and draw their own table to collect their data. [O2]
- Give groups different types of scaled graph paper for them to choose and **justify** a scale. [O1]
- Students will **present** their data in the form of a line graph to the class. [O2]

 © HarperCollins*Publishers* Limited 2016

Explain

- Each group will **explain** any relationship, emphasising any '… er' and '… er' finding (for example, the longer the heating, the higher the temperature) and **highlight** the change of state. [O3]
- Students **explain** what is happening to the particles between particular time intervals, for example, between 1 and 3 min, between 6 and 9 min. [O3]
- Students **explain** why the transition between the heating of one state of matter and the bond-breaking latent heating is curved and not a sharp angle, as many websites depict. [O3]
  - High demand: Students should **quantify** the gradient at any point in their graph. [O3]

Consolidate and apply

- Groups **predict** and then **plot** the cooling curve of paraffin wax, but extending the time by 10 min on each side. [O1, O2]
- Students answer questions from the Student Book, Section 1.10. [O3]

Extend

Ask students able to progress further to do the following:

- **Explain** how they might construct a similar graph for a salt solution, to explain why salt is added to roads in the winter. [O3]

Plenary suggestions

Students create their own checklist of things that they have learned this lesson. Swapping checklists in groups will enable some peer review, to help improve the lists.

Answers to questions

Worksheet 1.10, Section 2

The independent variables are:

a) shoe size
b) amount you eat
c) wind speed
d) change in pressure
e) energy supplied by heater.

 © HarperCollins*Publishers* Limited 2016

Lesson 1.11: Practical: Investigating specific heat capacity

Lesson overview

OCR Specification reference

Prac P5

Learning objectives

- Use theories to develop a hypothesis.
- Evaluate a method and suggest improvements.
- Perform calculations to support conclusions.

Learning outcomes

- [O1] Use theories about specific heat capacity to write a hypothesis for the experiment.
- [O2] Suggest improvements to the method for determining specific heat capacity.
- [O3] Calculate, using the specific heat capacity equation, to justify a conclusion.

Skills development

- Use scientific theories and explanations to develop hypotheses.
- Apply a knowledge of a range of techniques, instruments, apparatus and materials to select those appropriate to the experiment.
- Make and record observations and measurements using a range of apparatus and methods.
- Evaluate methods and suggest possible improvements and further investigations.

Maths focus

- Substitute numerical values into algebraic equations using appropriate units for physical quantities.
- Plot two variables from experimental or other data.
- Determine the slope and intercept of a linear graph.

Resources needed copper and aluminium blocks, each with two holes, 250 cm^3 beaker, taps, thermometer, petroleum jelly, 50 W, 12 V heater, 12 V power supply, insulation, ammeter, voltmeter, leads, stopwatch, balance (mass); Worksheet 1.11, Practical sheet 1.11, Technician's notes 1.11

Key vocabulary energy store, energy transfer, specific heat capacity

Teaching and learning

Engage

- Challenge students' thinking and understanding of how to determine the specific heat capacity of a substance by discussion. Suggested questions to consider: What is specific heat capacity? What is the equation? What must be measured to experimentally determine the specific heat capacity of a substance? [O1, O2]
- Explain that the class will be doing an experiment to determine the specific heat capacity of three different substances. Have pairs of students discuss what is meant by a hypothesis and what it should contain and then have them share their ideas with the class. [O1]

Challenge and develop

- Students, either individually or in small groups, write a hypothesis on the specific heat capacities of copper, aluminium and water. Choose three groups to share their hypotheses with the class and let the others comment on what was good in that hypothesis and how it could be improved. This should be used for all students, to improve their own hypotheses. It might be wise to have small samples of the three materials to hand, so that the students can hold them (certainly the metals) to help them formulate their hypotheses. [O1]

 © HarperCollins*Publishers* Limited 2016

- Carry out Practical 1.11. [O2, O3]
  - Low demand: Guide students through the process carefully.
  - Standard demand: Guide students through the calculations.
  - High demand: Students work independently.

Explain

- Discuss how the practical could have been improved, first in pairs, then groups of four, then groups of eight. Finally, someone in each group shares ideas for improvement with the class. Challenge students' thinking and understanding of the practical using the evaluation section of Practical sheet 1.11. [O1, O2, O3]

Consolidate and apply

- Distribute Worksheet 1.11 and proceed according to the students' ability.
  - Low demand: Guide students through tasks carefully. It might be necessary to help the students with the scales on the *x*- and *y*-axes when plotting the graph.
  - Standard and high demand: Work independently or in pairs or small groups. [O1, O2, O3]

Extend

- In small groups, students discuss why they think different substances have different specific heat capacities. (The easiest explanation, although not the whole story, is that substances with heavier atoms have fewer atoms per kilogram, so each atom gains more energy compared with other substances. Thus, atoms heat up more easily, so the substance has a lower specific heat capacity.)

Plenary suggestions

Ask me a question: Ask the students to write a question about something from the topic and then a mark scheme for the answer. Encourage them to come up with ones worth more than one or two marks and to try out their questions on one another.

Heads and tails: Ask each student to write a question about something from the topic on a coloured strip of paper and the answer on another colour. In groups of six to eight, hand out the strips so that each student gets a question and an answer. One student reads out his or her question. The student with the right answer then reads it out, followed by his or her question, and so on.

Answers to questions

Worksheet 1.11

1. 50 W
2.

Time (s)	Energy (J)	Temperature (°C)
0	0	18
60	3,000	18
120	6,000	20
180	9,000	22
240	12,000	24
300	15,000	26
360	18,000	28
420	21,000	30
480	24,000	32
540	27,000	34
600	30,000	36

3. graph with axes labelled, with units, scale leads to spread of points over at least half a page, sensible scale (not 3, 6, 7, 9), points plotted, line of best fit
4. gradient should be around 0.00067 (°C/J if you want to include gradient units)
5. 750 J/(kg °C)
6. Not all of the energy supplied goes into heating the metal; some is lost to surroundings, also to heating heater and thermometer.
7. More reliable since makes use of more data; rate of change of temperature is not constant and this isn't taken into account when using just one energy and change in temperature.

Practical sheet evaluation

1. Any sensible answer with explanation, consistent with how the results match the hypothesis.
2. Whichever matches the results but the most likely is: My values for specific heat capacity were all higher than the real values.
3. They varied by only small amounts.
4. So heat could conduct to the thermometer more easily.
5. So full volume of water was heated evenly *(or words to that effect)*.
6. It takes time for heat to transfer to thermometer.
7. By repeating the results.

 © HarperCollins*Publishers* Limited 2016

Lesson 1.12: Particle motion in gases

Lesson overview

OCR Specification reference

P1.3 a, b, c

Learning objectives

- Relate the temperature of a gas to the average kinetic energy of the particle.
- Explain how a gas has a pressure.
- Explain that changing the temperature of a gas held at constant volume changes its pressure.

Learning outcomes

- [O1] Describe the qualitative relationship between the speed of the particles and the temperature of the gas.
- [O2] Relate the importance of the wall of a container to the pressure of a gas.
- [O3] Explain how pressure can increase the temperature of a gas.

Skills development

- Think scientifically: analyse data.
- Work scientifically: interpret evidence.
- Learner development: ask questions.

Resources needed glass bottle, ice bath, small bowl, large coin, round-bottomed flask, coloured water, bung, glass tube, Bourdon gauge (or data logger pressure gauge), test tube, tubing, mini whiteboards; Worksheet 1.12, Practical sheet 1.12, Technician's notes 1.12

Digital resources projector, access to PhET website, digital cameras, some images or slow-motion videos of explosions (for example, popcorn popping, eggs in a microwave), videos of earth-based particle models and Tim Peake micro-gravity models

Key vocabulary kinetic, gas pressure, collide, randomly

Teaching and learning

Engage

- Demonstrate the inside-out balloon experiment and ask students to describe what happened (the effect can be accelerated using an ice bath or running cold tap). Repeat this with a volunteer and guide students to **explain** what happened by completing the sentence, "The sides of the balloon were pushed inwards as it got colder because …" to introduce the concepts of pressure and temperature. [O1, O2]
- Project some images of explosions – slow-motion videos can be useful here, for example, popcorn popping, eggs in a microwave – and ask students, paired by ability, to **describe** what was increasing inside the object and what was (trying to) stay the same. [O2]

Challenge and develop

- Conduct Practical 1.12, grouping students by ability. Students should describe the effect on the pressure of the gas (and hence on the coin), perhaps writing on mini whiteboards. An ice bath is a good way to cool the air inside the bottle. [O1, O2]
- Project a change of state from the interactive PhET website and ask student pairs to **explain** why the lid was blown apart when the temperature became too high.
- Students draw a bottle on a mini whiteboard. They then **draw** a series of sketches to **model** what happens, over time, to the particles inside the bottle and **take photos** of each sketch they make. [O3]

 © HarperCollins*Publishers* Limited 2016

Explain

- Each group **presents** their series of sketches to the class to explain why the coin jumped, focusing on 'collisions' between the particles themselves and the walls of the container. [O2]
- Ask students to **predict** what will happen when you demonstrate, with a Bourdon gauge or data logger pressure gauge, what happens to the air inside a stoppered test tube that has been immersed in ice. [O1, O2, O3]
- Students **plot** values of pressure against temperature in real time, following Practical sheet 1.12. [O1, O2, O3]
    - High demand: Students should **develop** the relationship mathematically and determine a value for the gradient of the graph. [O1]

Consolidate and apply

- Students **multiply** any corresponding points of pressure and temperature to **show** that the answer is always the same, using Worksheet 1.12. [O1]
- Students **compare** videos of earth-based particle models with a Tim Peake micro-gravity model. [O2]
- Students answer questions from Section 1.12 of the Student Book. [O1, O2, O3]

Extend

If time permits, ask students able to progress further to do the following:

- Warm the round-bottomed flask set-up described in Technician's notes 1.12. **Explain** (orally) what is happening to the particle motion now that there is no fixed volume. Focus on the collisions between the particles themselves and the walls of the container. [O2, O3]

Plenary suggestions

Play some **speed dating** by placing half the class as experts, the other as questioners and give them thirty seconds before rotating to another partner, asking questions based on this lesson. Any questions that are too difficult to be answered can be shared at the end of the lesson.

 © HarperCollins*Publishers* Limited 2016

Lesson 1.13: Increasing the pressure of a gas

Lesson overview

OCR Specification reference

P1.3d, e

Learning objectives

- Describe the relationship between the pressure and volume of a gas at constant temperature.
- Calculate the change in the pressure or volume of a gas held at constant temperature when either the pressure or volume is increased or decreased.
- Explain how doing work on a gas can increase its temperature.

Learning outcomes

- [O1] Predict how volume changes as the result of an increase or decrease of pressure in a gas.
- [O2] Understand that the volume of a gas is reduced by increasing the pressure for a system with constant temperature.
- [O3] Refer to particle collisions when describing how an increase in temperature is caused by mechanical work.

Skills development

- Think scientifically: analyse data.
- Work scientifically: present evidence.
- Learner development: collaborate effectively.

Maths focus

- Calculate an average.
- Draw graphs.

Resources needed 30 ml syringe, wooden blocks (optional), clamp, clamp stand, boss head, 0.5 kg masses, board, balloon, bucket (or waste-paper basket), lid, water pistol with pump; Worksheet 1.13, Practical sheet 1.13, Technician's notes 1.13

Digital resources projector, access to PhET website

Key vocabulary temperature, pressure, collide, compressed

Teaching and learning

Engage

- Ask: "Can one balloon support your bodyweight?" Give students twenty seconds' **thinking** time before demonstrating placing a balloon in a suitably sized bucket or waste-paper bin and covering it with a lid. Ask a volunteer to place a foot on the lid and slowly add his or her bodyweight (using the support of a hand or nearby wall). [O1, O2]
- Project (students could use their own computers or tablets) the simulation to alter pressure and volume (with the temperature parameter set constant). Ask students to **predict** how the rate of collisions with the walls of the container will change before starting. [O2]

Challenge and develop

- Demonstrate the investigation. Particular attention should be paid to how you will be asking students to load the masses and maintain good balance on the plunger (Practical sheet 1.13 and Technician's notes 1.13). On unloading, the board must be tapped to bring it to its equilibrium position. [O2]

 © HarperCollins*Publishers* Limited 2016

- Students should **collect** data and explain which is the dependent and which the independent variable for this investigation. Though the trend will be obvious, the quality of the data will rely on the quality and lubricant of the syringes that are used. [O1, O2]
- Students will **present** their data in the form of a graph. [O2]
    - High demand: Students will change their plots to **explore** an inversely proportional relationship. [O2]

Explain

- Each group will **state** and **explain** the relationship, emphasising an '… er' and '… er' finding (for example, the larg*er* the force or pressure, the small*er* the volume). [O2]
- Students **predict** whether the curve would touch either axis if it could be extrapolated. [O1, O2, O3]
    - High demand: Students should **quantify** the gradient of their graph. [O1]

Consolidate and apply

- Students **multiply** any corresponding points of pressure and volume to **show** that the answer is always the same, using Worksheet 1.13. [O1, O2]
- Use the aforementioned PhET simulation (with the temperature parameter not set to constant) to discuss, in **talk-triads**, the role of mechanical work in increasing temperature. Two students discuss this while a third listens. The listener is responsible for reporting the discussion to the class. [O3]

Extend

Ask students able to progress further to do the following:

- **Explain**, with relation to mechanical work, why the gas release from an aerosol can gets colder (due to the gas doing the work itself). [O2, O3]

Plenary suggestions

Squirt some water from a water pistol of the type that must be pumped beforehand. Ask students to draw a large circle in their books and write in it one thing they have learned from today's lesson, which explains the water pistol firing mechanism, and any question they would like to ask about something they are unsure of.

 © HarperCollins*Publishers* Limited 2016

Lesson 1.14: Pressure in a fluid

Lesson overview

OCR Specification reference

P1.3h, I, j

Learning objectives

- Explain how pressure acts in a fluid.
- Calculate pressure at different depths in a liquid (higher tier).
- Explain what causes upthrust (higher tier).

Learning outcomes

- [O1] State that pressure increases with depth.
- [O2] Calculate pressure.
- [O3] Explain why some objects float and others sink.

Maths focus

- Rearrange equations.
- Use the correct units for pressure in a fluid and density.
- Carry out a practical to obtain accurate results.

Resources needed tank of water or 2 litre pop bottle, equipment listed in Technician's notes; Worksheets 1.14.1, 1.14.2 and 1.14.3, Practical sheet 1.14, Technician's notes 1.14

Key vocabulary upthrust, incompressible, pressure, density, pascal (Pa)

Teaching and learning

Engage

Show students an image of a scuba diver and ask them why scuba divers must ascend slowly to avoid 'the bends'.

Challenge and develop

- To demonstrate the increasing pressure with depth, show students a water bottle with holes punched through at different depths. The angle at which the water comes out should then be linked to the pressure.
- The students could be asked to place their hands under the water as it emerges from the bottle; they should be able to feel the difference in pressure.
- Students could then return to the original question and improve their answers. [O1]

Explain

- Students should now be introduced to the idea of calculating the pressure exerted by different volumes of water. Show students a tank of water of known depth, and ask the students to think of ways to calculate the pressure at the bottom of the tank due to the water. They should discuss their ideas before feeding back, developing their ideas further. They might need prompting to think about factors that will affect the pressure, such as the force due to gravity. [O1, O2]
- The equation pressure = $h\rho g$ should be introduced before calculating the pressure at the bottom of the tank. [O1, O2]
- If no tank is available, a 2 litre pop bottle would be a suitable substitute.

Consolidate and apply

- Students should now investigate floating and sinking (Practical sheet 1.14). [O3]

 © HarperCollinsPublishers Limited 2016

- They should be told that the density of water is 1 g/cm^3. They should recall how to calculate the density of the objects.
- Students should be encouraged to state a rule (from their experimental results) to determine whether an object will float or sink in water. [O3]
- Issue each student with a pressure in a fluid worksheet at an appropriate level:
  - Low demand: Worksheet 1.14.1 [O1, O2, O3]
  - Standard demand: Worksheet 1.14.2 [O1, O2, O3]
  - High demand: Worksheet 1.14.3 [O1, O2, O3]

Extend

Students could be asked to explain how a hydraulic jack operates. To help, they could be given the equation $\frac{\text{large force}}{\text{large area}} = \frac{\text{small force}}{\text{small area}}$

Plenary suggestions

Ask me a question: Ask students to write a question about something from the topic and then a mark scheme for the answer. Encourage them to come up with questions worth more than one or two marks and to try their questions out on each other.

Where's the answer? As students enter the room, give them a card with a word written on it. At the end of the lesson, ask a question, and the students should say if they think they have the answer card and explain why.

Answers to questions

Worksheets 1.14.1, 1.14.2 and 1.14.3

Density (g/cm^3)	Mass (g)	Volume (cm^3)	Float or sink in water?
2	10	5	Sink
0.5	5	10	Float
0.2	4	20	Float
5	15	3	Sink

1. 40 000 Pa
2. 61 740 Pa
3. 4000 Pa
4. The same; because $p = h\rho g$, all quantities are the same in both situations so the pressure will be the same.

 © HarperCollinsPublishers Limited 2016

Lesson 1.15: Atmospheric pressure

Lesson overview

OCR Specification reference

P1.3f, g

Learning objectives

- Show that the atmosphere exerts a pressure.
- Explain variations in atmospheric pressure with height.
- Describe a simple model of the Earth's atmosphere and atmospheric pressure.

Learning outcomes

- [O1] Describe a demonstration that shows the effect of atmospheric pressure.
- [O2] Plot a graph of atmospheric pressure against height.
- [O3] Research the effect of changing air pressure on the body.

Maths focus

- Use the correct units for atmospheric pressure.

Resources needed drinks can, water, Bunsen burner, tray of cold water; Worksheets 1.15.1, 1.15.2 and 1.15.3

Digital resources image of oxygen tent such as those used by climbers when scaling Mount Everest, access to online research materials

Key vocabulary atmospheric pressure, barometer, atmosphere

Teaching and learning

Engage

- Show students an image of an oxygen tent such as those used by climbers when scaling Mount Everest. Students should discuss in pairs why these are used.
- Take feedback from a number of pairs.

Challenge and develop

- To show the high pressure the atmosphere exerts, demonstrate a drinks can collapsing when air is removed. To do this, take a drinks can and add around 10 cm^3 of water. Heat strongly over a Bunsen burner to ensure that all the air in the can is replaced by steam. Invert the can as quickly as possible into a tray of cold water to condense the steam (essentially creating a vacuum). At this point, the atmospheric pressure is now far greater than the pressure inside the can, resulting in the can collapsing. During the demonstration, be sure to discuss what is causing the pressure in different places. [O1]
- Follow this demonstration by asking the students to discuss why we are not crushed by air pressure.

Explain

- Give students data with which to plot a graph of height (*x*-axis) against air pressure (*y*-axis). Suitable data:

Height (m)	Air pressure (kPa)
0	100
2000	75
5000	50
10 000	25
15 000	10

 © HarperCollins*Publishers* Limited 2016

- The students should then return to the question from the start of the lesson and, given that Mount Everest is 8850 m high, they should estimate the air pressure at the top of the mountain. They should use this information to improve their answer to the original question. [O2]

Consolidate and apply

Issue each student with an atmospheric pressure worksheet at an appropriate level:

- Low demand: Worksheet 1.15.1 [O3]
- Standard demand: Worksheet 1.15.2 [O3]
- High demand: Worksheet 1.15.3 [O3]

Extend

- Students could be asked to explain how a mercury barometer works.
- Very strong candidates could be introduced to the barometric equation.

Plenary suggestions

Ideas hothouse: Ask students to work in pairs to list points about what they know about a particular idea. Then ask the pairs to join together into groups of four and then groups of six to eight to discuss this further and to come up with an agreed list of points. Ask one person from each group to report back to the class.

The big ideas: Ask students to write down three ideas they have learned during the topic. Then ask them to share their facts in groups and to compile a master list of facts, with the most important at the top. Ask for ideas to be shared and find out which other group(s) agreed.

 © HarperCollins*Publishers* Limited 2016

Lesson 1.16: Maths skills: Handling data

Lesson overview

OCR Specification reference

M2 c, f

Learning objectives

- Recognise the difference between mean, mode and median.
- Explain the use of tables and frequency tables.
- Explain when to use scatter diagrams, bar charts and histograms.

Learning outcomes

- [O1] State why data is shown in charts or graphs.
- [O2] Collect categoric and continuous data and draw suitable charts or graphs from their data.
- [O3] Distinguish between mean, mode and median and decide which to use in a given situation.

Skills development

- Collect and display data appropriately.

Maths focus

- Use the terms 'mean', 'mode' and 'median' correctly.
- Plot bar charts, histograms and scatter diagrams.

Resources needed a selection of tables, charts and graphs from the week's news, either cut from newspapers, or found online, at least four different types of ball (for example, tennis ball, table tennis ball, plastic ball, foam ball), metre rulers; Worksheets 1.16.1, 1.16.2 and 1.16.3

Digital resources selection of tables, charts and graphs from the week's news, either clipped from newspapers or found online

Key vocabulary continuous data, frequency table, scatter diagram, correlation, line of best fit, bar chart, histogram, independent variable, dependent variable, mean, median, mode, anomalous, outlier

Teaching and learning

Engage

- Show students a selection of tables, charts and graphs from the week's news, either clipped from a number of different newspapers or found online. Pose a series of questions to the students:
  - What do these show?
  - Why have they been drawn this way?
  - Why are tabulated data often also shown in a chart or a graph? [O1]

Challenge and develop

- The students should investigate the bouncing of a ball in two different ways:
  - Using different balls dropped from the same height and measuring the height of the first bounce. [O2]
  - Using a single ball, measuring the height of the first bounce when dropped from different starting heights. [O2]

Explain

- Remind students of the distinction between categoric and continuous data and how these types of data are displayed on charts and graphs.

 © HarperCollins*Publishers* Limited 2016

- The students should now take the data they have collected from their investigations to draw suitable charts or graphs. [O2]

Consolidate and apply

- Issue students with a Maths skills worksheet at an appropriate level. These worksheets include the consideration of the mean, mode and median.
  - Low demand: Worksheet 1.16.1
  - Standard demand: Worksheet 1.16.2
  - High demand: Worksheet 1.16.3 [O2, O3]

Extend

Ask students able to progress further to suggest (with reasons), in each of the following situations, which type of average would be the most appropriate to use:

- a retailer wanting to decide which products to restock (mode)
- a broadcaster wanting to investigate how many hours per day teenagers spend watching TV (mean)
- a group of entrepreneurs looking to start a business with a fixed salary budget, needing to decide how many staff they can employ and at what rates (median).

Plenary suggestions

What do I know? Ask students to each write down one thing about the Matter topic they are sure of, one thing they are unsure of and one thing they need to know more about, being specific. Ask them to work in groups of five or six to agree on group lists. Ask each group to say what they decided and agree as a class about what they are confident about, what they are less sure of and what things they want to know more about.

The big ideas: Ask students to write down three ideas they have learned during the Matter topic. Then ask them to share their facts in groups and to compile a master list of facts, with the most important at the top. Ask for ideas to be shared and find out which other group(s) agreed.

Heads and tails: Ask each student to write a question about something from the Matter topic so far on a coloured paper strip and the answer on another coloured strip. Hand out the strips to groups of six to eight students, so that each student gets a question and an answer. One student reads out his or her question. The student with the correct answer then reads it out, followed by his or her question, and so on.

Answers to questions

Worksheet 1.16.1

1. mean: £12 000; median: £10 000; mode: £10 000
2. mean: £16 000; median: £12 500; mode: £10 000
3. mean and median
4. mean: increased by £4000; median: increased by £2500

Worksheet 1.16.2

1. mean: £12 000; median: £10 000; mode: £10 000
2. mean: £16 000; median: £12 500; mode: £10 000
3. mean and median
4. mean: increased by £4000; median: increased by £2500
5. mode: hasn't changed; median: increased; mean: increased more

Worksheet 1.16.3

1. mean: £12 000; median: £10 000; mode: £10 000
2. mean: £16 000; median: £12 500; mode: £10 000
3. mean and median
4. mean: increased by £4000; median: increased by £2500
5. mode: hasn't changed; median: increased; mean: increased more
6. median: increased because a higher value has been added to the dataset; mean: increased more because the new value is an outlier at the higher end of the dataset

 © HarperCollins*Publishers* Limited 2016

When and how to use these pages: Check your progress, Worked example and End of chapter test

Check your progress

Check your progress is a summary of what students should know and be able to do when they have completed the chapter. Check your progress is organised in three columns to show how ideas and skills progress in sophistication. Students aiming for top grades need to have mastered all the skills and ideas articulated in the final column (shaded pink in the Student book).

Check your progress can be used for individual or class revision using any combination of the suggestions below:

- Ask students to construct a mind map linking the points in Check your progress
- Work through Check your progress as a class and note the points that need further discussion
- Ask the students to tick the boxes on the Check your progress worksheet (Teacher Pack CD). Any points they have not been confident to tick they should revisit in the Student Book.
- Ask students to do further research on the different points listed in Check your progress
- Students work in pairs and ask each other what points they think they can do and why they think they can do those, and not others

Worked example

The worked example talks students through a series of exam-style questions. Sample student answers are provided, which are annotated to show how they could be improved.

- Give students the Worked example worksheet (Teacher Pack CD). The annotation boxes on this are blank. Ask students to discuss and write their own improvements before reviewing the annotated Worked example in the Student Book. This can be done as an individual, group or class activity.

End of chapter test

The End of chapter test gives students the opportunity to practice answering the different types of questions that they will encounter in their final exams. You can use the Marking grid provided in this Teacher Pack or on the CD Rom to analyse results. This shows the Assessment Objective for each question, so you can review trends and see individual student and class performance in answering questions for the different Assessment Objectives and to highlight areas for improvement.

- Questions could be used as a test once you have completed the chapter
- Questions could be worked through as part of a revision lesson
- Ask Students to mark each other's work and then talk through the mark scheme provided
- As a class, make a list of questions that most students did not get right. Work through these as a class.

 © HarperCollins*Publishers* Limited 2016

Marking Grid for End of Chapter 1 Test

	Getting Started [Foundation Tier]								Going Further [Foundation and Higher Tiers]					More Challenging [Higher Tier]			Most demanding [Higher Tier]					
Student Name	Q. 1 (AO1) 1 mark	Q. 2 (AO1) 2 marks	Q. 3 (AO1) 1 mark	Q. 4 (AO1) 1 mark	Q. 5 (AO1) 1 mark	Q. 6 (AO2) 1 mark	Q. 7 (AO2) 2 marks	Q. 8 (AO2) 1 mark	Q. 9 (AO1) 1 mark	Q. 10 (AO1) 1 mark	Q. 11 (AO1) 2 marks	Q. 12 (AO2) 4 marks	Q. 13 (AO2) 2 marks	Q. 14 (AO1) 2 marks	Q. 15 (AO1) 2 marks	Q. 16 (AO2) 6 marks	Q. 17 (AO2) 2 marks	Q. 18 (AO2) 2 marks	Q. 19 (AO3) 6 marks	Q. 20 (AO3) 2 marks	**Total marks**	**Percentage**

 © HarperCollins*Publishers* Limited 2016

Check your progress

You should be able to:

Use density = mass/volume to calculate density	Use particle diagrams to communicate ideas about relative densities of different states Use the density equation to calculate mass and volume	Link the particle model for solids, liquids and gases with density values in terms of the arrangements of the atoms or molecules
Describe changes of state as physical changes	Describe how mass is conserved when substances change state Explain that changes of state are physical, not chemical, changes because the material recovers its original properties if the change is reversed	Explain that changes of state conserve mass
Describe how heating raises the temperature of a system	Describe that heating raises the temperature or changes the state of a system but not at the same time	Explain that internal energy is the total kinetic energy and potential energy of all the particles that make up a system
Describe the effect of an increase in temperature on the motion of the particles	Use the specific heat capacity equation to calculate the energy required to change the temperature of a certain mass of a substance	Use the specific heat capacity equation to calculate mass, specific heat capacity or temperature change
State that when an object changes state there is no change in temperature	Describe the latent heats of fusion and of vaporisation Use the equation $E = mL$	Use the particle model to explain why the latent heat of vaporisation is much larger than the latent heat of fusion
In the particle model the higher the temperature the faster the molecules move.	Use the particle model to explain the effect on temperature of increasing the pressure of a gas at constant volume	Describe that the temperature of a gas is related to the average kinetic energy of the molecules
Recall that gases can be compressed or expanded by pressure changes	Use the equation pV = constant to calculate the pressure or volume of a gas at constant temperature	Use the particle model to explain that increasing the volume of a gas, at constant temperature, can lead to a decrease in pressure
Describe how a fluid exerts a pressure on a surface Describe how pressure varies with depth in a fluid	Calculate pressure at any depth in a fluid and explain what causes atmospheric pressure	Explain how a partially (or totally) submerged object experiences upthrust and why atmospheric pressure decreases with height Describe the factors which influence floating and sinking

 © HarperCollins*Publishers* Limited 2016

Worked example

1 **Alex is heating a beaker of water by the rays of a sunlamp. The energy transferred was 21000 J. The time taken for the water to increase by 10 °C was 6000 seconds. Calculate the power supplied by the sunlamp.**

3.5

This is the correct numerical answer, but

a You have not given any units to the number. Always give the unit to the quantity.

b You have not shown any working. When working a calculation; write the equation, and show step by step how you do the calculation.

2 **Explain how raising the temperature of a gas, keeping the volume constant, increases the pressure exerted by the gas.**

The high temperature makes the molecules vibrate faster so there is a stronger force on the side of the container.

You are on the right lines. It is the force of the molecules hitting the side of the container that creates the pressure. With a gas, the molecules are no longer vibrating but they are free to move around at speed. Heating increases their kinetic energy.

3 **What does it mean to say that the changes of state are reversible?**

It means that the changes go both ways.

Correct, but you should give an example as well. Give an example of a change that goes both ways, e.g. water turns to steam and steam turns back to water.

4 **Explain what heating does to the energy stores of a system.**

Heating raises the total energy inside the system.

Yes it does, but again you should aim to give more specific information about the system. You need to mention the energy transfers that have taken place.

© HarperCollins*Publishers* Limited 2016

Forces: Introduction

When and how to use these pages

This unit will build on ideas the students have met before, such as:

- The average speed is the distance travelled divided by the time taken.
- If the speed of a car is increasing, it is accelerating.
- A journey can be represented by a distance–time graph.
- Forces cause acceleration.
- The bigger the force, the bigger the change in speed.
- Forces can be contact or non-contact.

Overview of the unit

In this unit, students will learn about what forces are and what they do, contact and non-contact forces, how force can cause acceleration (and deceleration), how motion can be calculated, Newton's three laws of motion, the turning effect of a force and pressure in solids, liquids and gases.

This unit offers a number of opportunities for students to work collaboratively to carry out investigations, draw conclusions from their investigations, perform calculations to model motion and refine their graph-plotting skills.

Obstacles to learning

Common misconceptions held by students:

- Keeping going requires a force.
- A lack of an idea about momentum – would you rather be hit by a bus or a bicycle? – their instinct may well be 'bicycle'.
- If an object is resting on a table, there are no forces acting on it.
- Friction occurs only between solid objects.
- Things fall naturally.
- Gravity affects only heavy things.

Practicals in this unit

In this unit, students will do the following practical work:

- investigate the motion of a ball
- investigate acceleration and displacement
- investigate Newton's first law
- investigate Newton's second law
- Practical: investigate the acceleration of an object
- simulate meteorites striking the Earth
- investigate and model power
- investigate the motion of an object in a circle
- understand the relationship between the force on a spring and its extension
- Practical: investigate the relationship between the force on a spring and its extension
- investigate transfers between gravitational energy and elastic potential energy
- investigate the link between mass and weight

 © HarperCollins*Publishers* Limited 2016

- investigate turning forces
- investigate linked gearwheels and their ratios.

	Lesson title	Overarching objectives
1	Scalars and vectors	• Define distance, displacement, speed, velocity and acceleration. • Recognise the difference between scalar and vector quantities. • State examples of scalar and vector quantities.
2	Speed	• Calculate speed using distance travelled divided by time taken. • Calculate speed from a distance–time graph. • Measure the gradient of a distance–time graph at any point.
3	Acceleration	• Describe acceleration. • Calculate acceleration. • [Higher tier] Explain motion in a circle.
4	Calculations of motion	• Describe uniform motion. • Use an equation for uniform motion. • Apply this equation to vertical motion.
5	Velocity–time graphs	• Draw velocity–time graphs. • Calculate acceleration using a velocity–time graph. • [Higher tier] Calculate displacement using a velocity–time graph.
6	Maths skills: Making estimates of calculations	• Estimate the results of simple calculations. • Round numbers to make an estimate. • Calculate order of magnitude.
7	Forces explain how objects interact	• Describe a force. • Recognise the difference between contact and non-contact forces. • State examples of scalar and vector quantities.
8	Forces and motion	• Understand what a force does. • Explain what happens to an object if all the forces acting on it cancel each other out. • Analyse how this applies to everyday situations.
9	Resultant forces	• Calculate the resultant of a number of forces. • Draw free-body diagrams to find resultant forces. • [Higher tier] Understand that a force can be resolved into two components acting at right angles to each other.
10	Forces and acceleration	• Explain what happens to the motion of an object when the resultant force is not zero. • Analyse situations in which a non-zero resultant force is acting. • Explain what inertia is.

 © HarperCollins*Publishers* Limited 2016

11	Momentum	• Explain what is meant by momentum. • Apply ideas about the rate of change of momentum to safety features in cars. • Use momentum calculations to predict what happens in a collision.
12	Practical: Investigating the acceleration of an object	• Plan an investigation to explore an idea. • Analyse results to identify patterns and draw conclusions. • Compare results with scientific theory.
13	Newton's third law	• Identify force pairs. • Understand and be able to apply Newton's third law.
14	Work done and energy transfer	• Understand what is meant by work done. • Explain the relationship between work done and force applied. • Identify the transfers between energy stores when work is done against friction.
15	Understanding power	• Define power. • Compare the rate of energy transfer by various machines and electrical appliances. • Calculate power.
16	Key concept: Forces and acceleration	• Recognise examples of balanced and unbalanced forces. • Apply ideas about speed and acceleration to explain sensations of movement. • Apply ideas about inertia and circular motion to explain braking and cornering.
17	Forces and energy in springs	• Explain why you need two forces to stretch a spring. • Describe the difference between elastic and inelastic deformation. • Calculate extension, compression and elastic potential energy.
18	Practical: Investigate the relationship between force and the extension of a spring	• Interpret readings to show patterns and trends. • Interpret graphs to form conclusions. • Apply the equation for a straight line to the graph.
19	Potential energy	• Consider what happens when a spring is stretched. • Describe what is meant by gravitational potential energy. • Calculate the energy stored by an object raised above ground level.
20	Heavy or massive?	• Identify the correct units for mass and weight. • Explain the difference between mass and weight. • Understand how weight is an effect of gravitational fields.
21	Moments	• Describe the turning effect of a force about a pivot. • Explain and use the principle of moments. • Explain what is meant by the centre of mass of an object.
22	Levers and gears	• Describe how levers and gears can be used to transmit the rotational effect of a force. • Explain how levers and gears transmit force.

 © HarperCollins*Publishers* Limited 2016

Lesson 2.1: Scalars and vectors

Lesson overview

OCR Specification reference

P2.1.2c, d

Learning objectives

- Define distance, displacement, speed, velocity and acceleration.
- Recognise the difference between scalar and vector quantities.
- State examples of scalar and vector quantities.

Learning outcomes

- [O1] Define distance, displacement, speed, velocity and acceleration.
- [O2] Match quantities to their correct SI unit.
- [O3] Identify both scalar and vector quantities.

Skills development

- Correct use of key terms and units.

Maths focus

- Use of correct units for quantities.

Resources needed map of the local town; Worksheets 2.1.1, 2.1.2 and 2.1.3

Key vocabulary distance, speed, displacement, velocity, acceleration, scalar, vector

Teaching and learning

Engage

- As the students enter the room, issue them with a small piece of paper. On one side of the paper, they should write what they understand the following terms to mean: distance, speed, acceleration and velocity. On the other side of the paper, they should write the units in which they think these quantities are measured. They may well come up with a variety, particularly for distance and speed, so the convention of using SI units could be introduced here. [O1, O2]

Challenge and develop

- Ask the students to share their definitions with a partner, before joining into groups of four to develop ideas further. Whilst they are working in a group of four, they should be asked the following question: "What is the difference between speed and velocity?". They should discuss any ideas they have between them, before feeding back to the class. [O1]
- If possible, issue the students with a map of the local town. Choose two locations on the map and ask them to work out the distance between the two points (i) 'as the crow flies' and (ii) following roads / paths. Once they have done this, introduce the idea of 'displacement' and 'distance' [O1, O3]

Explain

- Show the students an image of a car or a motorbike travelling round a roundabout. State that the circumference of the roundabout is 150 m. Ask the students (i) if the vehicle goes once round the roundabout, what is the distance travelled and the displacement? (ii) if the vehicle is travelling at a constant speed, how does the velocity differ from the speed? (iii) as acceleration is the change in velocity / time, is the vehicle accelerating? These questions should lead into a discussion about the difference between a scalar and a vector quantity. [O3]

 © HarperCollins*Publishers* Limited 2016

- Following this discussion, write a list of quantities (not necessarily related to forces) on the board and issue the students with two cards of different colours. As you read down the list of quantities, the students should use their cards to 'vote' for whether they think the quantity is a scalar or a vector quantity. If they are finding it difficult to decide, a useful question to pose is "Does the direction make a difference?" [O3]

Consolidate and apply

- The students should then be issued with the scalars and vectors worksheet suitable to their ability:
  - Low demand – Worksheet 2.1.1 [O1, O2, O3]
  - Standard demand – Worksheet 2.1.2 [O1, O2, O3]
  - High demand – Worksheet 2.1.3 [O1, O2, O3]

Extend

- Ask students to design a one-page summary sheet that could be given to students who are finding classifying quantities as scalars or vectors difficult. [O1, O3]

Plenary suggestions

Learning triangle: At the end of a lesson, ask students to draw a large triangle with a smaller inverted triangle that just fits inside it (so they have four triangles). Ask students to think back over the lesson and identify and write in the outer triangles:

- something they saw
- something they did
- something they discussed.

Ask students to write in the central triangle something they learned.

What do I know? Ask students to each write down one thing about the topic they are sure of, one thing they are unsure of and one thing they need to know more about, being specific. Ask them to work in groups of five or six to agree on group lists. Ask each group to say what they decided and agree as a class what they are confident about, what they are less sure of and what things they want to know more about.

Heads and tails: Ask each student to write a question about something from the lesson on a coloured paper strip and the answer to the question on another colour. In groups of six to eight, hand out the strips so that each student gets a question and an answer. One student reads out their question. The student with the right answer then reads it out, followed by their question, and so on.

Answers to questions

Worksheets 2.1.1, 2.1.2 and 2.1.3

Quantities and units

Distance – m; speed – m/s; velocity – m/s; acceleration – m/s^2; displacement – m

Scalars or vectors?

Energy – S; Force – V; Temperature – S; Mass – S; Acceleration – V; Distance – S

Representing vectors

1. Suggest a 1 cm arrow pointing due East.
2. Suggest a 2 cm arrow pointing due South.
3. Suggest a 1.5 cm arrow at 35°.
4. Suggest a 10 cm arrow at 65° (*Worksheet 2.1.3 only*).

 © HarperCollins*Publishers* Limited 2016

Lesson 2.2: Speed

Lesson overview

OCR Specification reference

P2.1a, b, c, e (part), g

Learning objectives

- Calculate speed using distance travelled divided by time taken.
- Calculate speed from a distance–time graph.
- Measure the gradient of a distance–time graph at any point.

Learning outcomes

- [O1] Plot a distance–time graph.
- [O2] Calculate speed when given distance and time.
- [O3] Use a distance–time graph to calculate speed at different points in a journey.

Skills development

- Manipulate an equation with three variables.
- Plot a set of data onto a graph.
- Use a graph to calculate additional information.

Maths focus

- Plot data.
- Calculate the gradient from a graph.

Resources needed apparatus listed in Technician's notes, graph paper; Worksheets 2.2.1, 2.2.2 and 2.2.3, Practical sheet 2.2, Technician's notes 2.2

Key vocabulary distance, time, speed, gradient, average speed, distance–time graph, tangent

Teaching and learning

Engage

The Hare and the Tortoise

- **Pair talk:** Issue students with a series of cards illustrating a variety of objects and animals. The students should arrange these cards in order of speed, from slowest to fastest. Examples could include a beetle, a person walking, a person riding a bicycle, a double-decker bus, a bird flying, a cheetah and a missile. [O2]

Challenge and develop

Investigating motion

- Practical 2.2: Students could generate distance–time data by rolling a tennis ball across the room or down the corridor. [O1]
- They should record the data for the journey in a shared table.

Explain

- Students should now plot the distance–time data onto a graph. [O1]
- A discussion could follow: why do we show data on a graph as opposed to leaving it in a table?
- Ask students: which part of the graph represents where the journey was the fastest? What about the slowest? How did you decide? [O3]

 © HarperCollins*Publishers* Limited 2016

Consolidate and apply

Representing motion

- **Pair talk:** Bridging question: What do we mean by 'speed'?
- Remind students of speed = distance ÷ time (examples could be given here, if necessary) and discuss how this relates to how they decided from their graph which parts of the journey were the fastest and which were the slowest. [O2]
- Students should then calculate the speed at different points in the journey from their graph. [O3]
- Students should then complete one of the following three worksheets to consolidate this work:
  - Low demand: Worksheet 2.2.1 (label the parts of a distance–time graph given a story) [O1]
  - Standard demand: Worksheet 2.2.2 (plot a distance–time graph given a story and a table of data and calculate the speed for different parts of the journey) [O1]
  - High demand: Worksheet 2.2.3 (plot a distance–time graph from a story and calculate the speed for different parts of the journey). [O1]
- Consider how a speed camera works.
- Discuss in pairs, then combine pairs into fours, then appoint a spokesperson: "How does a speed camera work, given that it cannot measure speed, it can only take photographs?" [O2]

Extend

Students could plot a distance–time graph for their journeys to school. They could then use their graphs to calculate the speed during the different parts of their journey. [O1, O3]

Plenary suggestions

Ask me a question: Ask students to write a question about something from the topic and then a mark scheme for the answer. Encourage them to come up with questions worth more than one or two marks and to try their questions out on each other.

Hot seat: Ask each student to think of a question, using material from the topic. Select someone to put in the hot seat; ask students to ask their questions and say at the end whether the answer is correct or incorrect.

Answers to questions

Worksheets 2.2.1, 2.2.2 and 2.2.3

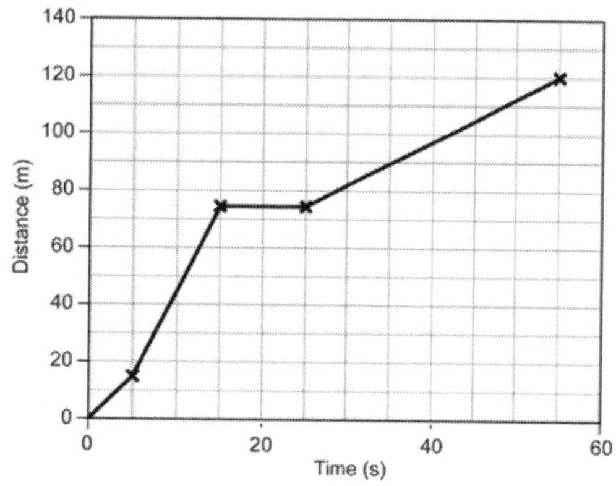

Data for graph

Time interval	Speed
0–5 s	3 m/s
5–15 s	6 m/s
15–25 s	0 m/s
25–55 s	1.5 m/s

 © HarperCollins*Publishers* Limited 2016

Lesson 2.3: Acceleration

Lesson overview

OCR Specification reference

P2.1h (part)

Learning objectives

- Describe acceleration.
- Calculate acceleration.
- [Higher tier] Explain motion in a circle.

Learning outcomes

- [O1] Link velocity to acceleration.
- [O2] Calculate acceleration from given data.
- [O3, higher tier] Explain why an object travelling at constant speed in a circle is accelerating.

Skills development

- Model Galileo's experiment.
- Calculate acceleration given initial and final velocities and time taken.
- Convert from kilometres to metres and from minutes to seconds.

Maths focus

- Manipulate an equation with three variables.
- Convert units.

Resources needed tennis balls, scrunched-up paper; Worksheets 2.3.1, 2.3.2 and 2.3.3

Key vocabulary gravity, air resistance, drag, acceleration, deceleration

Teaching and learning

Engage

Demonstration or pairs activity: Model the Galileo experiment by dropping a tennis ball and some scrunched-up paper, which should be as close as possible to the same volume as the ball. What can students deduce about the acceleration of the two objects? [O1]

Challenge and develop

Open out the piece of paper and drop the ball and the paper again. Ask students to identify what has changed about the situation and what effect this has on the velocity of the piece of paper. If the acceleration of the two objects in the first situation was the same, is it still the same in this situation? [Clearly not.] What must acceleration depend on [velocity] and what must it be independent of [mass]? [O1]

Explain

- Using the change of velocity of the piece of paper in both situations, introduce the equation for calculating acceleration as the change in velocity divided by the time taken. Be sure to highlight that the initial velocity is zero. [O1, O2]
- [Higher tier only] When the equation for acceleration has been established, ask the students why, if a car is travelling at a constant speed around a roundabout, it is accelerating. The students should be able to recall that velocity is a vector quantity and so, as the direction is constantly changing, the acceleration must be changing, despite the constant speed. [O3]

Consolidate and apply

Issue students an acceleration worksheet at an appropriate level:

 © HarperCollins*Publishers* Limited 2016

- Low demand: Worksheet 2.3.1 (the three equations are given) [O1, O2]
- Standard demand: Worksheet 2.3.2 (the equation acceleration = change in velocity ÷ time is given) [O1, O2]
- High demand: Worksheet 2.3.3 (the equation acceleration = change in velocity ÷ time is given, and mixed units are used) [O1, O2, O3]

Extend

Students could be asked to suggest, using ideas about acceleration (and deceleration), why there is a cushioned buffer at the end of the 60 m sprint at indoor athletics competitions.

Plenary suggestions

The big ideas: Ask students to write down three ideas they have learned during the last three lessons. Then ask them to share their facts in groups and to compile a master list of facts, with the most important at the top. Ask for ideas to be shared and find out which other group(s) agreed.

Ideas hothouse: Ask students to work in pairs to list points about what they know about the topic. Then ask the pairs to join together into groups of four and then groups of six to eight to discuss this further and to come up with an agreed list of points. Ask one person from each group to report back to the class.

Answers to questions

Worksheet 2.3.1

1. 2.0 m/s^2
2. 4.14 m/s^2
3. 10.3 m/s^2

Worksheets 2.3.2 and 2.3.3

1. 2.0 m/s^2
2. 7 s
3. 27.8 m/s

 © HarperCollins*Publishers* Limited 2016

Lesson 2.4: Calculations of motion

Lesson overview

OCR Specification reference

P2.1h (part)

Learning objectives

- Describe uniform motion.
- Use an equation for uniform motion.
- Apply this equation to vertical motion.

Learning outcomes

- [O1] Calculate acceleration and displacement from a velocity–time graph.
- [O2] Use a velocity–time graph to derive two equations of motion.
- [O3] Apply the equations of motion to situations including vertical motion.

Skills development

- Manipulate an equation.
- Plot a set of data onto a graph.
- Use a graph to calculate additional information.

Maths focus

- Plot data.
- Calculate the gradient from a graph.
- Calculate the area under a graph.

Resources needed graph paper; Worksheets 2.4.1, 2.4.2, 2.4.3, 2.4.4, 2.4.5 and 2.4.6

Key vocabulary velocity, time, speed, gradient, area, acceleration, velocity–time graph, uniform motion, vertical

Teaching and learning

Engage

- Ask students to plot a sketch velocity–time graph of the following situation: 'A car passes Grace, travelling at a speed of 10 m/s. Next, 15 s later, the car passes Bradley, travelling at a speed of 20 m/s. The car is speeding up evenly.' [O1]
- On the velocity axis, the students should label the initial velocity '*u*' and the final velocity '*v*'.

Challenge and develop

- The students should now answer the questions "How far did the car travel?" and "What was the acceleration of the car?" by calculating the area under the graph and the gradient, respectively. [O1]
- Discuss with students how they can tell from a velocity–time graph if the acceleration is constant [the graph gives a straight line with a uniform gradient]. Introduce the term 'uniform'.

Explain

The students should now be given a calculations of motion worksheet at an appropriate level:

- Low demand: Worksheet 2.4.1 [O2]
- Standard demand: Worksheet 2.4.2 [O2]
- High demand: Worksheet 2.4.3 [O2]

 © HarperCollins*Publishers* Limited 2016

Consolidate and apply

Students should then complete one of the following three worksheets to consolidate this work:

- Low demand: Worksheet 2.4.4 (applying the equations of motion) [O3]
- Standard demand: Worksheet 2.4.5 (applying and rearranging the equations of motion) [O3]
- High demand: Worksheet 2.4.6 (applying and rearranging the equations of motion to situations including vertical motion) [O3]

Extend

- Given $s = ut + \frac{1}{2}at^2$, $v = u + at$ and $t = \frac{(v-u)}{a}$, students could be asked to prove that $v^2 = u^2 + 2as$.
- Alternatively, students could be given some data for a rocket launch and asked to calculate the time taken for the rocket to reach orbit. [O2, O3]

Plenary suggestions

Hot seat: Ask each student to think of a question, using material from the topic. Select someone to put in the hot seat; ask students to ask their questions and say at the end whether the answer is correct or incorrect.

What do I know? Ask students to each write down one thing about the topic they are sure of, one thing they are unsure of and one thing they need to know more about, being specific. Ask them to work in groups of five or six to agree on group lists; ask each group to say what they decided and agree as a class about what they are confident about, what they are less sure of and what things they want to know more about.

Answers to questions

Worksheets 2.4.1 and 2.4.2

Part 1

1. $v - u$
2. $a = \frac{(v-u)}{t}$
3. $v = u + at$

Part 2

1. area of rectangle = ut
2. area of triangle = $\frac{1}{2}t(v-u)$
3. total area = $ut + \frac{1}{2}t(v-u)$
4. $s = ut + \frac{1}{2}t(at)$
5. $s = ut + \frac{1}{2}at^2$

Worksheet 2.4.3

Part 1

1. $\frac{(v-u)}{t}$
2. $v = u + at$

Part 2

1. area of rectangle = ut

 © HarperCollins*Publishers* Limited 2016

2. area of triangle = $\frac{1}{2}t(v-u)$
3. total area = $ut+\frac{1}{2}t(v-u)$
4. $s=ut+\frac{1}{2}t(at)$
5. $s=ut+\frac{1}{2}at^2$

Worksheet 2.4.4

1. 15 m/s
2. 40 m/s
3. 24 m/s
4. 120 m
5. 250 m
6. Yes, it has travelled 4.5 m
7. 16.25 m/s
8. 54.8 m/s
9. 20 m/s

Worksheet 2.4.5

1. 15 m/s
2. 40 m/s
3. 2 m/s^2
4. 120 m
5. 250 m
6. Yes, it has travelled 4.5 m
7. 16.25 m/s
8. 54.8 m/s
9. 5 m

Worksheet 2.4.6

1. 15 m/s
2. 40 m/s
3. 2 m/s^2
4. 120 m
5. 250 m
6. Yes, it has travelled 4.5 m
7. 16.25 m/s
8. 54.8 m/s
9. 5 m
10. 3.6 m
11. 0.45 m (45 cm)
12. 14 m/s

1\. 15 m/s
2\. 40 m/s
3\. 2 m/s^s
1\. 120 m
2\. 250 m
3\. Yes, it has travelled 4.5 m
1\. 16.24 m/s
2\. 5 m

1\. 15 m/s
2\. 40 m/s
3\. 2 m/s^s
1\. 120 m
2\. 250 m
3\. Yes, it has travelled 4.5 m
1\. 16.24 m/s
2\. 5 m
1\. 3.6 m
2\. 0.45 m / 45 cm

 © HarperCollins*Publishers* Limited 2016

Lesson 2.5: Velocity–time graphs

OCR Specification reference

P2.1f

Lesson overview

Learning objectives

- Draw velocity–time graphs.
- Calculate acceleration using a velocity–time graph.
- [Higher tier] Calculate displacement using a velocity–time graph.

Learning outcomes

- [O1] Review the meaning of the terms 'velocity' and 'acceleration'.
- [O2] Generate velocity–time data to produce a graph.
- [O3] Calculate displacement and acceleration when given velocity and time.

Skills development

- Manipulate an equation with three variables.
- Plot a set of data onto a graph.
- Use a graph to calculate additional information.

Maths focus

- Plot data.
- Calculate the gradient from a graph.
- [Higher tier] Calculate the area under a graph.

Resources needed ticker timer, ticker tape, glue, scissors, graph paper; Worksheets 2.5.1, 2.5.2 and 2.5.3, Practical sheet 2.5, Technician's notes 2.5

Digital resources ticker-timer simulator (e.g. http://www.phy.ntnu.edu.tw/ntnujava/index.php?topic=2602.0)

Key vocabulary velocity, time, speed, gradient, area, acceleration, velocity–time graph, sketch graph, rate of change, displacement

Teaching and learning

Engage

Pair talk: Ask: "Does constant acceleration mean constant velocity?" Let students discuss the question in pairs before taking ideas from a few different pairs. [O1]

Challenge and develop

- Students should generate velocity–time data using a ticker timer (Practical sheet 2.5). [O1]
- If ticker timers are not available, generate the data using a ticker-timer simulator.

Explain

- Students should now generate the velocity–time graph by cutting out the ticker-timer strips and sticking them in place. [O2]
- A discussion could follow: why do we show data on a graph as opposed to leaving it in a table?
- Talk the students through the process of calculating the gradient (and therefore the acceleration) and the area under the line (and therefore the displacement) of the velocity–time graph generated in the practical task. [O3]

© HarperCollins*Publishers* Limited 2016

Consolidate and apply

- Students should then complete a worksheet at an appropriate level to consolidate this work:
  - Low demand: Worksheet 2.5.1 (plot a velocity–time graph given a story and a table of data) [O1, O2]
  - Standard demand: Worksheet 2.5.2 (plot a velocity–time graph given a story and a table of data and calculate the acceleration and displacement for different parts of the journey) [O2, O3]
  - High demand: Worksheet 2.5.3 (plot a velocity–time graph from a story and calculate the acceleration and displacement for different parts of the journey) [O2, O3]
- Return to the original question from the start of the lesson and ask students to describe the difference between velocity and acceleration. [O1]

Extend

Students could plot a velocity–time graph for a journey that includes reversing. [O2, O3]

Plenary suggestions

Learning triangle: At the end of the lesson, ask students to draw a large triangle with a smaller inverted triangle that just fits inside it (so they have four triangles). Ask students to think back over the lesson and identify and write in the outer triangles:

- something they saw
- something they did
- something they discussed.

Ask students to write in the central triangle something they learned.

Heads and tails: Ask each student to write a question about something from the topic on a coloured paper strip and the answer on another coloured strip. Hand out the strips to groups of six to eight students, so that each student gets a question and an answer. One student reads out his or her question. The student with the correct answer then reads it out, followed by his or her question, and so on.

Answers to questions

Worksheets 2.5.1, 2.5.2 and 2.5.3

Velocity–time graph

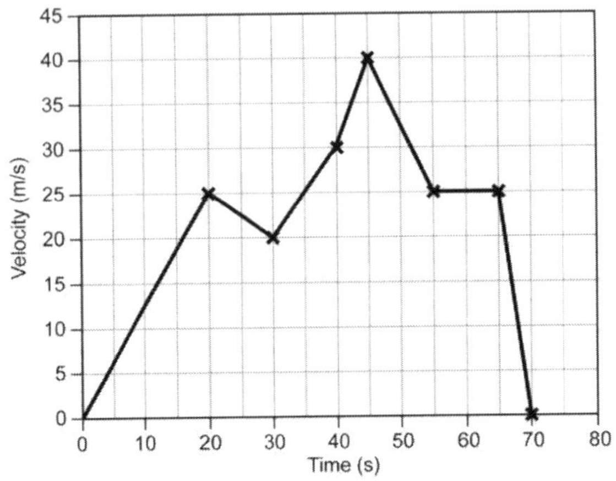

Acceleration

1. 1.25 m/s^2
2. −0.5 m/s^2
3. 1 m/s^2
4. 2 m/s^2
5. −1.5 m/s^2
6. 0 m/s^2
7. −5 m/s^2

 © HarperCollins*Publishers* Limited 2016

Lesson 2.6: Maths skills: Making estimates of calculations

Lesson overview

OCR Specification reference

PM2.3i to vi

Learning objectives

- Estimate the results of simple calculations.
- Round numbers to make an estimate.
- Calculate order of magnitude.

Learning outcomes

- [O1] Make an estimate, having rounded numbers.
- [O2] Calculate an order of magnitude.

Skills development

- Make sensible estimates.

Maths focus

- Round numbers to make an estimate.

Resources needed Worksheets 2.6.1, 2.6.2 and 2.6.3

Key vocabulary estimate, round, order of magnitude

Teaching and learning

Engage

Ask students to estimate how much change they would get if they paid for a packet of crisps costing 63p, a sandwich costing £2.25 and a bottle of water costing 97p using a £10 note. They must not use a calculator. Ask the students to discuss in pairs how they made their estimate. [O1]

Challenge and develop

Ask students to think about how they could estimate the speed of the traffic travelling along a road. Let them discuss this in pairs, before joining into groups of four to develop their ideas further. Take feedback from some groups. [O1]

Explain

Students should now answer questions 3, 4 and 5 from the Student Book.

Consolidate and apply

Students should then complete a worksheet at an appropriate level to consolidate this work:

- Low demand: Worksheet 2.6.1 [O1, O2]
- Standard demand: Worksheet 2.6.2 [O1, O2]
- High demand: Worksheet 2.6.3 [O1, O2]

Extend

Ask students to estimate the volume of air in the classroom.

Plenary suggestions

Hot seat: Ask each student to think of a question, using material from the topic. Select someone to put in the hot seat; ask students to ask their questions and say at the end whether the answer is correct or incorrect.

 © HarperCollins*Publishers* Limited 2016

What do I know? Ask students to each write down one thing about the topic they are sure of, one thing they are unsure of and one thing they need to know more about, being specific. Ask them to work in groups of five or six to agree on group lists; ask each group to say what they decided and agree as a class on what they are confident about, what they are less sure of and what things they want to know more about.

The big ideas: Ask students to write down three ideas they have learned during the topic. Then ask them to share their facts in groups and to compile a master list of facts, with the most important at the top. Ask for ideas to be shared, and find out which other group(s) agreed.

Answers to questions

Worksheets 2.6.1, 2.6.2 and 2.6.3

1. Depends on the book chosen.
2. £50
3. ≈ -4 m/s^2
4. $10^6 \div 10^1 = 10^5$ s

 © HarperCollins*Publishers* Limited 2016

Lesson 2.7: Forces explain how objects interact

Lesson overview

OCR Specification reference

P2.2d, e

Learning objectives

- Describe a force.
- Recognise the difference between contact and non-contact forces.
- State examples of scalar and vector quantities.

Learning outcomes

- [O1] Identify forces in everyday situations.
- [O2] Categorise forces as either contact or non-contact forces.
- [O3] Identify both scalar and vector quantities.

Skills development

- Correct use of key terms.

Resources needed Worksheets 2.7.1, 2.7.2 and 2.7.3

Key vocabulary push, pull, contact force, non-contact force, displacement, velocity, scalar, vector

Teaching and learning

Engage

- As a reminder of the knowledge they should have from KS3 Science, show the students an image of a floating catamaran and ask them to discuss what is keeping it afloat. Take ideas from several pairs, before agreeing on an answer as a class. [O1]
- Alternatively, show the students an image of a racing car and ask them to identify the forces acting on the car. An extension to this would be to ask students how the forces on the car change as the car speeds up. [O1]

Challenge and develop

Issue the students with a copy of the forces worksheet appropriate to their level:

- Low demand: Worksheet 2.7.1 (names of forces and a description of the forces are given) [O1]
- Standard demand: Worksheet 2.7.2 (names of forces are given) [O1]
- High demand: Worksheet 2.7.3 [O1]

Explain

- Ask the students to rub their hands together and state what it feels like. Use this to introduce the idea of a contact force. [O2]
- Issue each pair of students with a pair of bar magnets and ask them to investigate the effect of holding opposite poles and like poles together. Use this to introduce the idea of a non-contact force. [O2]
- The students should then be encouraged to return to their worksheet and, in a second coloured pen, annotate the diagrams to show whether the forces they have labelled are examples of contact forces or non-contact forces. [O2]
- As a way of introducing the idea of scalar and vector quantities, set a scene for the students, such as two cars driving down a very narrow lane at 30 miles per hour. Pose the question: "What will happen?" and allow

 © HarperCollins*Publishers* Limited 2016

the students to discuss this, before taking some ideas. The answer, of course, is that they don't know. If the cars are travelling in the same direction, they will simply carry on their journey. If they are travelling in opposite directions however, they will crash into each other. Ask the students to discuss other situations where considering the direction as well as the magnitude of a quantity is important. Force is likely to form part of their discussion. (Images from the worksheet could be used to prompt discussion.) [O3]

- Alternatively, students could be asked to map their journey to school and home again and the idea of the difference between distance and displacement could be introduced here. [O3]

Consolidate and apply

- Once the students have the idea of the definition of a scalar and a vector quantity, issue students with two cards of different colours. One colour should represent a scalar quantity and the other colour should represent a scalar quantity. Read down a list of quantities and the students should use their cards to vote whether they think the quantity is a scalar or a vector. Quantities might include: length (s), speed (s), volume (s), temperature (s), energy (s), power (s), displacement (v), velocity (v), acceleration (v), momentum (v), force (v), weight (v). [O3]
- The students could then be put into groups of three or four and given an A3 picture of a city scene. They should be asked to annotate the picture with all the forces they can see in action in the picture, along with the scalar and vector quantities. For example, there might be a moving car in the picture. The forces labelled might be friction, weight, thrust, air resistance, etc. and the scalar quantities could be the length of the car, its speed and the kinetic energy it has. The vector quantities might be the car's velocity, its acceleration and its weight. Once the students have annotated their picture, they should pass it on to the next group, where they should then add any further ideas, before passing it to a third group. The image should then be returned to the original group. [O1, O3]

Extend

- Ask students to research the term 'resultant vector' and give three everyday examples of vector addition. [O3]

Plenary suggestions

Silent animation: Show students a short video clip without sound and ask them to say what they think is happening. They could either suggest the ideas or do a 'voice over' for it.

Heads and tails: Ask each student to write a question about something from the lesson on a coloured paper strip and the answer to the question on another colour. In groups of six to eight, hand out the strips so that each student gets a question and an answer. One student reads out their question. The student with the right answer then reads it out, followed by their question, and so on.

 © HarperCollins*Publishers* Limited 2016

Lesson 2.8: Forces and motion

Lesson overview

OCR Specification reference

P2.2d, e

Learning objectives

- Understand what a force does.
- Explain what happens to an object if all the forces acting on it cancel each other out.
- Analyse how this applies to everyday situations.

Learning outcomes

- [O1] Identify situations where the forces on an object are balanced and situations where the forces are unbalanced.
- [O2] State Newton's first law.
- [O3] Give everyday examples of Newton's first law.

Skills development

- Interpret information from diagrams.

Resources needed marbles, ramps; Worksheets 2.8.1, 2.8.2 and 2.8.3, Practical sheet 2.8, Technician's notes 2.8

Digital resources http://practicalphysics.iopconfs.org/galileos-thought-experiment.html

Key vocabulary balanced force, resultant force, equilibrium, Newton's first law

Teaching and learning

Engage

Show students a number of different situations, such as a book resting on a table, a tennis ball **accelerating** along a table, a weight hanging on the end of a length of string, a raindrop travelling through the air. Issue each student with two cards of different colours. The students should use the cards to vote whether the forces in the situation are balanced or unbalanced. [O1, O2]

Challenge and develop

- **Pair talk:** Ask the students: "Why do things stop?" Let them discuss their ideas in pairs.
- **Pairs to fours:** Each pair should then join with another pair to explain and compare ideas before feeding back to the class. [O1]

Explain

- Use the ideas from the discussion to introduce Newton's first law. Students could then investigate Newton's first law (Practical sheet 2.8). [O2, O3]
- Alternatively, introduce Galileo's thought experiment (http://practicalphysics.iopconfs.org/galileos-thought-experiment.html) and ask the students to work in groups of four to plan how they would carry out this experiment and to predict the results. [O2, O3]

Consolidate and apply

Students should then complete a forces and motion worksheet at an appropriate level to consolidate this work:

- Low demand: Worksheet 2.8.1 [O2, O3]
- Standard demand: Worksheet 2.8.2 [O2, O3]
- High demand: Worksheet 2.8.3 [O2, O3]

 © HarperCollins*Publishers* Limited 2016

Extend

Set up a debate, with the students split into two 'teams'. One team is given the quote 'You need a force to make something go', and the other team is given the quote 'You need a force to make something stop'. Allow the students to debate, using ideas learned in the lesson. [O2]

Plenary suggestions

Ideas hothouse: Ask students to work in pairs to list points about what they know about the topic. Then ask the pairs to join together into groups of four and then groups of six to eight to discuss this further and to come up with an agreed list of points. Ask one person from each group to report back to the class.

Learning triangle: At the end of the lesson, ask students to draw a large triangle with a smaller inverted triangle that just fits inside it (so they have four triangles). Ask students to think back over the lesson and identify and write in the outer triangles:

- something they saw
- something they did
- something they discussed.

Ask students to write in the central triangle something they learned.

 © HarperCollins*Publishers* Limited 2016

Lesson 2.9: Resultant forces

Lesson overview

OCR Specification reference

P2.2e (part), g, h (part)

Learning objectives

- Calculate the resultant of a number of forces.
- Draw free-body diagrams to find resultant forces.
- [Higher tier] Understand that a force can be resolved into two components acting at right angles to each other.

Learning outcomes

- [O1] Draw scaled arrows on images to calculate resultant forces.
- [O2] Draw free-body diagrams and find the resultant force from the diagrams.
- [Higher tier, O3] Use scale vector diagrams to resolve forces.

Skills development

- Interpret information from diagrams.

Maths focus

- Draw arrows to scale.
- Calculate resultant forces by scaled drawing.
- Measure angles using a protractor.

Resources needed a copy of an image of a tug of war for each student, an image of a car with 'thrust' and 'friction' labelled along with their magnitudes; Worksheets 2.9.1, 2.9.2 and 2.9.3

Key vocabulary free-body diagram, balanced forces, unbalanced forces, resultant force, resolving a force, components of a force, magnitude

Teaching and learning

Engage

Ask students to discuss the following question: "Despite having fallen a very long way, raindrops do not leave holes in the pavement when they fall. Why not?"

Challenge and develop

- Issue each student with an image of a tug of war. Ask them to add arrows to the image to represent the forces. Encourage them to draw the arrows to scale – the longer the arrow, the larger the force. [O1]
- **Pair talk:** Ask students to discuss why, despite everyone in the picture pulling really hard, nobody is moving. Take ideas from different groups before introducing the term 'resultant force'. [O1, O2]
- Now tell students: "Fred joins the team pulling to the left. How does this change the forces in the picture? What will happen? (i.e. which team will win the tug of war and why?)" [O1]

Explain

- Demonstrate two situations to the students: one in which forces are balanced, and one in which they are not. An example might be a book resting on a table and then the book on the table after it has been pushed.
- Ask students to sketch each situation and draw force arrows to scale for both situations. Be sure to emphasise that a force is not needed to keep a moving object moving, and so the gravity pulling the book down, the table pushing upwards on the book and the friction opposing the direction of motion are the only forces acting in the second situation. The students should easily be able to identify in which situation the forces are balanced and in which situation they are unbalanced. [O1, O2]

 © HarperCollinsPublishers Limited 2016

- Show the students an image of a car, with 'thrust' and 'friction' labelled and with a magnitude given for each of the forces. Ask the students to calculate the resultant force on the car and the direction in which the car will travel. [O1, O2]
- [Higher tier] Explain to students how to determine the resultant force of two forces that are perpendicular to each other using a scale drawing. The example outlined in Figure 2.25 in the Student Book could be used. Encourage students to determine both the magnitude and the direction (using a protractor). Follow this by explaining how a resultant force can be resolved into its component forces, again using a scale diagram. [O3]

Consolidate and apply

Pupils should then complete a resultant forces worksheet at an appropriate level to consolidate this work:

- Low demand: Worksheet 2.9.1 (labelling diagrams with scaled force arrows, calculating resultant forces and constructing free-body diagrams) [O2]
- Standard demand: Worksheet 2.9.2 (labelling diagrams with scaled force arrows, calculating resultant forces and constructing free-body diagrams) [O2]
- High demand: Worksheet 2.9.3 (labelling diagrams with scaled force arrows, calculating resultant forces, calculating missing forces when given resultant forces, constructing free-body diagrams and resolving forces using scale drawings) [O2, O3]

Extend

Ask students to design an instruction sheet for another student in their class to explain how to resolve forces into perpendicular components using scale diagrams.

Plenary suggestions

Ask me a question: Ask students to write a question about something from the topic and then a mark scheme for the answer. Encourage them to come up with questions worth more than one or two marks and to try their questions out on each other.

Freeze frame: Ask students to create a 'freeze frame' of an idea in the topic. This involves three or four students arranging themselves as a static image and other students suggesting what it represents.

 © HarperCollins*Publishers* Limited 2016

Lesson 2.10: Forces and acceleration

OCR Specification reference

P2.2i, j

Lesson overview

Learning objectives

- Explain what happens to the motion of an object when the resultant force is not zero.
- Analyse situations in which a non-zero resultant force is acting.
- Explain what inertia is.

Learning outcomes

- [O1] Draw force diagrams for objects where the combination of forces acting on them is not zero.
- [O2] State Newton's second law.
- [O3] Give everyday examples of Newton's second law.

Skills development

- Interpret information from diagrams.
- Investigate a phenomenon practically.
- Analyse practical results and comment on the accuracy of the data.

Maths focus

- Plot a graph.
- Calculate a gradient from a graph.

Resources needed equipment listed in Technician's notes; Practical sheets 2.10.1, 2.10.2 and 2.10.3, Technician's notes 2.10

Digital resources
http://www.thephysicsaviary.com/Physics/Programs/Labs/MeasuringPhotogateAccelerationPrelab/index.html

Key vocabulary resultant force, Newton's second law, acceleration, inertia, inertial mass, gravitational mass

Teaching and learning

Engage

Ask students to draw a force diagram to represent the forces acting on a car accelerating *slowly* away from a set of traffic lights. Ask the students to say whether the forces they have labelled are balanced or not. In which direction is the resultant force acting? [O1]

Challenge and develop

- **Pair talk:** Ask students to discuss how their diagram would change if the car were accelerating away from the lights more quickly. How would they show this on their force diagram? How would it change the magnitude of the resultant force? [O1, O3]
- Collect the students' ideas before asking them to suggest what the magnitude of the resultant force depends on, using their diagrams to help. They should realise that the magnitude of the resultant force depends on the acceleration of the object (in this case the car). [O1, O3]
- Introduce $F = ma$ and give an example of calculating force when given the mass and acceleration, calculating mass when given the force and acceleration, and calculating acceleration when given force and mass. [O2]

Explain

- Students should investigate Newton's second law by carrying out the practical at the appropriate level:

 © HarperCollins*Publishers* Limited 2016

- Low demand: Practical sheet 2.10.1 [O2]
- Standard demand: Practical sheet 2.10.2 [O2]
- High demand: Practical sheet 2.10.3 [O2]

- Alternatively, a simulation such as http://www.thephysicsaviary.com/Physics/Programs/Labs/MeasuringPhotogateAccelerationPrelab/index.html could be used to investigate Newton's second law. [O2]

Consolidate and apply

- Demonstrate the idea of inertia: hang two cans on two pieces of string of equal length. Fill one can with sand (or rice), and leave the second can empty. Ask a student to apply enough force to start the empty can moving. Ask the student to apply the same force (as best as possible) to attempt to move the filled can. What do the students notice?
- Introduce the term 'inertia' as a measure of how difficult it is to change the velocity of the object. [O3]
- Ask the students to use the idea of inertia to explain why lorries have a speed limit of 40 mph on single carriageway roads, but cars have a speed limit of 60 mph on the same roads. Introduce the term 'inertial mass' as a measure of how difficult it is to change the velocity of an object, and show that inertial mass can be substituted into $F = ma$ so that the equation becomes force = inertial mass × acceleration. [O3]

Extend

Students should be encouraged to research why seat belts in modern cars might be called inertia belts. [O3]

Plenary suggestions

Hot seat: Ask each student to think of a question, using material from the lesson. Select someone to put in the hot seat; ask students to ask their questions and say at the end whether the answer is correct or incorrect.

Heads and tails: Ask each student to write a question about something from the topic on a coloured paper strip and the answer on another coloured strip. Hand out the strips to groups of six to eight students, so that each student gets a question and an answer. One student reads out his or her question. The student with the correct answer then reads it out, followed by his or her question, and so on.

© HarperCollins*Publishers* Limited 2016

Lesson 2.11: Momentum [higher tier]

OCR Specification reference

P2.2k

Lesson overview

Learning objectives

- Explain what is meant by momentum.
- Apply ideas about the rate of change of momentum to safety features in cars.
- Use momentum calculations to predict what happens in a collision.

Learning outcomes

- [O1] Define momentum.
- [O2] Explain how safety features in cars use ideas about the rate of change of momentum.
- [O3] Calculate momentum.

Skills development

- Relate mathematical concepts to real-life situations.

Maths focus

- Rearrange equations.

Resources needed two trolleys (ideally of equal mass), balance (to measure the mass of the trolleys); Worksheets 2.11.1, 2.11.2 and 2.11.3

Digital resources video clip, e.g. https://www.youtube.com/watch?v=d7iYZPp2zYY of a crash test with and without a seat belt and airbag

Key vocabulary momentum, airbag, seat belt, conservation, crumple zones, conservation of momentum

Teaching and learning

Engage

- Show students a video clip, e.g. https://www.youtube.com/watch?v=d7iYZPp2zYY, of a crash test with and without a seat belt and airbag.
- **Pair talk:** Ask: "What things affect the degree of impact in a crash?" Take feedback from some student pairs. Students should identify mass and velocity (or at least speed) as two factors that will affect the degree of impact. [O1]

Challenge and develop

- Introduce the term 'momentum', as Newton did at the beginning of *Principia*, as 'the quantity of motion' and state that it is equal to the product of an object's mass and its velocity. [O1]
- Link this to the next task by stating that car manufacturers try to reduce the rate of change of momentum by increasing the time taken to reduce velocity, i.e. features are added to cars to slow down the occupants. [O2]

Explain

- Remind students of the equation for Newton's second law of motion, $F = ma$, and ask students to substitute acceleration as change in velocity ÷ time to lead to the equation $F = \frac{m(v-u)}{t}$.
- Ask students to consider how this equation could be useful for car safety feature manufacturers. They should discuss their ideas in pairs, before joining together in fours to share ideas. [O2]

 © HarperCollins*Publishers* Limited 2016

- Take feedback from several groups of students, before concluding that, to reduce the force on the car's occupants, the change in momentum should be as small as possible and the time taken to reduce the velocity should be as long as possible. [O2]
- Students should then answer question 3 from the Student Book. Alternatively, students could consider, using ideas about momentum, how a crumple zone in a car improves safety. [O2]

Consolidate and apply

- Demonstrate to the students two trolleys (ideally of equal mass) colliding. Have one trolley stationary (trolley A) and the other trolley colliding into it (trolley B).
- Ask the students to watch the demonstration in pairs. One student should watch trolley A and the other trolley B. Students should each note down what happens to the trolley that they were watching.
- Use this demonstration to introduce the concept of conservation of momentum, $m_1u_1 + m_2u_2 = (m_1 + m_2)v$. [O3]
- Model an example calculation, perhaps finding the common final velocity of two trolleys that have collided. [O3]
- The students should now answer questions 6 and 7 in the Student Book. Alternatively, the students could be given a momentum worksheet at an appropriate level (all higher tier):
  - Low demand: Worksheet 2.11.1 [O3]
  - Standard demand: Worksheet 2.11.2 [O3]
  - High demand: Worksheet 2.11.3. [O3]

Extend

- Students could be given the following problem:
  A seat belt stretches a little on impact. Use ideas about momentum to explain why. [O2]

Plenary suggestions

What do I know? Ask students to each write down one thing about the topic they are sure of, one thing they are unsure of and one thing they need to know more about, being specific. Ask them to work in groups of five or six to agree on group lists; ask each group to say what they decided and agree as a class about what they are confident about, what they are less sure of and what things they want to know more about.

Heads and tails: Ask each student to write a question (or calculation) on a coloured paper strip and the answer on another coloured strip. Hand out the strips to groups of six to eight students, so that each student gets a question and an answer. One student reads out his or her question. The student with the correct answer then reads it out, followed by his or her question, and so on.

Ideas hothouse: Ask students to work in pairs to list points about what they know about momentum. Then ask the pairs to join together into groups of four and then groups of six to eight to discuss this further and to come up with an agreed list of points. Ask one person from each group to report back to the class.

Answers to questions

Worksheet 2.11.1

1. a quantity with both magnitude and direction
2. mass × velocity (the quantity of motion)
3. 0 kg m/s
4. 22 500 kg m/s
5. 75 000 kg m/s
6. 2 m/s
7. 40 kg
8. $F = \frac{m(v-u)}{t}$

Worksheet 1.2

1. vector (as velocity is a vector quantity)
2. kg m/s
3. zero
4. 22,500 kg m/s
5. 2 m/s
6. 40 kg
7. $F = \frac{m(v-u)}{t}$
8. 2.2 s

 © HarperCollins*Publishers* Limited 2016

Worksheet 2.11.2

1. vector (because velocity is a vector quantity)
2. kg m/s
3. 0 kg m/s
4. 22 500 kg m/s
5. 2 m/s
6. 40 kg
7. $F = \frac{m(v-u)}{t}$
8. 2.2 s

Worksheet 2.11.3

1. vector (because velocity is a vector quantity)
2. kg m/s
3. 22 500 kg m/s
4. 2 m/s
5. 40 kg
6. $F = \frac{m(v-u)}{t}$
7. 2.2 s
8. 4 m/s

 © HarperCollins*Publishers* Limited 2016

Lesson 2.12: Practical: Investigating the acceleration of an object

Lesson overview

OCR Specification reference

P2.2i

Learning objectives

- Plan an investigation to explore an idea.
- Analyse results to identify patterns and draw conclusions.
- Compare results with scientific theory.

Learning outcomes

- [O1] Identify independent, dependent and, therefore, control variables.
- [O2] Plan a valid investigation.
- [O3] Analyse results from a practical investigation.

Skills development

- Carry out fair and accurate investigations.
- Analyse data collected from experiments.
- Draw sensible conclusions from experimental data.

Maths focus

- Calculate acceleration.

Resources needed apparatus listed in Technician's notes; Practical sheet 2.12, Technician's notes 2.12

Digital resources picture of a head-up display in a car or acceleration app for a smart phone

Key vocabulary acceleration, force, mass, weight, gravitational field strength, newtons, kilograms, Newton's second law, direct proportion, inverse proportion

Teaching and learning

Engage

- Show students a picture of a head-up display in a car and ask them to discuss what it shows.
- Alternatively, show them an acceleration app for a smart phone and ask them to discuss how it works.
- Issue the students with their challenge: to investigate the relationship between the force on an object and its acceleration.
- Students should be encouraged to think of different ways in which this relationship could be investigated safely in the lab. [O2]
- Ideas might include: use of ticker timers with different masses on the end or a trolley and a light gate.
- Ask students to plan their investigation, considering the following factors carefully: [O1, O2]
  - safety
  - practicability
  - control variables
  - independent variable
  - dependent variable

 © HarperCollins*Publishers* Limited 2016

- the results table format
- the graph they will generate from their independent and dependent variables.

Explain

Ask students to investigate the relationship between the force on an object and its acceleration (Practical sheet 2.12). [O1, O2]

Consolidate and apply

- Students should then analyse the results they have collected by plotting the graph of force (on the *y*-axis; note: this is different from the usual 'plot the independent variable on the *x*-axis' advice) and acceleration (on the *x*-axis). [O3]
- Plotting the graph in this way will enable the mass, *m*, to be found by calculating the gradient of the line.
- If students have been issued with different masses, the graphs can be compared.

Extend

- Ask students to discuss and suggest how a Nintendo Wii controller works.
- Alternatively, students could return to the head-up display or acceleration app from the start of the lesson and develop their ideas about how they work, having completed their investigation.

Plenary suggestions

What do I know? Ask students to each write down one thing about the topic they are sure of, one thing they are unsure of and one thing they need to know more about, being specific. Ask them to work in groups of five or six to agree on group lists; ask each group to say what they decided and agree as a class about what they are confident about, what they are less sure of and what things they want to know more about.

Hot seat: Ask each student to think of a question, using material from the topic. Select someone to put in the hot seat; ask students to ask their questions and say at the end whether the answer is correct or incorrect.

Ideas hothouse: Ask students to work in pairs to list points about what they know about a particular idea. Then ask the pairs to join together into groups of four and then groups of six to eight to discuss this further and to come up with an agreed list of points. Ask one person from each group to report back to the class.

 © HarperCollins*Publishers* Limited 2016

Lesson 2.13: Newton's third law

Lesson overview

OCR Specification reference

P2.2p

Learning objectives

- Identify force pairs.
- Understand and be able to apply Newton's third law.

Learning outcomes

- [O1] Label force pairs.
- [O2] State Newton's third law.
- [O3] Identify Newton's third law pairs of forces.

Resources needed Worksheets 2.13.1, 2.13.2 and 2.13.3

Digital resources clip of royal gun salute, e.g. https://www.theguardian.com/uk-news/video/2014/apr/21/queens-birthday-marked-royal-gun-salute-video, video clip for demonstration of Newton's third lawhttp://www.theguardian.com/uk-news/video/2014/apr/21/queens-birthday-marked-royal-gun-salute

Key vocabulary Newton's third law, action, reaction, equal, opposite, force pair, weight

Teaching and learning

Engage

Show students a clip of a royal gun salute and ask: "Why do the soldiers not stand behind the gun when it is fired?"

Challenge and develop

- **Demonstration:** If possible, have two students sitting on two wheeled office chairs about 5 m apart. Give the students a rope to hold between them and ask one of the students to pull the other towards him or her. The other students should observe what happens. [O1]
- Lead a group discussion centred on the observations, leading to the introduction of Newton's third law. Be sure to emphasise that the equal and opposite forces are acting on different objects, because if they were working on the same object, the resultant force would be zero and nothing would ever move. [O1, O2]
- Alternatively, show a video clip of the same situation and follow it with the discussion. [O1, O2]

Explain

Issue students with a Newton's third law worksheet at an appropriate level:

- Low demand: Worksheet 2.13.1 [O1, O3]
- Standard demand: Worksheet 2.13.2 [O1, O3]
- High demand: Worksheet 2.13.3 [O1, O3]

Consolidate and apply

Students should then be given time to consolidate Newton's three laws of motion by designing a poster for display in the Science block outlining why the ideas in the three laws are so important to us.

Extend

Students could be given a picture of a book resting on a table labelled with the weight of the book and the force of the table pushing up on the book. The students should then explain why this is *not* an example of a third law pair, despite the forces being equal and opposite.

 © HarperCollins*Publishers* Limited 2016

Plenary suggestions

Ask me a question: Ask students to write a question about something from the topic and then a mark scheme for the answer. Encourage them to come up with questions worth more than one or two marks and to try their questions out on each other.

Silent animation: Show students a short video clip without sound and ask them to say what they think is happening. They could suggest ideas or provide a 'voice over'.

Answers to questions

Worksheet 2.13.1

1. If object A exerts a force on object B, then object B exerts an equal but opposite force on object A.
2. a) an arrow showing the wall pushing back on the person
   b) an arrow showing the quayside pushing back on the person
   c) an arrow showing the recoil of the gun
   d) an arrow showing the other ice skater pushing on the first ice skater
3. The bullet has less mass and so travels faster.

Worksheets 2.13.2 and 2.13.3

1. If object A exerts a force on object B, then object B exerts an equal but opposite force on object A.
2. a) an arrow showing the person pushing on the wall and an arrow showing the wall pushing back on the person
   b) an arrow showing the person pushing against the quayside and an arrow showing the quayside pushing back on the person
   c) an arrow showing the thrust of the bullet and an arrow showing the recoil of the gun
   d) an arrow showing the first ice skater pushing on the second ice skater and an arrow showing the second ice skater pushing on the first ice skater
3. The bullet has less mass and so travels faster.

 © HarperCollins*Publishers* Limited 2016

Lesson 2.14: Work done and energy transfer

Lesson overview

OCR Specification reference

P2.2m, n

Learning objectives

- Understand what is meant by work done.
- Explain the relationship between work done and force applied.
- Identify the transfers between energy stores when work is done against friction.

Learning outcomes

- [O1] Define work done.
- [O2] Calculate, using the work done equation.
- [O3] Link kinetic energy and work done, including using calculations.

Skills development

- Use a variety of models, such as representational, spatial, descriptive, computational and mathematical, to solve problems, make predictions and develop scientific explanations and understanding of familiar and unfamiliar facts.
- Make and record observations and measurements using a range of apparatus and methods.
- Recognise the importance of scientific quantities and understand how they are determined.

Maths focus

- Recognise and use expressions in standard form.
- Find arithmetic means.
- Change the subject of an equation.
- Substitute numerical values into algebraic equations using appropriate units for physical quantities.
- Solve simple algebraic equations.

Resources needed sand in a tray, metre rule, ball bearings, smoother, mass balance; Worksheets 2.14.1, 2.14.2 and 2.14.3, Practical sheets 2.14.1 and 2.14.2, Technician's notes 2.14

Key vocabulary work, energy transfer, kinetic energy, force

Teaching and learning

Engage

- As the students enter the room, give them a small piece of paper and ask them to give their best answer to the following question: what is energy?
- **Pair talk:** Ask students to share ideas with a partner, adapting their answer if they think it is appropriate.
- **Pairs to fours:** Pairs should join together to finalise their answer. Take feedback from a number of groups before developing the idea that energy is the ability to do work. Discuss what we mean by doing work after showing the work done video.
- Play the work done video showing a 'plane pull' by four-times World's Strongest Man Žydrūnas Savickas.
- After the video, ask the students the following questions: Why is the plane pull such hard work? Why couldn't many people do this? What factors would make it harder or easier? What energy transfers were there during the plane pull? [O1, O2, O3]
- Show the work done animation(s), especially, 'work' and 'work against friction'.
- Class discussion on work done leading to the definition (energy transfer when a force causes an object to move) and the equation: work done = force × distance or $W = Fs$. Put equation into the triangle. Note: students will be expected to remember this equation. [O1, O2]

 © HarperCollins*Publishers* Limited 2016

- Issue students with a piece of paper with questions related to the 'work done and kinetic energy' videos appropriate to their ability:
  - Low demand: What caused the truck to stop? Why was there sand at the end of the dragster runway?
  - Standard demand: As low demand, plus: What factors would make the vehicle take a longer or shorter distance to stop?
  - High demand: As standard demand, plus: What information do you need to know to work out the braking distance of a vehicle?
- Watch both 'Work done and kinetic energy' videos. Use the information from the videos to point out that in braking or accelerating: work done = change in kinetic energy. [O1, O2, O3]

Challenge and develop

- Watch the meteorite strike simulation. Discuss what this video shows. How are potential energy, kinetic energy and work done involved here? Note: It is estimated that about 500 meteorites hit the surface of the Earth each year (obviously not of the size that wiped out the dinosaurs!). [O1, O3]
- Carry out the practical.
  - Low demand: Use Practical sheet 2.14.2, in which it is deemed that students will find conversion between grams and kilograms as well as centimetres (and millimetres) and metres too difficult. Students need to be guided through the calculations carefully.
  - Standard demand: Use Practical sheet 2.14.1. Go through conversions between grams and kilograms as well as between centimetres (and millimetres) and metres.
  - High demand: Use Practical sheet 2.14.1 independently. [O1, O2, O3]
- Discuss how the practical could have been improved so that it more closely simulates meteorites striking the Earth – first in pairs, then groups of four, then groups of eight. Finally, someone from each group shares ideas for improvement with the class. [O1, O2, O3]

Explain

- Challenge students' thinking and understanding of work done from the practical activities using the evaluation sections of the practical sheet (these sections are identical). [O1, O2, O3]

Consolidate and apply

Distribute the worksheets, according to ability.

- Low demand: Guide students through tasks carefully.
- Standard and high demand: Work independently or in pairs or small groups. [O1, O2, O3]

Extend

Ask students able to progress further to do the following:

- Suggest why a large meteorite such as the one that killed the dinosaurs would do so much damage. [O1, O3]

Answers to questions

Worksheet question 1 (2.14.1 – a–c; 2.14.2 – a–d; 2.14.3 – a–e)
a) the greater the braking force, the shorter the braking distance b) 75 000 J c) 2000 N d) 18.3 m e) 2.1×10^{19} N
Worksheet question 2 (2.14.1 – a–c; 2.14.2 – a–c; 2.14.3 – a–d)
a) more kinetic energy; more work needs to be done to stop player in same distance; more force required b) more kinetic energy; more work needs to be done to stop ball in same distance; more force required or a greater stopping distance c) (i) 360 000 J (ii) 0 J (iii) 360 000 J (iv) 360 000 J (or 360 000 Nm) (v) 4800 N d) (i) 2106 J (ii) 416 J (iii) 1690 J (iv) 1690 J (or 1690 Nm)
Practical sheet evaluation
1. Any sensible answer, including: go into ground, depth of crater depends on kinetic energy, some sand is thrown out.
2. The higher the kinetic energy of impact, the deeper the crater.
3. As it falls, gravitational potential energy is transferred into kinetic energy (or ball bearings don't reach high speeds, so little heat loss).
4. Air resistance leads to some energy transfer to heat.
5. work done = change in kinetic energy
6. The greater the energy transfer, the greater the work done on the ball bearing.
7. transferred to heat energy and as sound, and to kinetic energy of sand
8. Any sensible suggestion, including comparing meteorites (ball bearings) of same surface area but different mass.

 © HarperCollins*Publishers* Limited 2016

Lesson 2.15: Understanding power

Lesson overview

OCR Specification reference

P2.2o

Learning objectives

- Define power.
- Compare the rate of energy transfer by various machines and electrical appliances.
- Calculate power.

Learning outcomes

- [O1] Define power.
- [O2] Use data on different machines and electrical appliances to compare powers.
- [O3] Calculate, using the power equation.

Skills development

- Explain everyday and technological applications of science; evaluate associated personal, social, economic and environmental implications; make decisions based on the evaluation of evidence and arguments.
- Make and record observations and measurements using a range of apparatus and methods.
- Use scientific vocabulary, terminology and definitions.

Maths focus

- Recognise and use expressions in standard form.
- Change the subject of an equation.
- Substitute numerical values into algebraic equations using appropriate units for physical quantities.

Resources needed bathroom-scales-style force meter, force meter, masses (100 g to 1 kg), stopwatch, metre rule or tape measure, flight of stairs; Worksheet 2.15, Practical sheets 2.15.1 and 2.15.2, Technician's notes 2.15.1 and 2.15.2

Digital resources power video

Key vocabulary power, energy transfer

Teaching and learning

Engage

- Watch the power video of NASA's Space Launch System, the next generation in rockets, which are designed to eventually take human beings to Mars. Power at lift-off will be roughly 4.8×10^{10} W or 48 000 000 000 W; compare that with 60 W in a typical filament bulb or 11 W in a typical energy-saving light bulb.
- After the video clip, ask the students to discuss what it means if one thing has a higher power than another. [O1, O2]
- Lead a class discussion on the definition of power and its equation – use questioning to lead students to 'power is defined as the rate at which energy is transferred or the rate at which work is done' and the equation:

  power = energy transferred ÷ time, or power = work done ÷ time
- Lead a discussion on the watt (if tempted, at regular intervals use the really bad joke: 'Watt is the unit of power' and when they answer, tell them you were just making a statement rather than asking a question). [O1, O2, O3]

 © HarperCollinsPublishers Limited 2016

Challenge and develop

- Carry out Practical 2.15.1 or Practical 2.15.2, depending on ability:
  - Low demand: Use either Practical sheet 2.15.1 or 2.15.2. Guide students through the process carefully.
  - Standard demand: Use either Practical sheet 2.15.1 or 2.15.2. Ensure that students understand how to use the force meter and how to perform the calculations.
  - High demand: Use either Practical sheet 2.15.1 or 2.15.2. Students work independently. [O1, O2, O3]
- Discuss how the practical could have been improved, first in pairs, then groups of four, then groups of eight. Finally, someone in each group shares ideas for improvement with the class. [O1, O2, O3]

Explain

- Challenge students' thinking and understanding of power from the practical activities using the evaluation sections of the practical sheets (these sections are identical). [O1, O2, O3]

Consolidate and apply

- The students should now complete the task on the understanding power worksheet. [O1, O2, O3]

Extend

- Each student or student pair discusses what it would be like at take-off to be an astronaut on the NASA SLS rocket, with explanations. [O1, O2]

Plenary suggestions

Hot seat: Ask each student to think up a question, using material from the topic. Select someone to put in the hot seat. Ask students to ask their question and say at the end whether the answer is correct or incorrect.

What do I know? Ask students to each write down one thing about the topic they are sure of, one thing they are unsure of and one thing they need to know more about, being specific. Ask them to work in groups of five or six to agree on group lists. Ask each group to say what they decided and agree as a class about what they are confident about, what they are less sure of and what things they want to know more about.

Answers to questions

Practical sheet evaluation

1. Power depends on both work done and time taken.
2. watt, W
3. Any sensible answer, including: depends on how hard they tried, when they got tired, miscounts, mistiming.
4. Any sensible answer, including: strength, stamina, how much they tried, different weights
5. Any sensible answer, including: only move a vertical height (there was some horizontal movement.)

 © HarperCollins*Publishers* Limited 2016

Lesson 2.16: Key concept: Forces and acceleration

Lesson overview

OCR Specification reference

KC

Learning objectives

- Recognise examples of balanced and unbalanced forces.
- Apply ideas about speed and acceleration to explain sensations of movement.
- Apply ideas about inertia and circular motion to explain braking and cornering.

Learning outcomes

- [O1] Name forces in a picture.
- [O2] Recall Newton's first and second laws and identify where they are in action.
- [O3] Carry out a practical based on circular motion.

Skills development

- Use key terms correctly.

Maths focus

- Apply an equation with three variables.

Resources needed picture of a theme park showing several rides, equipment listed in Technician's notes; Worksheets 2.16.1, 2.16.2 and 2.16.3, Practical sheet 2.16, Technician's notes 2.16

Digital resources video clip of a theme park ride, e.g. https://www.youtube.com/watch?v=XIOK73JNV3A

Key vocabulary force, Newton's first law, Newton's second law, circular motion, speed, acceleration, balanced, inertia, velocity

Teaching and learning

Engage

Show students a video clip of a theme park ride (e.g. https://www.youtube.com/watch?v=XIOK73JNV3A) and ask them to write down the names of all the forces they see in the clip. [O1]

Challenge and develop

- Issue pairs of students with a picture of a theme park showing several rides. They should work in pairs to label, in one colour, all the forces they can see in the picture. [O1]
- In a second colour, they should annotate places where they see Newton's first law in action. [O2]
- In a third colour, they should annotate places where they see Newton's second law in action. [O2]
- Next, students should pass their pictures to another pair, who should add anything they think their peers have missed.
- The pictures should be passed to one further pair, who should add any ideas, before returning the picture to the original pair. [O1, O2]

Explain

Ask students to investigate motion in a circle (to model the motion at the top of the Stealth ride they saw in the video clip at the start of the lesson), as outlined on Practical sheet 2.16.

Consolidate and apply

Issue each student with a forces and acceleration worksheet at an appropriate level:

 © HarperCollins*Publishers* Limited 2016

- Low demand: Worksheet 2.16.1 [O1, O2, O3]
- Standard demand: Worksheet 2.16.2 [O1, O2, O3]
- High demand: Worksheet 2.16.3 [O1, O2, O3]

Extend

Ask students to return to the picture of the theme park from the start of the lesson and to label all the Newton's third law force pairs.

Plenary suggestions

Learning triangle: At the end of the lesson, ask students to draw a large triangle with a smaller inverted triangle that just fits inside it (so they have four triangles). Ask students to think back over the lesson and identify and write in the outer triangles:

- something they saw
- something they did
- something they discussed.

Ask students to write in the central triangle something they learned.

Ask me a question: Ask students to write a question about something from the topic and then a mark scheme for the answer. Encourage them to come up with questions worth more than one or two marks and to try their questions out on each other.

Ideas hothouse: Ask students to work in pairs to list points about what they know about a particular idea. Then ask the pairs to join together into groups of four and then groups of six to eight to discuss this further and to come up with an agreed list of points. Ask one person from each group to report back to the class.

Answers to questions

Worksheets 2.16.1, 2.16.2 ad 2.16.3

2. a) m/s^2 b) 150 N c) 50 kg d) 15 m/s^2
3. A scalar quantity is one that just has a magnitude. Examples include energy and time.
4. A vector quantity has both magnitude and direction. Examples include acceleration and velocity.
5. No, the acceleration is not constant, because the velocity is constantly changing (acceleration = change in velocity ÷ time)
6. Acceleration is independent of mass, and the acceleration due to gravity (g) is the same for both objects (10 m/s^2).

 © HarperCollins*Publishers* Limited 2016

Lesson 2.17: Forces and energy in springs

Lesson overview

OCR Specification reference

P2.3a, b

Learning objectives

- Explain why you need two forces to stretch a spring.
- Describe the difference between elastic and inelastic deformation.
- Calculate extension, compression and elastic potential energy.

Learning outcomes

- [O1] Identify the forces acting on a spring when it is stretched.
- [O2] State the meaning of the terms 'elastic deformation' and 'inelastic deformation'.
- [O3] Calculate extension and elastic potential energy.

Skills development

- Carry out a valid experiment.
- Analyse experimental results to draw sensible conclusions.

Maths focus

- Multiply numbers.
- Recognise a proportional relationship.

Resources needed equipment listed in Technician's notes; Worksheets 2.17.1 and 2.17.2, Practical sheet 2.17, Technician's notes 2.17

Digital resources image of a skyscraper

Key vocabulary elastic deformation, inelastic deformation, extension, compression, limit of proportionality, linear, non-linear, spring constant, elastic potential energy

Teaching and learning

Engage

Show students an image of a skyscraper. Ask: "How might the architects ensure that the building doesn't shake too much in an earthquake?" Allow the students to discuss any design features they think might be present, before telling them that skyscrapers are often mounted on (stiff) springs (that are strong enough to withstand the weight of the building).

Challenge and develop

Demonstrate an expendable spring being stretched and ask students to identify the forces acting in the demonstration. Ask why the spring doesn't just accelerate towards the ground (as the masses on their own would if they were dropped). [O1]

Explain

Students should carry out Practical 2.17 to investigate the stretching of a spring. [O1, O2]

Consolidate and apply

Issue each student with a forces and energy in springs worksheet at an appropriate level:

- Low demand: Worksheet 2.17.1 [O2, O3]
- Standard and high demand: Worksheet 2.17.2 [O2, O3]

 © HarperCollins*Publishers* Limited 2016

Extend

Ask students to give an example of an everyday situation where a spring with a) a high spring constant and b) a low spring constant should be used.

Plenary suggestions

The big ideas: Ask students to write down three ideas they have learned during the topic. Then ask them to share their facts in groups and to compile a master list of facts, with the most important at the top. Ask for ideas to be shared, and find out which other group(s) agreed.

What do I know? Ask students to each write down one thing about the topic they are sure of, one thing they are unsure of and one thing they need to know more about, being specific. Ask them to work in groups of five or six to agree on group lists; ask each group to say what they decided and agree as a class what they are confident about, what they are less sure of and what things they want to know more about.

Hot seat: Ask each student to think of a question, using material from the topic. Select someone to put in the hot seat; ask students to ask their questions and say at the end whether the answer is correct or incorrect.

Answers to questions

Worksheets 2.17.1 and 2.17.2

1.

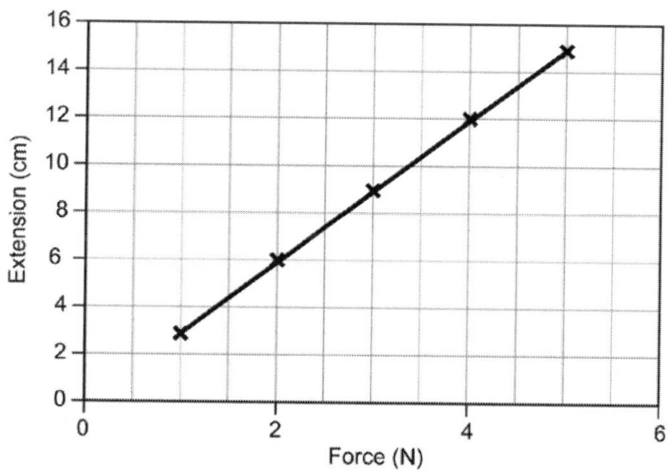

2. The gradient should be approximately 3.
4. 0.0438 J

Worksheet 2.17.1

5. 0.075 J

Worksheet 2.17.2

5. 0.258 m

 © HarperCollins*Publishers* Limited 2016

Lesson 2.18: Practical: Investigate the relationship between force and the extension of a spring

Lesson overview

OCR Specification reference

P2.3c, d, e

Learning objectives

- Interpret readings to show patterns and trends.
- Interpret graphs to form conclusions.
- Apply the equation for a straight line to the graph.

Learning outcomes

- [O1] Identify independent, dependent and therefore control variables.
- [O2] Plan a valid investigation.
- [O3] Analyse results from a practical investigation.

Skills development

- Carry out fair and accurate investigations.
- Analyse data collected from experiments.
- Draw sensible conclusions from experimental data.

Maths focus

- Draw a graph.
- Calculate the gradient of a graph.

Resources needed apparatus listed in Technician's notes; Practical sheet 2.18, Technician's notes 2.18

Digital resources an image of some mechanical weighing scales that are used in supermarkets

Key vocabulary Hooke's law, spring, stretching, measure, spring constant, extension, prediction, newtons, anomaly, mean

Teaching and learning

Engage

Show students an image of some mechanical weighing scales that are used in supermarkets. Ask the students to discuss how they think they work. Take feedback from a number of pairs to develop ideas.

Challenge and develop

- Issue students with their challenge: to investigate the relationship between the force on a spring and the extension.
- Students should be encouraged to think of different ways in which the relationship could be investigated safely in the lab. [O2]
- Ideas might include: elastic bands with masses hung on, a slinky spring with masses hung on, a piece of bungee cord (such as those used for securing loads in vehicles) with masses hung on, as well as the more obvious choice of a small spring with masses hung on.
- Ask students to plan their investigation, considering the following factors carefully: [O1, O2]
  - safety
  - practicability

 © HarperCollins*Publishers* Limited 2016

- control variables
- independent variable
- dependent variable
- the results table format
- the graph they will generate from their independent and dependent variables.

Explain

Ask students to investigate the relationship between the force on a spring and the extension (Practical sheet 2.18). [O1, O2]

Consolidate and apply

- Students should then analyse the results they have collected by plotting the graph of force (on the *y*-axis; note: this is different from the usual 'plot the independent variable on the *x*-axis' advice) and extension (on the *x*-axis). [O3]
- Plotting the graph in this way will enable the spring constant, *k*, to be found by calculating the gradient of the line.
- If students have been issued with 'different' springs, the spring constants can then be compared.

Extend

Ask students to consider the elastic limit of the spring – what happens after this? They could then be asked to categorise some everyday items as either plastic or elastic based on the material's behaviour.

Plenary suggestions

Heads and tails: Ask each student to write a question about something from the topic on a coloured paper strip and the answer on another coloured strip. Hand out the strips to groups of six to eight students, so that each student gets a question and an answer. One student reads out his or her question. The student with the correct answer then reads it out, followed by his or her question, and so on.

Where's the answer? As students enter the room, give them a card with a word written on it. At the end of the lesson, ask a question, and the students should say if they think they have the answer card and explain why.

What do I know? Ask students to each write down one thing about the topic they are sure of, one thing they are unsure of and one thing they need to know more about, being specific. Ask them to work in groups of five or six to agree on group lists; ask each group to say what they decided and agree as a class what they are confident about, what they are less sure of and what things they want to know more about.

 © HarperCollins*Publishers* Limited 2016

Lesson 2.19: Potential energy

Lesson overview

OCR Specification reference

P2.3f

Learning objectives

- Consider what happens when a spring is stretched.
- Describe what is meant by gravitational potential energy.
- Calculate the energy stored by an object raised above ground level.

Learning outcomes

- [O1] Describe different types of energy store, including elastic potential energy and gravitational potential energy.
- [O2] Calculate, using the elastic potential energy equation.
- [O3] Calculate, using the gravitational potential energy equation.

Skills development

- Evaluate methods and suggest possible improvements and further investigations.
- Carry out and represent mathematical and statistical analysis.
- Use SI units (for example, grams, metres, joules) and IUPAC chemical nomenclature unless inappropriate.

Maths focus

- Change the subject of an equation.
- Substitute numerical values into algebraic equations using appropriate units for physical quantities.
- Solve simple algebraic equations.

Resources needed clamp stands, clamps, bosses, springs of known spring constant, 100 g mass holders, 100 g masses, metre rules, clear sticky tape, card, scissors; Worksheets 2.19.1, 2.19.2 and 2.19.3, Practical sheet 2.19, Technician's notes 2.19

Digital resources elastic potential energy video, gravitational potential energy video

Key vocabulary energy store, energy transfer, elastic potential energy, gravitational potential energy, gravitational field strength

Teaching and learning

Engage

- Watch the elastic potential energy video showing archers shooting arrows. Challenge students' thinking and understanding of the energy involved. You could ask: What energy does the arrow have when moving? Where does this energy come from? What type of energy does the bow have when the bowstring is pulled back? How can the energy be increased? Why would different bows contain different amounts of energy?
- Lead a discussion on the energy being stored in the bow until released by the archer – all stored energies are called potential energies. Then introduce the equation for elastic potential energy: $E_e = ½ ke^2$. Formula triangles are trickier when there are more than three quantities, but show that it is possible with E_e at the top. Emphasise that to determine e, students should use the triangle to find e^2, then take the square root. Note: students will always be given this equation in examinations on the physics equation sheet. [O1, O2]
- Watch the gravitational potential energy video showing pole-vaulters. Challenge students' thinking and understanding of the energy involved. You could ask: What energy does the pole-vaulter have at the top of motion? Where does this energy come from? What type of energy does the pole have when it is bent? How would this be different for a heavier athlete? How would this be different on different planets?
- Re-emphasise the idea of stored energies as potential energies. Then introduce the equation for gravitational potential energy: $E_p = mgh$. Formula triangles are trickier with more than three quantities, but show that it is possible with E_p at the top. Note: students are expected to remember this equation in examinations. [O1, O3]

Challenge and develop

- The students should then carry out Practical 2.19, where they will investigate transfers between gravitational energy and elastic potential energy.

 © HarperCollins*Publishers* Limited 2016

- After the practical is done, discuss how the practical could have been improved, first in pairs, then groups of four, then groups of eight. Finally, someone from each group shares ideas for improvement with the class. [O1, O2, O3]

Explain

- Challenge students' thinking and understanding of elastic potential energy and gravitational potential energy from the practical activities using the evaluation section of Practical sheet 2.19. [O1, O2, O3]

Consolidate and apply

The students should now be given the potential energy worksheet suitable to their ability:

- Low demand – Worksheet 2.19.1 [O1, O2, O3]
- Standard demand – Worksheet 2.19.2 [O1, O2, O3]
- High demand – Worksheet 2.19.3 [O1, O2, O3]

Extend

Ask students able to progress further to do the following:

- Each student or student pair suggests how to produce a catapult that would fire an object to the greatest height. Discuss how you could test this. [O1, O2, O3]

Plenary suggestions

What's in the picture? Have a picture of an object such as a catapult concealed by rectangles, each of which can be removed in turn. Ask for suggestions as to which tile should be removed and what the partially revealed graphic shows. Encourage speculation and inference; ask for suggestions as to which tile should go next and continue until a full image has been produced.

Hot seat: Ask each student to think up a question, using material from the lesson. Select someone to put in the hot seat. Ask students to ask their question and say at the end whether the aswer is correct or incorrect.

Answers to questions

Worksheet question 1 (2.19.1 – a, b, d, e and f; 2.19.2 – a, c, d, f and g; 2.19.3 – a, c, d, g and h)

a) stored energy

b) any sensible answers, including: bows, wind-up toys, car suspensions

c) increase extension, higher spring constant

d) $E_e = \frac{1}{2} ke^2$

e)

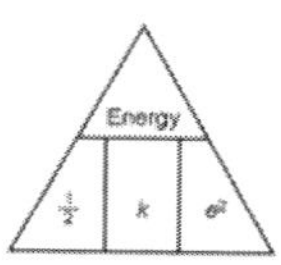

f) 147 J

g) 20 N/m

h) 0.20 m

Worksheet question 2 (2.19.1 – a, b, d, e and g; 2.19.2 – a, c, d, f and g; 2.19.3 – a, c, d, g and h)

a) any sensible answers, above ground (technically anything not at infinity; values would be negative but this isn't A-level)

b) increase height, greater mass

c) different value of g (gravitational field strength)

d) $E_p = mgh$

e)

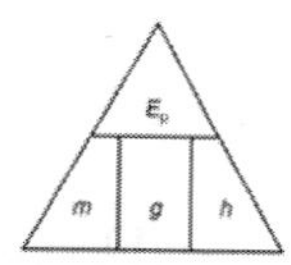

f) 3289 J; she would have had to jump higher than the bar to clear it

g) 100 m

h) 0.16 kg

Worksheet question 3 (2.19.1 – a and b; 2.19.2 – a, b and c; 2.19.3 – a, b, c and d)

a) 62.5 J

b) 62.5 J

c) 312.5 m

d) air resistance

Practical sheet evaluation

1. The greater the mass, the greater the loss in gravitational potential energy.
2. The greater the extension, the greater the gain in elastic potential energy.
3. Probably B, with the explanation that there is a difference due to air resistance; if students' results show A, then allow, with the explanation that as the mass falls, E_p converts to E_e.
4. Any sensible answers, including accounting for mass of spring and determining whether k changes over range of extension used.

 © HarperCollins*Publishers* Limited 2016

Lesson 2.20: Heavy or massive?

Lesson overview

OCR Specification reference

P2.3g, h, I, j

Learning objectives

- Identify the correct units for mass and weight.
- Explain the difference between mass and weight.
- Understand how weight is an effect of gravitational fields.

Learning outcomes

- [O1] Define mass and weight.
- [O2] Link weight to gravitational field strength.
- [O3] Calculate weight when given mass and gravitational field strength.

Skills development

- Carry out a valid experiment.
- Analyse experimental results to draw sensible conclusions.
- Use the correct units for mass and weight.

Resources needed a textbook, equipment listed in Technician's notes; Worksheets 2.20.1, 2.20.2 and 2.20.3, Practical sheet 2.20, Technician's notes 2.20

Key vocabulary mass, weight, gravity, gravitational field strength, newtons, kilograms

Teaching and learning

Engage

Pass a textbook round for each student to hold. The students should write down what they think the book ***weighs***. Take some suggestions from the students (without saying whether they are correct or not). At this point, students are likely to give an estimated ***mass***.

Challenge and develop

- **Pair talk:** Ask: "Would the book weigh the same if we took it to the Moon?" Introduce the idea that **weight** depends on gravitational field strength. [O1, O2]
- Pass the book round again. Ask students to suggest its weight and its mass. Following this, ask the students to suggest the units that weight and mass would be measured in, given that the unit for gravitational field strength is N/kg. [O1, O2]

Explain

- Students should use Practical sheet 2.20 to investigate $W = mg$. They will take a number of different objects and measure the masses (on a balance) and the weights (on a force meter). [O1]
- When students have collected the data, they should plot a graph of weight (on the *y*-axis) and mass (on the *x*-axis). They should then calculate the gradient of the line, which will be equal to the gravitational field strength. [O2]

Consolidate and apply

Issue each student with a worksheet on mass and weight at an appropriate level:

- Low demand: Worksheet 2.20.1 [O2, O3]
- Standard demand: Worksheet 2.20.2 [O2, O3]

 © HarperCollins*Publishers* Limited 2016

- High demand: Worksheet 2.20.3 [O2, O3]

Extend

An astronaut in deep space is at one end of her spacecraft and wants to move to the other end. She pushes against the inside of the craft with a force of 25 N, which causes her to accelerate at 0.5 m/s^2. What is her mass?

Plenary suggestions

Learning triangle: At the end of the lesson, ask students to draw a large triangle with a smaller inverted triangle that just fits inside it (so they have four triangles). Ask students to think back over the lesson and identify and write in the outer triangles:

- something they saw
- something they did
- something they discussed.

Ask students to write in the central triangle something they learned.

Heads and tails: Ask each student to write a question about something from the lesson on a coloured paper strip and the answer on another coloured strip. Hand out the strips to groups of six to eight students, so that each student gets a question and an answer. One student reads out his or her question. The student with the correct answer then reads it out, followed by his or her question, and so on.

Answers to questions

Worksheets 2.20.1, 2.20.2 and 2.20.3

1. the amount of matter in an object
2. the force on an object due to gravity
3. a) kilograms; b) newtons
4. 20 N
5. 1 kg
6. 550 N
7. 1.6 N/kg

 © HarperCollins*Publishers* Limited 2016

Lesson 2.21: Moments

Lesson overview

OCR Specification reference

P2.3k, l

Learning objectives

- Describe the turning effect of a force about a pivot.
- Explain and use the principle of moments.
- Explain what is meant by the centre of mass of an object.

Learning outcomes

- [O1] Define the term 'moment'.
- [O2] Carry out a practical investigation into the principle of moments.
- [O3] Determine the centre of mass of a metre rule.

Skills development

- Carry out a valid experiment.
- Analyse experimental results to draw sensible conclusions.
- Use the correct unit for moment.

Maths focus

- Multiply numbers.

Resources needed apparatus listed in Technician's notes; Worksheets 2.21.1, 2.21.2 and 2.21.3, Practical sheet 2.21, Technician's notes 2.21

Key vocabulary moment, clockwise, anticlockwise, centre of mass, centre of gravity, pivot, newton-metre

Teaching and learning

Engage

- **Pair talk:** Why is the handle of a door placed a long way from the hinge?

Challenge and develop

- Open the door. Ask a student to try to close it by pushing near the hinge. What does the student have to do to be able to close it? Ask the student to try to push the door further away from the hinge, keeping the force as constant as possible. [O1]
- From this, introduce the equation 'moment = force × distance' by linking the constant force but increased distance leading to a greater turning effect, known as a moment. [O1]

Explain

- The students should carry out Practical 2.21, to investigate the principle of moments. [O1, O2]
- Following the practical, students should be given a metre rule and, using what they have found out from their investigation, they should find the point on the ruler where it balances on their finger. Identify this point as the centre of mass of the ruler and define it as the point through which all the weight of the object seems to act. [O2, O3]

Consolidate and apply

Issue each student with a moments worksheet at an appropriate level:

 © HarperCollins*Publishers* Limited 2016

- Low demand: Worksheet 2.21.1 [O2, O3]
- Standard demand: Worksheet 2.21.2 [O2, O3]
- High demand: Worksheet 2.21.3 [O2, O3]

Extend

Why does a boat become less stable if you stand up in it? [O1]

Plenary suggestions

The big ideas: Ask students to write down three ideas they have learned during the topic. Then ask them to share their facts in groups and to compile a master list of facts, with the most important at the top. Ask for ideas to be shared and find out which other group(s) agreed.

Hot seat: Ask each student to think of a question, using material from the lesson. Select someone to put in the hot seat; ask students to ask their questions and say at the end whether the answer is correct or incorrect.

Answers to questions

Worksheets 2.21.1 and 2.21.2

4. The moment will increase.
5. The moment will increase.
6. a) 10 Nm b) 70 Nm
7. a) 3 m b) 2.3 m
8. a) 10 N b) 11 N
9. 1 m
10. 0.5 m

Worksheet 2.21.3

4. The moment will increase.
5. The moment will increase.
6. a) 10 Nm (1000 Ncm) b) 70 Nm (7000 Ncm)
7. a) 3 cm b) 2.3 cm
8. a) 100 N b) 110 N
9. 1 m
10. 0.5 m

 © HarperCollins*Publishers* Limited 2016

Lesson 2.22: Levers and gears

Lesson overview

OCR Specification reference

P2.3m

Learning objectives

- Describe how levers and gears can be used to transmit the rotational effect of a force.
- Explain how levers and gears transmit force.

Learning outcomes

- [O1] State the purpose of levers and gears.
- [O2] Distinguish between levers acting as force multipliers and levers acting as distance multipliers.
- [O3] State how gears transmit force.

Resources needed image of ancient Egyptians using levers, pictures of different levers used in everyday life, pictures of different everyday items that use gears, equipment listed in Technician's notes; Worksheets 2.22.1, 2.22.2 and 2.22.3, Practical sheet 2.22, Technician's notes 2.22

Digital resources http://s1114.photobucket.com/user/Torrisholme-Mick/media/WhiskKC.jpg.html

Key vocabulary lever, force multiplier, distance multiplier, pivot, fulcrum, gear

Teaching and learning

Engage

Pair talk: How were the pyramids in Egypt built when there was very little technology at the time? How were the heavy objects moved? Students could be shown an image of the ancient Egyptians using levers, to provide a visual prompt.

Challenge and develop

- Show students a number of pictures of levers being used in everyday life, such as using a screwdriver to remove the lid from a paint can, a pair of scissors, a wheelbarrow, a pair of pliers, an elbow joint. Ask them to discuss and then suggest how these work as simple machines. [O1]
- As the discussion develops, introduce the idea that levers can act as **force multipliers** (where the output force is much greater than the input force) or as **distance multipliers** (where the output force moves further than the input force). Ask students to categorise the levers in the pictures as either force multipliers or distance multipliers. [O1, O2]

Explain

- Show students an image of a hand-held egg whisk (e.g. http://s1114.photobucket.com/user/Torrisholme-Mick/media/WhiskKC.jpg.html) and ask them to discuss in pairs how they think it works.
- Take feedback from several groups to develop ideas. [O3]
- Show students a number of pictures of different everyday items that use gears, e.g. a hand-crank pencil sharpener, a bicycle, a can opener, a clock with hands, a wind-up toy. [O3]
- Ask students to suggest what the items have in common, in the way that they function. [O3]
- The students should then be given a number of plastic gearwheels of differing sizes to investigate their motion (Practical sheet 2.22). [O3]

Consolidate and apply

Issue each student with a levers and gears worksheet at an appropriate level:

 © HarperCollins*Publishers* Limited 2016

- Low demand: Worksheet 2.22.1 [O1, O2]
- Standard demand: Worksheet 2.22.2 [O1, O2]
- High demand: Worksheet 2.22.3 [O1, O2]

Extend

Students could be asked to show how a claw hammer used to remove a nail from a piece of wood is an example of a force multiplier. [O2]

Plenary suggestions

Ask me a question: Ask students to write a question about something from the topic and then a mark scheme for the answer. Encourage them to come up with questions worth more than one or two marks and to try their questions out on each other.

Where's the answer? As students enter the room, give them a card with a word written on it. At the end of the lesson, ask a question, and the students should say if they think they have the answer card and explain why.

Answers to questions

Worksheets 2.22.1, 2.22.2 and 2.22.3

Type 1 levers: oars on a boat, pliers, see-saw

Type 2 levers: crowbar, nutcracker, wheelbarrow

Type 3 levers: tweezers, stapler, cooking tongs

 © HarperCollins*Publishers* Limited 2016

When and how to use these pages: Check your progress, Worked example and End of chapter test

Check your progress

Check your progress is a summary of what students should know and be able to do when they have completed the chapter. Check your progress is organised in three columns to show how ideas and skills progress in sophistication. Students aiming for top grades need to have mastered all the skills and ideas articulated in the final column (shaded pink in the Student book).

Check your progress can be used for individual or class revision using any combination of the suggestions below:

- Ask students to construct a mind map linking the points in Check your progress
- Work through Check your progress as a class and note the points that need further discussion
- Ask the students to tick the boxes on the Check your progress worksheet (Teacher Pack CD). Any points they have not been confident to tick they should revisit in the Student Book.
- Ask students to do further research on the different points listed in Check your progress
- Students work in pairs and ask each other what points they think they can do and why they think they can do those, and not others

Worked example

The worked example talks students through a series of exam-style questions. Sample student answers are provided, which are annotated to show how they could be improved.

- Give students the Worked example worksheet (Teacher Pack CD). The annotation boxes on this are blank. Ask students to discuss and write their own improvements before reviewing the annotated Worked example in the Student Book. This can be done as an individual, group or class activity.

End of chapter test

The End of chapter test gives students the opportunity to practice answering the different types of questions that they will encounter in their final exams. You can use the Marking grid provided in this Teacher Pack or on the CD Rom to analyse results. This shows the Assessment Objective for each question, so you can review trends and see individual student and class performance in answering questions for the different Assessment Objectives and to highlight areas for improvement.

- Questions could be used as a test once you have completed the chapter
- Questions could be worked through as part of a revision lesson
- Ask Students to mark each other's work and then talk through the mark scheme provided
- As a class, make a list of questions that most students did not get right. Work through these as a class.

 © HarperCollins*Publishers* Limited 2016

Marking Grid for End of Chapter 2 Test

	Getting Started [Foundation Tier]								**Going Further [Foundation and Higher Tiers]**					**More Challenging [Higher Tier]**			**Most demanding [Higher Tier]**			**Total marks**	**Percentage**
Student Name	Q. 1 (AO1) 1 mark	Q. 2 (AO1) 1 mark	Q. 3 (AO1) 2 marks	Q. 4 (AO1) 1 mark	Q. 5 (AO2) 1 mark	Q. 6 (AO1) 1 mark	Q. 7 (AO2) 1 mark	Q. 8 (AO2) 3 marks	Q. 9 (AO1) 1 mark	Q. 10 (AO1) 1 mark	Q. 11 (AO2) 2 marks	Q. 12 (AO2) 4 marks	Q. 13 (AO1) 1 mark	Q. 14 (AO2) 1 mark	Q. 15 (AO1) 2 marks	Q. 16 (AO2) 6 marks	Q. 17 (AO1) 2 marks	Q. 18 (AO2) 2 marks	Q. 19 (AO3) 8 marks		

 © HarperCollins*Publishers* Limited 2016

Check your progress

You should be able to:

Know that forces are vectors and have magnitude and direction	Explain the difference between contact and non-contact forces	Represent vector quantities by arrows
Understand that average speed = distance/time	Know that acceleration is the rate at which speed changes	Calculate acceleration from change in velocity/time taken
Explain the significance of the gradient of a distance–time graph	Interpret a journey represented on a distance–time graph	Determine the instantaneous speed from the tangent to a distance–time graph of an accelerating object
Explain the significance of the gradient of a velocity–time graph	Interpret a journey represented on a velocity–time graph	Determine total distance travelled from a velocity–time graph
Recall the equation for uniform motion	Apply the equation for uniform motion	Rearrange the equation for uniform motion
Draw a free-body diagram to represent forces acting on an object	Calculate the resultant force acting on an object	Determine the components of a force
Apply Newton's first law to a stationary object and an object moving in a straight line at a constant speed	Link Newton's first law to the idea of a zero resultant force	Explain what is meant by inertia
State Newton's second law and recall the equation $F = ma$	Use $F = ma$ to determine force, mass or acceleration	Explain what is meant by inertial mass
Recognise that weight and mass are not the same	Explain the difference between weight and mass	Relate the ideas of weight and mass to Newton's second law
State Newton's third law	Apply Newton's third law to simple equilibrium situations	Explain how Newton's third law applies
Explain what is meant by momentum	Relate ideas about forces and kinetic energy, and to rate of change of momentum	Apply the principle of conservation of momentum to collisions
Explain that a moment is the turning effect of a force	Calculate the size and direction of a moment Explain how gears and levers transmit the rotational effect of a force	Apply the idea of moments to contexts such as the balancing of a seesaw
recognise that when a force moves an object along the line of action of the force, work is being done state that various devices do work and, in doing so, transfer energy	calculate the work done by a force from the size of the force and the distance moved	use the equation for work done to solve problems, including changing the subject of the equation

 © HarperCollins*Publishers* Limited 2016

Worked example

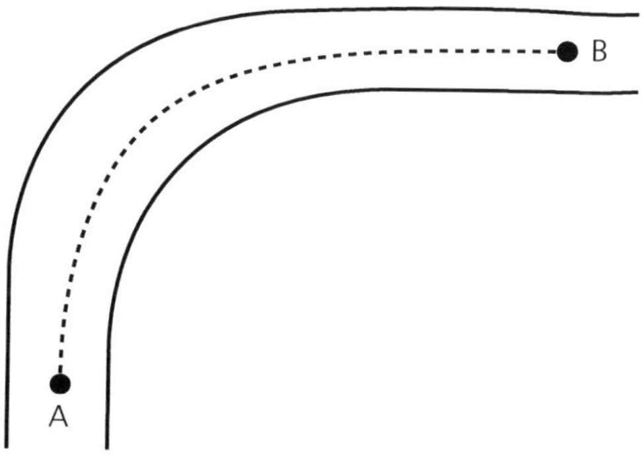

The diagram shows a car travelling round a bend at a constant speed of 20 m/s. It took the car 4.5 s to move from point A to B.

1. **Explain why the distance travelled by the car from A to B is greater than the car's displacement at B.**

   The car's displacement is the distance from A to B in a straight line. The car travelled a further distance as it was going round the curve.

   This is correct. You should also note that distance is a scalar quantity but displacement is a vector quantity with size and direction.

2. **Calculate the distance the car travelled from A to B.**

   $v = \frac{s}{t}$ so $s = \frac{v}{t}$

   $s = \frac{20}{4.5}$

   $s = 4.44$ m

   The student correctly remembered the equation $v = \frac{s}{t}$ but failed to rearrange it correctly as $s = v \times t$ to give the answer 90 m.

3. **Explain why the car is accelerating as it travels round the bend at a constant speed.**

   Acceleration is a change in velocity and the car's velocity is changing as it goes round the bend.

   Velocity is speed in a given direction. As the car goes round the corner, its speed stays the same but the direction is constantly changing, so its velocity is constantly changing which is acceleration.

4. **Explain what is meant by stopping distance for a car and what factors might affect this.**

   Stopping distance is the distance the car takes while braking and the distance travelled while thinking to apply the brakes. Ice on the road will cause the brakes not to work and increase the stopping distance.

   You should aim to give at least two factors that affect thinking distance and braking distance. Thinking distance can be affected by tiredness, drugs and alcohol. Braking distance can be affected by adverse road conditions such as rain or ice, and worn tyres or brakes.

© HarperCollins*Publishers* Limited 2016

Electricity: Introduction

When and how to use these pages

This unit will build on ideas the students have met before, such as:

- There are two kinds of electric charge – positive and negative: like charges repel and unlike charges attract.
- An electric current is due to the flow of charge.
- The current in a circuit can be controlled by resistors.
- Cells provide a source of electrical energy; batteries are a number of cells joined together.
- The electrical supply in a home is alternating current (ac).
- Mains electricity has a high potential difference.
- Fuses switch off the current if a fault occurs.

Overview of the unit

In this unit, students will learn about static and current electricity. They will investigate different types of circuit and learn some of the key features of each type of circuit. They will distinguish between current and potential difference and investigate factors that affect resistance in a circuit. They will discover how electricity is transmitted to homes and the features of mains electricity. They will investigate power and energy transfers and calculate power.

This unit offers a number of opportunities for students to investigate phenomena through practical work, work collaboratively with peers and critically evaluate evidence before drawing conclusions.

Obstacles to learning

Students may need extra guidance with the following common misconceptions:

- confusion between the terms 'current' and 'potential difference'
- current, energy and voltage get 'used up'
- voltage passes through things
- a cell and a battery are the same thing
- batteries 'die'.

Practicals in this unit

In this unit students will do the following practical work:

- research static electricity
- plot a 'field'
- build circuits to determine the resistance of a fixed resistor and a filament lamp
- Practical: investigate the characteristics of a thermistor, a light-dependent resistor and a diode
- test insulators and conductors
- investigate the properties of thermistors, light-dependent resistors (LDRs) and diodes
- determine the energy transferred by and power ratings of a range of household appliances
- investigate the power rating of different domestic appliances
- Practical: investigate factors that affect resistance in a circuit
- investigate current, potential difference and resistance in series and parallel circuits
- use a circuit to determine the resistance of a fixed resistor

 © HarperCollins*Publishers* Limited 2016

	Lesson title	Overarching objectives
1	Static electricity	• Describe how insulating materials can become charged. • Know that there are two kinds of electric charge. • Explain these observations in terms of electron transfer.
2	Electric fields and currents	• Explain what an electric field is. • Draw an electric field pattern for a charged sphere. • Use the idea of an electric field to explain electrostatic attraction and sparking. • Recall that an electric current is a rate of flow of electric charge.
3	Electric circuits and potential difference	• Know circuit symbols. • Recall that current (I) depends on resistance (R) and potential difference (V). • Explain how an electric current passes round a circuit.
4	Series and parallel circuits	• Recognise series and parallel circuits. • Describe the changes in the current and potential difference in series and parallel circuits.
5	Resistance	• Set up a circuit to investigate resistance. • Investigate the changing resistance of a filament lamp. • Compare the properties of a resistor and a filament lamp.
6	Practical: Use circuit diagrams to set up and check appropriate circuits to investigate the factors affecting the resistance	• Use a circuit to determine resistance. • Gather valid data to use in calculations. • Apply the circuit to determine the resistance of different components.
7	Investigating circuits	• Classify materials as either conducting or insulating. • Use series circuits to test components and make measurements. • Carry out calculations on series circuits.
8	Control circuits	• Use a thermistor and a light-dependent resistor (LDR). • Investigate the properties of thermistors, LDRs and diodes.
9	Power and energy transfers	• Describe the energy transfers in different domestic appliances. • Describe power as a rate of energy transfer. • Calculate the energy transferred.

 © HarperCollins*Publishers* Limited 2016

10	Calculating power	• Calculate power. • Use power equations to solve problems. • Consider power ratings and changes in stored energy.
11	Investigating series and parallel circuits	• Use a circuit to determine resistance. • Gather valid data for use in calculations. • Apply the circuit to determine the resistance of combinations of components.
12	What's the difference between potential difference and current?	• Understand and be able to apply the concepts of current and potential difference. • Use these concepts to explain various situations.
13	Using algebra in electric circuit calculations	• Solve algebraic equations including using appropriate substitutions, numerical values and units. • Change the subject of an equation. • Use the symbols $=, <, \ll, \gg, >, \propto, \sim$.
14	Maths skills: Using formulae and understanding graphs	• Recognise how algebraic equations define the relationships between variables. • Solve simple algebraic equations by substituting numerical values. • Describe relationships expressed in graphical form.

© HarperCollins*Publishers* Limited 2016

Lesson 3.1: Static electricity

Lesson overview

OCR Specification reference

P3.1a, b, c

Learning objectives

- Describe how insulating materials can become charged.
- Know that there are two kinds of electric charge.
- Explain these observations in terms of electron transfer.

Learning outcomes

- [O1] Describe how insulating materials can become charged.
- [O2] Recall that there are two types of charge and that like charges repel and unlike charges attract.
- [O3] Explain how a person can get an electric shock and explain static electricity in terms of electric fields.

Skills development

- Carry out focused research and describe findings to peers.

Resources needed Van de Graaff generator, polythene rods, Perspex rods, dusters, small pieces of paper, balloons, watch glasses; Worksheet 3.1

Key vocabulary attract, conductor, electron, insulator, repel

Teaching and learning

Engage

- Issue the students with a sheet of scrap paper and ask them to write what they understand by the following terms: 'a conductor', 'an insulator', 'attraction' and 'repulsion'. Follow this by asking students to give an example for each of these terms.
- Alternatively, the words and definitions of the words could be written on separate cards and the students could be asked to match them up.

Challenge and develop

Show the students a range of demonstrations to illustrate the effect of charging an insulator, for example:

- Charge a polythene rod with a duster and use it to pick up small pieces of paper (circles of paper cut with a hole puncher would be ideal). [O1, O2]
- Rub a balloon on a jumper and either stick it to the wall or hold it close to someone's hair. [O1, O2]
- Charge a polythene rod and hold it close to (but not touching) a very thin stream of water from the tap; watch the water 'bend'. [O1, O2]
- Arrange two watch glasses on top of one another and rest a Perspex rod on the top of one of them; charging a second Perspex rod with a duster. If the charged rod is held next to the rod resting on the watch glasses, the watch glass on top will rotate due to the repulsion. [O1, O2]

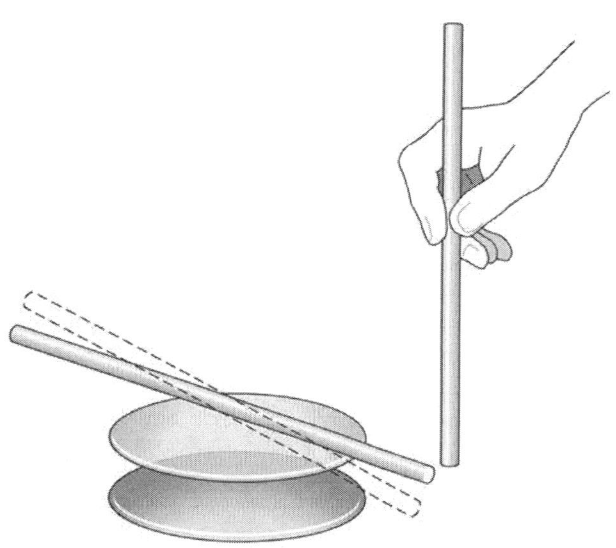

 © HarperCollinsPublishers Limited 2016

Explain

Jigsaw: The students split into five groups; each group should be given a situation involving static electricity and access to research materials. Students should use the research materials to find out about their given situation. Students should be given a situation appropriate to their level:

- Low demand: situations 1 and 2 [O1, O2, O3]
- Standard demand: situation 3 [O1, O2, O3]
- High demand: situations 4 and 5 [O1, O2, O3]

The situations suggested are:

1. rubber used on aircraft tyres
2. anti-static flooring in operating theatres
3. cloud-to-ground lightning
4. photocopiers
5. paint spraying.

Consolidate and apply

- The students should then regroup into groups of five so that there is one person from each original group in each new group. [O1, O2, O3]
- Students should use the table on Worksheet 3.1 as a 'script' for sharing their research findings between the members of the group so that they all have the same information by the end of the feedback session. [O1, O2, O3]

Extend

Ask students able to progress further to do one of the following:

- Explain how a Van de Graaff generator works, using ideas they have met in the lesson. It might be useful to show them a Van de Graaff generator if one is available, or a video clip of one if the equipment is not available. [O1, O2]
- Alternatively, the students could be asked to explain why static charges do not build up on a conductor.

Plenary suggestions

Ideas hothouse: Ask students to work in pairs to list three points about what they know or have learned about static electricity from the lesson today. Then ask the pairs to join together into groups of four and then groups of six to eight to discuss this further and to come up with an agreed list of points. Ask one person from each group to report back to the class.

Hot seat: Ask each student to think of a question, using material from the topic. Select someone to put in the hot seat. Ask students to ask their questions and say at the end whether the answer is correct or incorrect.

 © HarperCollins*Publishers* Limited 2016

Lesson 3.2: Electric fields and currents

Lesson overview

Learning objectives

- Explain what an electric field is.
- Draw an electric field pattern for a charged sphere.
- Use the idea of an electric field to explain electrostatic attraction and sparking.
- Recall that an electric current is a rate of flow of electric charge.

Learning outcomes

- [O1] State what the lines on a field diagram show.
- [O2] Compare magnetic and electric fields.
-
- [O3] Use ideas about electric fields to explain how a lightning conductor works.

Skills development

Summarise information clearly.

Maths focus

Resources needed bar magnets, iron filings, plotting compasses; Worksheets 3.2.1, 3.2.2 and 3.2.3

Key vocabulary electric field, spark, field lines, electron

Teaching and learning

Engage

- Show the students a video clip of the effect of an electric field on semolina seeds in castor oil (suggested YouTube search 'semolina electric field'). It might be a good idea to play the clip without any sound to begin with and ask the students to write a commentary.
- **Pair talk:** Ask the students to discuss what they think is happening in the video clip.
- **Pairs to fours:** Ask pairs of students to join together into groups of four to develop their ideas about what they think could be happening before taking feedback from some of the groups.
- The clip could then be played again, this time with sound, and the students could see how close their commentaries were. [O1]

Challenge and develop

- Issue the students with a bar magnet (ideally wrapped in a plastic bag to prevent iron filings coming into contact with the magnet) and a container of iron filings. Ask the students to place the bar magnet under a sheet of A4 plain paper and carefully sprinkle the iron filings over the magnet. Ask the students to describe the pattern they see. [O1, O2]
- Then ask the students to return the iron filings to the container before placing the bar magnet on top of the piece of paper. They should then draw around the magnet in pencil. Taking a plotting compass, they should place the compass at different points around the magnet and mark on the direction of the arrow on the compass. Once they have plotted about 25–30 points, they should join up the plots and mark on the direction of the magnetic field. [O1]
- Ask students to explain what they think the lines are showing.

Explain

- Now issue the students with the electric fields worksheet appropriate to their level. This worksheet will require them to compare magnetic and electric fields.

 © HarperCollins*Publishers* Limited 2016

- Low demand: Worksheet 3.2.1 [O1, O2]
- Standard demand: Worksheet 3.2.2 [O1, O2]
- High demand: Worksheet 3.2.3 [O1, O2]

- The students should then be given a chance to share their ideas, before the correct answers are given.

Consolidate and apply

- The students should carefully read the sections in the Student Book on 'Sparking' and 'Electric current and charge'. [O3]
- Once they have read it, they should summarise the ideas in each section in no more than 30 words for each section. [O3]
- Alternatively, students could be shown a video clip of a Faraday cage or Richard Hammond struck by lightning in a car (suggested YouTube search: 'Richard Hammond Faraday cage') and then could be asked to explain what they have seen using ideas about electric fields they have met in the lesson. [O3]

Extend

Ask students able to progress further to do the following:

- Explain how a lightning conductor works, using ideas about electric fields. [O3]

Plenary suggestions

Learning triangle: At the end of the lesson, ask students to draw a large triangle with a smaller inverted triangle that just fits inside it (so they have four triangles). In the outer three ask them to think back over the lesson and identify (and write in the respective triangles):

- something they saw
- something they did
- something they discussed.

Ask them to add in the central triangle something they learned.

Ask me a question: Ask students to write a question about something from the lesson and then devise a mark scheme for the answer. Encourage them to come up with questions worth more than one or two marks and to try their questions on one another.

Answers to questions

Worksheets 3.2.1, 3.2.2 and 3.2.3

Field	Force between	Attraction between	Repulsion between	Direction of arrows	Difference
Magnetic	poles	opposite poles	like poles	from north to south	cannot have N without S
Electric	charges or charged particles	opposite charges	like charges	from positive to negative	can have separate positive and negative charges

 © HarperCollins*Publishers* Limited 2016

Lesson 3.3: Electric circuits and potential difference

Lesson overview

OCR specification reference

P3.1e, f, g

Learning objectives

- Know circuit symbols.
- Recall that current (*I*) depends on resistance (*R*) and potential difference (*V*).
- Explain how an electric current passes round a circuit.

Learning outcomes

- [O1] Recall that an electric current is a flow of electrical charge and is measured in amperes *(A)*.
- [O2] Remember that charge is measured in coulombs *(C)* and recall and use the equation $Q = It$.
- [O3] Explain the concept that current is the rate of flow of charge. Rearrange and apply the equation $Q = It$.

Skills development

Use a model to explain a concept.

Maths focus

Use and rearrange an equation with three quantities.

Resources needed Worksheets 3.3.1, 3.3.2, 3.3.3 and 3.3.4

Key vocabulary potential difference, parallel, resistance, voltmeter, coloumb, charge, current

Teaching and learning

Engage

- Issue the students with Worksheet 3.3.1. The students should match the component symbol with the component name. [O1]
- Alternatively, a set of dominoes could be made with the component name and a different circuit symbol on each, and students could be asked to match the dominoes. [O1]

Challenge and develop

- Develop ideas that:
  1. current is a measure of how many electrons are moving (past a point (per second))*
  2. 'voltage' is a measure of how much energy each electron is transferring / transfers*
  3. in developing a mental model, our ideas have to match what actually happens in real circuits.

*Point out that these are not completely correct definitions, but at this stage avoid concepts of 'charge' and 'coulomb' in our model. [O1, O3]

- To do this, build a student circuit as follows:
  1. Seat a student (at the front) – they are a cell.
  2. Seat a second student in a suitable place (there will need to be a convenient 'conducting path' to be a lamp).
  3. Question – "Is the lamp shining?" Answer – no, as there is no energy getting to the bulb. Give the 'cell' a pile of paper (scrap A4 torn to A6 size is suitable). Q – "Is the lamp shining now?" A – no; energy is not transferred as there is no pathway and no mechanism to transfer the energy. The cell cannot 'throw' the energy since in the real world, lamps do not light when you put a cell nearby.

 © HarperCollins*Publishers* Limited 2016

4. Introduce a student to be an electron – they should collect one piece of paper from the cell, follow the conducting path and transfer the paper to the lamp.
5. Q – "What does the bulb do now?" A – it transfers energy to heat and light (the student acting as the lamp should tear the paper in two) and emits the energy (throw the paper up into the air (safely)). Q – "Why do we need a complete circuit?" A – so the electron can continue round and collect more energy.
6. Introduce additional electrons – with the rules that: 1 – electrons cannot overtake each other (because they repel); 2 – each time we run the circuit, electrons should remember their starting position and stop in the same place when we 'switch off'.
7. Introduce a second cell – electrons collect one piece of paper (energy) from each cell. Run the circuit. Q – "What effect will this have on the bulb?" A – it will be brighter – more energy.
8. Introduce a second bulb in series. Run the circuit (add more electrons as you like!) – electrons should transfer one piece of paper to each lamp. Q – "How is the brightness of each lamp affected?" A – it returns to its original brightness.
9. Now introduce the idea of current. Station three students around the circuit (these will be ammeters) as follows:
   - one at the 'start' next to the cells
   - one between the bulbs
   - one at the 'end' next to the cells.

   Their job is to count how many 'electrons' pass them when the circuit runs (electrons need to stop where they start!). Run the circuit. Compare the 'measurements' – they should all be the same.
10. Now introduce the idea of 'voltage' (to save distinguishing between potential difference and emf at this stage). Station the students in three places (these will be voltmeters; leave the 'ammeters' in place):
    - one opposite the cells
    - one opposite each lamp.

    Their job is to count the pieces of paper given out / torn up respectively.

    Run the circuit. "How do the totals compare?" (leads to the energy supplied being shared between the lamps) Divide the 'voltmeter' totals by the 'ammeter' totals (leads to the J/C idea and is needed for the parallel circuit version to add up correctly).
11. Parallel version: in this version, again run with two 'cells' in the battery. Place the two 'lamps' so that parallel paths are available in the room. This time, when the 'electrons' reach the junction they should take the path the electron in front of them didn't take (i.e. alternating). With care, they should join back into the same order when the branches combine. 'Ammeters' and 'voltmeters' run as before, but will need four of each for a simple parallel circuit of two lamps.
12. Resistance: the concept of resistance within the wires can be discussed by asking for the effect of placing chairs around the conducting pathway (care needed if this is actually tried!).

- The conclusions to measurements made can be compared with those made in real circuits next lesson, leading to Kirchoff's rules.

Explain

- Using the ideas from the model they have just experienced, the students should now be asked to come up with their own models for remembering the difference between electric current, potential difference and resistance. [O1, O3]
- **Pairs to fours:** Once they have come up with their models, they should share their ideas in a pair, before joining with another pair and sharing their ideas further. During this activity, they should discuss and improve their models following questioning or feedback from their peers. [O1, O3]

 © HarperCollins*Publishers* Limited 2016

Consolidate and apply

- The students should be issued with the electric current worksheet suitable to their ability:
  - Low demand: Worksheet 3.3.2
  - Standard demand: Worksheet 3.3.3
  - High demand: Worksheet 3.3.4

Extend

Ask students able to progress further to do the following:

- Describe what superconductors are and explain how they work. They should also suggest some uses for superconductors.

Plenary suggestions

Learning triangle: At the end of a lesson ask students to draw a large triangle with a smaller inverted triangle that just fits inside it (so they have four triangles). In the outer three ask them to think back over the lesson and identify (and write in the respective triangles):

- something they saw
- something they did
- something they discussed.

Ask students to write in the central triangle something they learned.

Heads and tails: Ask each student to write a question about something from the lesson on a coloured paper strip and the answer on another colour. In groups of six to eight, hand out the strips so that each student gets a question and an answer. One student reads out their question. The student with the right answer then reads it out, followed by their question. This could take the form of calculations using $Q = It$ and $V = IR$.

What do I know? Ask students to each write down one thing about the topic they are sure of, one thing they are unsure of and one thing they need to know more about, being specific. Ask them to work in groups of five or six to agree on group lists, ask each group to say what they decided and agree as a class what they are confident about, what they are less sure of and what things they want to know more about.

Answers to questions

Worksheets 3.3.2 and 3.3.3

1. 180 C
2. 120 C
3. 10 A
4. 12 s
5. 18 V
6. 75 V
7. 5 A
8. 4 Ω

Worksheet 3.3.4

1. 180 C
2. 360 C
3. 5 A
4. 12 s
5. 18 V
6. 0.075 V
7. 0.05 A
8. 4 Ω

 © HarperCollins*Publishers* Limited 2016

Lesson 3.4: Series and parallel circuits

Lesson overview

OCR Specification reference

P3.2a, i and j (part)

Learning objectives

- Recognise series and parallel circuits.
- Describe the changes in the current and potential difference in series and parallel circuits.

Learning outcomes

- [O1] Draw and recognise series and parallel circuits. Compare the brightness of lamps connected in series and parallel.
- [O2] Recall that the current in a series circuit is always the same and that the total current in a parallel circuit is the sum of the currents through each branch.
- [O3] Recall and apply the equation $V = IR$ and, for a series circuit, $R_{total} = R_1 + R_2$.

Skills development

- Follow circuit symbol conventions.
- Use circuit symbols to construct a physical circuit from which measurements can be made.

Maths focus

- Use and rearrange an equation with three quantities.

Resources needed cells, leads, lamps, ammeters, voltmeters, resistors – typically 1 to 10 Ω; Worksheets 3.4.1, 3.4.2 and 3.4.3

Key vocabulary series circuit, parallel circuit, current, potential difference, resistance

Teaching and learning

Engage

- Ask students whether the appliances in their houses are connected in series or parallel circuits. Ask students to justify their answers. [O1]
- Give the students a few minutes to think about their answers before sharing ideas with the group. [O1]

Challenge and develop

- The students should begin by building a simple series circuit with two cells and two lamps. They should measure the current either side of the lamps and between the lamps, and the potential difference across each lamp. They should record their results in a suitable table. The series circuit diagram in Figure 3.12 on page 108 of the Student Book could be shown if the students need a diagram to work from.
- Compare the measurements with those from the model in Lesson 3.3.
- Students should then build a simple parallel circuit with two cells and two lamps. They should measure the current before the split, next to each lamp and after the split, and the potential difference across each lamp. The parallel circuit diagram in Figure 3.12 could be shown if the students need a diagram to work from. Again, students should record their measurements in a suitable table before comparing their measurements with those generated from the model last lesson. [O1, O2]

 © HarperCollins*Publishers* Limited 2016

Explain

- The students should then build a series circuit with two cells and two resistors and measure the current before, between and after the resistors, and the potential difference across each resistor, recording their measurements in a table. Figure 3.13 on page 108 in the Student Book could be used as a reference.
- Finally, the students should build a parallel circuit with two cells and two resistors and measure the current before the split and next to each resistor and the potential difference across each resistor, recording their measurements in a table. Figure 3.14 on page 109 in the Student Book could be used as a reference if necessary. [O1, O2]

Consolidate and apply

Issue each student with a series and parallel circuits worksheet at an appropriate level:

- Low demand: Worksheet 3.4.1 [O3]
- Standard demand: Worksheet 3.4.2 [O3]
- High demand: Worksheet 3.4.3 [O3]

Extend

Ask students able to progress further to do the following:

- Research Kirchhoff's laws.

Plenary suggestions

Learning triangle: At the end of the lesson, ask students to draw a large triangle with a smaller inverted triangle that just fits inside it (so they have four triangles). Ask students to think back over the lesson and identify and write in the outer triangles:

- something they saw
- something they did
- something they discussed.

Ask students to write in the central triangle something they learned.

What do I know? Ask students to each write down one thing about the topic they are sure of, one thing they are unsure of and one thing they need to know more about, being specific. Ask them to work in groups of five or six to agree on group lists; ask each group to say what they decided and agree as a class about what they are confident about, what they are less sure of and what things they want to know more about.

Answers to questions

Worksheets 3.4.1 and 3.4.2

1. 6 V
2. 72 V
3. 120 V
4. 3 Ω
5. 0.83 Ω
6. 0.4 Ω
7. (a) 0.4 A (b) 30 Ω

Worksheet 3.4.2

7. (c) 8 V

Worksheet 3.4.3

1. 6 V
2. 3 Ω
3. 0.83 Ω
4. 0.4 Ω
5. (a) 0.4 A (b) 30 Ω (c) 8 V
6. (a) 0.6 A (b) 12 V (c) smaller than 10 Ω as R_{total} is always smaller than the smallest resistance of any component.

 © HarperCollinsPublishers Limited 2016

Lesson 3.5: Resistance

Lesson overview

OCR specification reference

P3.2d, e

Learning objectives

- Set up a circuit to investigate resistance.
- Investigate the changing resistance of a filament lamp.
- Compare the properties of a resistor and a filament lamp.

Learning outcomes

- [O1] Set up a circuit to investigate the relationship between *V, I* and *R* for a fixed resistor.
- [O2] Draw *I-V* graphs for a fixed resistor.
- [O3] Draw an *I-V* graph for a filament lamp.

Skills development

- Follow circuit symbol conventions.
- Use circuit symbols to construct a physical circuit from which measurements can be made.

Maths focus

- Plot a graph of current against potential difference and interpret the line produced.

Resources needed cells, leads, piece of copper wire, filament lamp, ammeters, voltmeters, variable resistors; Practical sheet 3.5, Technician's notes 3.5

Key vocabulary filament lamp,ohmic, non-ohmic, resistance, potential difference, current

Teaching and learning

Engage

- Show the students an image of a filament lamp and an energy saving light bulb.
- Ask the students to list as many similarities and differences between the two as they can.
- Ask the students to suggest some advantages to using the energy saving bulbs. Can they think of any disadvantages?

Challenge and develop

- The students should begin by building a circuit to measure the resistance of a length of copper wire (see Practical sheet 3.5, part 1). [O1]
- They should take at least 6 pairs of readings before plotting a graph of current (on the y-axis) against potential difference (on the y-axis). [O1, O2]
- They should then use the gradient of the line to calculate the resistance of the copper wire (this will be equal to 1/gradient). [O1, O2]

Explain

- Explain to the students that if a component gives a straight line through the origin, then the two quantities plotted are directly proportional to one another, and in this case, the component obeys Ohm's law and is known as an ohmic resistor. Remind the students of Ohm's law and ask them to explain how the constant of proportionality is reached from the graph.

 © HarperCollins*Publishers* Limited 2016

Consolidate and apply

- The students should now use the same circuit, but this time replace the copper wire with a filament lamp. (see Practical sheet 3.5, part 2). [O3]
- Again, the students should take at least 6 pairs of readings before plotting a graph of current (on the y-axis) against potential difference (on the x-axis). [O3]
- They should draw a curved line of best fit through their points. They could then be asked to calculate the gradient of the graph at different values for potential difference and use the value for the gradient to work out the resistance at each potential difference (this will be given by 1/gradient). [O3]

Extend

Ask students able to progress further to do the following:

- Suggest why the resistance of a filament lamp increases as the temperature increases. They should use ideas about particles in their explanation

Plenary suggestions

Hot seat: Ask each student to think up a question, using material from the lesson. Select someone to put in the hot seat, ask pupils to ask their question and say at the end whether the answer is correct or incorrect.

Heads and tails: Ask each student to write a question about something from the topic on a coloured strip of paper and the answer on another colour. In groups of six to eight, hand out the strips so that each student gets a question and an answer. One student reads out their question. The student with the right answer then reads it out, followed by their question, and so on.

 © HarperCollins*Publishers* Limited 2016

Lesson 3.6: Practical: Use circuit diagrams to set up and check appropriate circuits to investigate factors affecting the resistance

Lesson overview

Learning objectives

- Use a circuit to determine resistance.
- Gather valid data to use in calculations.
- Apply the circuit to determine the resistance of different components.

Learning outcomes

- [O1] Gather valid data.
- [O2] Plot data on a graph.
- [O3] Use a graph to draw conclusions.

Skills development

- Understand the scientific process.
- Use graphs to analyse data.

Maths focus

- Plot and then evaluate graphs.

Resources needed Practical sheet 3.6, Technician's notes 3.6

Key vocabulary Ohm's law, filament lamp, diode, resistor, ohmic conductor, independent variable, dependent variable

Teaching and learning

Engage

Show the students an image of an LED and ask them a) what it is and b) what it is used for. You might like to show them a remote control and tell them that there are two diodes in operation – one that emits visible light to let you know that the signal is being sent and one that emits a signal at an invisible wavelength.

Challenge and develop

- Split the class into three different groups and ask them to use the test circuit to generate I-V data for three components. One group should generate data for the filament lamp, one for the diode and one for the resistor at constant temperature. Details of how to do this are on Practical sheet 3.6. [O1]
- Once students have collected the data, they should share it and plot the I-V graphs for each of the three components. [O1, O2]

Explain

- Ask the students to recall Ohm's law from earlier in the unit. If necessary, give them some time to go back and look it up in either their notes or a suitable textbook. [O3]
- Draw the students' attention to the fact that Ohm's law states that the potential difference is proportional to the current, and that the proportionality constant is the resistance.
- Ask the students to state how a graph can be used to show that one factor is proportional to another (straight line graph that goes through the origin).

Consolidate and apply

- The students should use the graphs they drew earlier in the lesson to decide which of the components they tested obeyed Ohm's law and which didn't, giving a reason for their answer. [O3]

 © HarperCollins*Publishers* Limited 2016

- The students should then answer the questions from the Student Book appropriate to their ability:
  - Low demand: questions 1, 2 and 4
  - Standard demand: questions 1, 3, 5 and 6
  - High demand: questions 2, 4, 5 and 6

Extend

Ask students able to progress further to do the following:

- Research the structure, function and uses of semiconductors.

Plenary suggestions

The big ideas: Ask students to write down three ideas they learned during the topic. Then ask them to share their facts in groups and to compile a master list of facts, with the most important at the top. Ask for ideas to be shared and find out which other group(s) agreed.

Ideas hothouse: Ask students to work in pairs to list points about what they know about a particular idea. Then ask the pairs to join together into groups of four and then groups of six to eight to discuss this further and to come up with an agreed list of points. Ask one person from each group to report back to the class.

 © HarperCollins*Publishers* Limited 2016

Lesson 3.7: Investigating circuits

Lesson overview

OCR Specification reference

P3.2j (part)

Learning objectives

- Classify materials as either conducting or insulating.
- Use series circuits to test components and make measurements.
- Carry out calculations on series circuits.

Learning outcomes

- [O1] Set up a test circuit and make measurements from the test circuit.
- [O2] Recall that the current through a component depends on the resistance of the component and the potential difference across it.
- [O3] Recall and apply the equation $V = IR$ and for a series circuit $R_{total} = R_1 + R_2$.

Skills development

- Follow circuit symbol conventions.
- Use circuit symbols to construct a physical circuit from which measurements can be made.

Maths focus

- Use and rearrange an equation with three quantities.

Resources needed cells, leads, lamps and buzzers, ammeters, voltmeters, variable resistors, range of materials to test (see Practical sheet 3.7); Worksheets 3.7.1, 3.7.2 and 3.7.3, Practical sheet 3.7, Technician's notes 3.7

Key vocabulary series circuit, current, potential difference, resistance, ammeter, equivalent resistance

Teaching and learning

Engage

- Ask the students to write a list of as many conducting and insulating materials as they can. [O1]
- Ask the students to share their lists with someone else, adding to their lists as appropriate. [O1]
- Finally, ask them to suggest how they can test whether they have the materials in the correct columns. [O1]

Challenge and develop

- The students should begin by building a test circuit with two cells, a buzzer and two terminals. They should then take a range of different materials, as suggested in Practical sheet 3.7, and use the circuit to test whether the materials are conductors or insulators.
- Students could then amend their lists from the start of the lesson to reflect their findings from the practical task.

Explain

- Students should then build a test circuit to calculate the resistance of a lamp. They should measure the current and the potential difference and use their measurements to calculate the resistance. [O2]

Consolidate and apply

Issue each student with an investigating circuits worksheet at an appropriate level:

- Low demand: Worksheet 3.7.1 [O2, O3]

 © HarperCollins*Publishers* Limited 2016

- Standard demand: Worksheet 3.7.2 [O2, O3]
- High demand: Worksheet 3.7.3 [O2, O3]

Extend

Ask students able to progress further to do the following:

- Research and explain what equivalent resistance is.

Plenary suggestions

Learning triangle: At the end of the lesson, ask students to draw a large triangle with a smaller inverted triangle that just fits inside it (so they have four triangles). Ask students to think back over the lesson and identify and write in the outer triangles:

- something they saw
- something they did
- something they discussed.

Ask students to write in the central triangle something they learned.

What do I know? Ask students to each write down one thing about the topic they are sure of, one thing they are unsure of and one thing they need to know more about, being specific. Ask them to work in groups of five or six to agree on group lists; ask each group to say what they decided and agree as a class about what they are confident about, what they are less sure of and what things they want to know more about.

Answers to questions

Worksheets 3.7.1, 3.7.2 and 3.7.3

1. b) 0.5 A
2. 20 Ω
3. b) 9 V

 © HarperCollins*Publishers* Limited 2016

Lesson 3.8: Control circuits

Lesson overview

OCR Specification reference

P3.2 k (part)

Learning objectives

- Use a thermistor and light-dependent resistor (LDR).
- Investigate the properties of thermistors, LDRs and diodes.

Learning outcomes

- [O1] State the main properties of a diode, thermistor and light-dependent resistor (LDR).
- [O2] Describe the behaviour of a thermistor and LDR in terms of changes to their resistance.
- [O3] Describe applications of diodes, thermistors and LDRs and explain their uses.

Skills development

- Follow circuit symbol conventions.
- Use circuit symbols to construct a physical circuit from which measurements can be made.

Maths focus

- Display experimental data in a suitable table.
- Plot graphs and draw lines of best fit from data.

Resources needed cells, thermistors, diodes, LDRs, ammeters, voltmeters, connecting leads and crocodile clips, variable resistors; Worksheet 3.8, Practical sheet 3.8, Technician's notes 3.8

Digital resources an image of an illuminated streetlight

Key vocabulary thermistor, light-dependent resistor (LDR), sensors, diode, light-emitting diode (LED)

Teaching and learning

Engage

- Show the students an image of an illuminated streetlight and pose the question: "How does the street light know when to switch on?" [O1, O3]
- Ask students to discuss their ideas with a partner, before joining pairs together into groups of four for them to develop their ideas further. [O1, O3]
- Take feedback from a number of groups, developing ideas as appropriate.

Challenge and develop

Issue Worksheet 3.8. Students should use their knowledge and, if available, the Student Book to suggest which sensor component should be used for each of the uses listed. [O3]

Explain

- Students should then investigate the current–potential difference characteristics for a thermistor, a diode and an LDR (Practical sheet 3.8). [O2]
- The students should plot *I–V* graphs for the thermistor and the diode. [O2]

Consolidate and apply

Students should now answer the questions from the Student Book appropriate for their level:

- Low demand: questions 1, 2 and 3 [O1, O2, O3]
- Standard demand: questions 3, 4 and 5 [O1, O2, O3]

 © HarperCollins*Publishers* Limited 2016

- High demand: questions 5, 6 and 7 [O1, O2, O3]

Extend

Ask students able to progress further to do the following:

- Suggest some uses of diodes, other than in LEDs.

Plenary suggestions

Learning triangle: At the end of the lesson, ask students to draw a large triangle with a smaller inverted triangle that just fits inside it (so they have four triangles). Ask students to think back over the lesson and identify and write in the outer triangles:

- something they saw
- something they did
- something they discussed.

Ask students to write in the central triangle something they learned.

Heads and tails: Ask each student to write a question about something from the lesson on a coloured paper strip and the answer on another coloured strip. Hand out the strips to groups of six to eight students, so that each student gets a question and an answer. One student reads out his or her question. The student with the correct answer then reads it out, followed by his or her question, and so on.

Answers to questions

Worksheet 3.8

LEDs: lighting in traffic lights and emergency vehicle lighting

LDRs: camera shutter control and street lighting

Thermistors: fire alarms and preventing a fish tank from becoming too cold

 © HarperCollins*Publishers* Limited 2016

Lesson 3.9: Power and energy transfers

Lesson overview

OCR Specification reference

P3.2 k (part)

Learning objectives

- Describe the energy transfers in different domestic appliances.
- Describe power as a rate of energy transfer.
- Calculate the energy transferred.

Learning outcomes

- [O1] Understand that everyday electrical appliances bring about energy transfers.
- [O2] Recall and use the equation energy transferred, $E = Pt$.
- [O3] Recall and apply the equation energy transferred, $E = QV$.

Maths focus

- Use and rearrange an equation with three quantities.

Resources needed range of domestic appliances (for example, hair dryer, electric drill, kettle, desk lamp, radio, microwave), power meters, stop clocks; Worksheets 3.9.1, 3.9.2 and 3.9.3, Practical sheet 3.9, Technician's notes 3.9

Key vocabulary power, energy transfer, joules, watts, power rating

Teaching and learning

Engage

- Show the students a range of domestic appliances and ask them to suggest the energy transfers occurring when each appliance is turned on. Suggested appliances include: a hair dryer, an electric drill, a kettle, a desk lamp, a radio and a microwave.
- If necessary, the students could be reminded of the work on energy stores carried out at Key Stage 3. [O1]

Challenge and develop

- Students should then use a power meter to collect data from each appliance (Practical sheet 3.9). [O1]

Explain

- Students should use the data gathered to calculate the electrical energy transferred by each device, by applying the equation:

  electrical energy transferred = power × time [O1, O2]
- Students should then rank the appliances from highest to lowest in terms of the amount of energy they transfer. [O1]

Consolidate and apply

Issue each student with a power and energy transfers worksheet at an appropriate level:

- Low demand: Worksheet 3.9.1 [O1, O2, O3]
- Standard demand: Worksheet 3.9.2 [O1, O2, O3]
- High demand: Worksheet 3.9.3 [O1, O2, O3]

Extend

Ask students able to progress further to do the following:

 © HarperCollins*Publishers* Limited 2016

- Use the equations $E = QV$ and $I = Q \div t$ to show that $E = VIt$. They could follow this by using the equations $P = VI$ and $E = Pt$ to confirm that $E = VIt$.

Plenary suggestions

The big ideas: Ask students to write down three ideas they have learned during the topic so far. Then ask them to share their facts in groups and to compile a master list of facts, with the most important at the top. Ask for ideas to be shared and find out which other group(s) agreed.

Heads and tails: Ask each student to write a question about something from the topic on a coloured paper strip and the answer on another coloured strip. One student reads out his or her question. The student with the correct answer then reads it out, followed by his or her question, and so on.

Hot seat: Ask each student to think of a question, using material from the topic. Select someone to put in the hot seat; ask students to ask their questions and say at the end whether the answer is correct or incorrect.

Answers to questions

Worksheets 3.9.1 to 3.9.3

1. (a) 600 s
   (b) 900 000 J (900 kJ)
2. (a) 2500 W
   (b) 2700 s
   (c) 6 750 000 J (6750 kJ)
3. 6900 J
4. 50 C

Worksheet 3.9.3

5. 40 V

 © HarperCollins*Publishers* Limited 2016

Lesson 3.10: Calculating power

Lesson overview

OCR Specification reference

P3.2 l (part)

Learning objectives

- Calculate power.
- Use power equations to solve problems.
- Consider power ratings and changes in stored energy.

Learning outcomes

- [O1] Recall that power is measured in watts (W) and that 1 kW equals 1000 W.
- [O2] Recall and use the equation $P = V \times I$.
- [O3] Recall and apply the equation $P = I^2R$.

Maths focus

- Use and rearrange an equation with three quantities.

Resources needed range of domestic appliances, power meters, stop clocks; Worksheets 3.10.1, 3.10.2 and 3.10.3, Practical sheet 3.10, Technician's notes 3.10

Key vocabulary power, energy transfer, joules, watts, kilowatts, power rating

Teaching and learning

Engage

- Show the students a range of images of different appliances with their power ratings. Give some power ratings in watts and some in kilowatts. Ask the students to order the images from the appliance with the highest power rating to the appliance with the lowest power rating. If necessary, remind students that 1 kW = 1000 W. [O1]
- Alternatively, ask the students to write down their understanding of the word 'power', given the work they carried out in Lesson 3.9. Take feedback from a number of students, developing ideas as necessary. [O1]

Challenge and develop

- Students should then use a power meter to collect data from each appliance (Practical sheet 3.10). [O1]

Explain

- Students should use the data gathered to calculate the power rating of each device, by applying the equation:

  power = potential difference × current [O1, O2]
- They should then calculate the resistance for each appliance (using the equation $R = V \div I$) before using the equation $P = I^2R$ to calculate the power rating of each device. They should notice that both equations give the same answer. [O1, O2, O3]

Consolidate and apply

Issue each student with a calculating power worksheet at an appropriate level:

- Low demand: Worksheet 3.10.1 [O1, O2, O3]
- Standard demand: Worksheet 3.10.2 [O1, O2, O3]
- High demand: Worksheet 3.10.3 [O1, O2, O3]

 © HarperCollins*Publishers* Limited 2016

Extend

Ask students able to progress further to do the following:

- Use the equations $V = IR$ and $P = (IR) \times I$ to show that $P = I^2R$.

Plenary suggestions

The big ideas: Ask students to write down three ideas they have learned during the topic. Then ask them to share their facts in groups and to compile a master list of facts, with the most important at the top. Ask for ideas to be shared and find out which other group(s) agreed.

Ideas hothouse: Ask students to work in pairs to list three points about what they know about a particular idea. Then ask the pairs to join together into groups of four and then groups of six to eight to discuss this further and to come up with an agreed list of points. Ask one person from each group to report back to the class.

Answers to questions

Worksheet 3.10.1

1. (a) 2760 W
   (b) 2.76 kW
2. 10 A
3. power: watts; current: amperes; resistance: ohms
4. (a) 1150 W
   (b) 46 Ω
5. 23 Ω

Worksheet 3.10.2

1. (a) 2760 W
   (b) 2.76 kW
2. 10 A
3. (a) 1150 W
   (b) 46 Ω
4. 23 Ω
5. 12.25 A

Worksheet 3.10.3

1. 10 A
2. (a) 1150 W
   (b) 46 Ω
3. 12.25 A
4. (a) 500 g
   (b) 500 × 4.2 × 80 = 168 000 J
   (c) 168 000 ÷ 2000 = 84 s (1 min 24 s)

 © HarperCollinsPublishers Limited 2016

Lesson 3.11: Practical: Investigating series and parallel circuits

Lesson overview

OCR Specification reference

P3.2h, k (part)

Learning objectives

- Use a circuit to determine resistance.
- Gather valid data for use in calculations.
- Apply the circuit to determine the resistance of combinations of components.

Learning outcomes

- [O1] Investigate the effect on resistance of the length of the wire.
- [O2] Investigate the effect on resistance of connecting resistors in series.
- [O3] Investigate the effect on resistance of connecting resistors in parallel.

Skills development

- Follow circuit symbol conventions.
- Use circuit symbols to construct a physical circuit from which measurements can be made.

Maths focus

- Display experimental data in a suitable table.

Resources needed pieces of resistance wire attached to a metre rule, ammeters, voltmeters, connecting leads and crocodile clips, variable resistors, 0–12 V dc power packs; Worksheet 3.11, Practical sheets 3.11.1, 3.11.2 and 3.11.3, Technician's notes 3.11

Digital resources images of a range of objects that use thermistors, such as digital thermometers, car engines, microwaves and rechargeable batteries

Key vocabulary resistance, potential difference, current, series, parallel

Teaching and learning

Engage

Show the students images of a range of objects that use thermistors, such as digital thermometers, car engines, microwaves and rechargeable batteries. State that thermistors are components whose resistance changes with temperature (temperature goes up, resistance goes down). Ask the students why the objects in the images contain thermistors.

Challenge and develop

- Students should be reminded of the equation linking potential difference, current and resistance ($V = I \times R$), using example calculations if this is necessary for the students in the class.
- Students should be asked to draw a standard test circuit, using conventional circuit symbols.

Explain

The students should be split into three groups, ideally based on ability:

 © HarperCollins*Publishers* Limited 2016

- Low demand (group 1): these students should carry out Practical 3.11.1, investigating the effect of changing the length of a wire on its resistance. [O1]
- Standard demand (group 2): these students should carry out Practical 3.11.2, investigating the effect on the total resistance of connecting components in series. [O2]
- High demand (group 3): these students should carry out Practical 3.11.3, investigating the effect on the total resistance of adding components in parallel. [O3]

Consolidate and apply

- Students should now be regrouped into groups of three so that there is one student from each of the practical groups in each group.
- The students should report their findings to the rest of the group. Worksheet 3.11 can be used to help collect the feedback from each student. [O1, O2, O3]

Extend

Ask students able to progress further to do the following:

- Research Kirchhoff's laws and apply these rules to the findings from the investigations carried out in Practicals 3.11.2 and 3.11.3.

Plenary suggestions

Heads and tails: Ask each student to write a question about something from the lesson on a coloured paper strip and the answer on another coloured strip. Hand out the strips to groups of six to eight students, so that each student gets a question and an answer. One student reads out his or her question. The student with the correct answer then reads it out, followed by his or her question, and so on.

Ask me a question: Ask students to write a question about something from the lesson and then a mark scheme for the answer. Encourage them to come up with questions worth more than one or two marks and to try their questions out on each other.

 © HarperCollins*Publishers* Limited 2016

Lesson 3.12: Key concept: What's the difference between potential difference and current?

Lesson overview

OCR Specification reference

P3.2 c, i

Learning objectives

- Understand and be able to apply the concepts of current and potential difference.
- Use these concepts to explain various situations.

Learning outcomes

- [O1] Recall the model of current and potential difference.
- [O2] Construct series and parallel circuits and measure the current and potential difference in these circuits.
- [O3] Consider the definition of current and potential difference.

Maths focus

- Use and rearrange an equation with three quantities.

Resources needed cells, connecting leads, lamps, ammeters, voltmeters; Worksheet 3.12, Practical sheets 3.12.1, 3.12.2 and 3.12.3, Technician's notes 3.12

Digital resources access to online research sources about current, potential difference and electrical safety

Key vocabulary potential difference, current, charge, resistance, power, energy transfer

Teaching and learning

Engage

Remind the students of the model in Lesson 3.3, in which the concepts of current and potential difference were introduced through setting up a 'circuit' using students as electrons, lamps, ammeters and voltmeters, and pieces of paper as charge. [O1]

Challenge and develop

Issue students with a practical sheet at an appropriate level:

- Low demand: Practical sheet 3.12.1, investigating the current and potential difference in series circuits. [O2]
- Standard demand: Practical sheet 3.12.2, investigating the current and potential difference in both series and parallel circuits. [O2]
- High demand: Practical sheet 3.12.3, investigating the current, potential difference and resistance in series and parallel circuits. [O2]

Explain

- Following the practical investigation, students should write a definition for both 'current' and 'potential difference'. They should be encouraged to share their definitions in groups, developing their own definitions as a result of the discussions. [O3]

Consolidate and apply

- Students should then research whether it is current, potential difference or both that makes electricity potentially dangerous. Worksheet 3.12 gives an outline for this research task. [O3]
- Alternatively, students could be asked to produce a guide for householders on 'domestic electricity', including definitions of 'current', 'potential difference', 'resistance', 'power' and 'energy transferred'. [O3]

 © HarperCollins*Publishers* Limited 2016

Extend

Ask students able to progress further to do the following:

- Suggest how a thunderstorm can produce such high potential differences (up to 500 000 V).

Plenary suggestions

Learning triangle: At the end of the lesson, ask students to draw a large triangle with a smaller inverted triangle that just fits inside it (so they have four triangles). Ask students to think back over the lesson and identify and write in the outer triangles:

- something they saw
- something they did
- something they discussed.

Ask students to write in the central triangle something they learned.

What do I know? Ask students to each write down one thing about the topic they are sure of, one thing they are unsure of and one thing they need to know more about, being specific. Ask them to work in groups of five or six to agree on group lists; ask each group to say what they decided and agree as a class about what they are confident about, what they are less sure of and what things they want to know more about.

 © HarperCollins*Publishers* Limited 2016

Lesson 3.13: Maths skills: Using algebra in electric circuit calculations

Lesson overview

Learning objectives

- Solve algebraic equations including using appropriate substitutions, numerical values and units.
- Change the subject of an equation.
- Use the symbols =, <, <<, >>, >, $\propto$, ~.

Learning outcomes

- [O1] Substitute numerical values and units into an equation.
- [O2] Rearrange equations to change the subject.
- [O3] Use conventional algebraic symbols.

Skills development

- Follow algebraic symbol conventions.

Maths focus

- Substitute values into and rearrange equations.

Resources needed mini whiteboards and whiteboard pens; Worksheets 3.13.1, 3.13.2 and 3.13.3

Key vocabulary proportional, subject of an equation, SI units, substitute

Teaching and learning

Engage

- As the students enter the room, issue them with a small piece of paper. They should write down what they understand by the term 'electric charge' on one side of the paper, and on the other, they should write down the equation they should use in order to calculate electric charge.

  They should use a different coloured pen to label the equation with the correct units for each quantity in the equation. [O1]

Challenge and develop

- The students should be reminded that the equation to calculate electric charge is $Q = It$ and they should use this equation to answer question 1 on page 126 of the Student Book. They could do this on a mini whiteboard to allow assessment of the answers given by the students.
- They should be encouraged to write out the equation first, check the units in the question (here they will need to convert from minutes to seconds) and then to substitute the values into the equation. [O1]

Explain

- The students should then be encouraged to put the equation $Q = It$ into an equation triangle (encourage them to place the equation they know to be correct in) and then to list the three rearrangements of the equation to make Q, I and t the subject of the equation. [O2]
- You may like to return to the example of speed = distance / time from Chapter 2 (Forces) in order to remind the students how to construct equation triangles. [O2]

Consolidate and apply

The students should then be issued with the maths skills worksheet appropriate to their ability:

- Low demand: worksheet 3.13.1 [O1, O2, O3]
- Standard demand: worksheet 3.13.2 [O1, O2, O3]
- High demand: worksheet 3.13.3 [O1, O2, O3]

 © HarperCollins*Publishers* Limited 2016

Extend

Ask students able to progress further to do the following:

- Design a single-sided summary sheet instructing their peers on how to rearrange equations with more than three quantities, and how to combine equations. [O1, O2, O3]

Plenary suggestions

Learning triangle: At the end of a lesson ask students to draw a large triangle with a smaller inverted triangle that just fits inside it (so they have four triangles). In the outer three ask them to think back over the lesson and identify (and write in the respective triangles):

- something they saw
- something they did
- something they discussed.

Ask them to write in the central triangle something they learned.

Heads and tails: Ask each student to write a question about something from the lesson on a coloured paper strip and the answer on a different coloured strip. In groups of six to eight, hand out the strips so that each student gets a question and an answer. One student reads out their question. The student with the correct answer then reads it out, followed by their question. This could take the form of calculations using and rearranging $Q = It$ and $V = IR$.

What do I know? Ask students to each write down one thing about the topic they are sure of, one thing they are unsure of and one thing they need to know more about, being specific. Ask them to work in groups of five or six to agree on group lists, ask each group to say what they decided and agree as a class what they are confident about, what they are less sure of and what things they want to know more about.

Answers to questions

Worksheet 3.13.1

1. Charge – coulombs; current – amperes/amps; time – seconds
2. 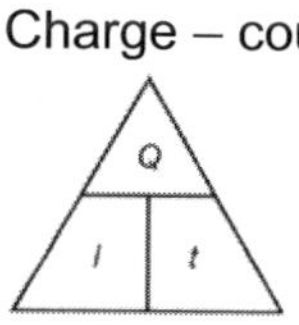

3. I = Q/t
4. 2 A
5. potential difference – volts; current – amperes/amps; resistance – ohms
6. 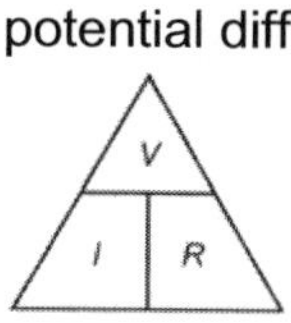

7. R = V/I
8. 2 Ω

Worksheet 3.13.2 and 3.13.3

1. Charge – coulombs; current – amperes/amps; time – seconds
2. 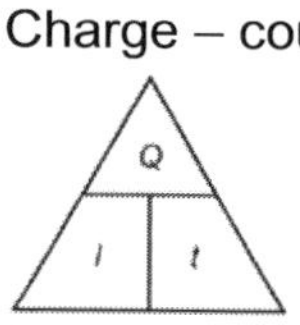

3. 2 A
4. 4 s
5. potential difference – volts; current – amperes/amps; resistance – ohms
6. 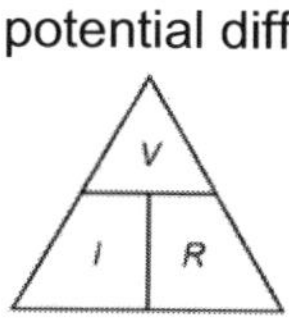

7. 2 Ω
8. correct substitution of V/R for I in Q = It

 © HarperCollinsPublishers Limited 2016

Lesson 3.14: Maths skills: Using formulae and understanding graphs

Lesson overview

Learning objectives

- Recognise how algebraic equations define the relationships between variables.
- Solve simple algebraic equations by substituting numerical values.
- Describe relationships expressed in graphical form.

Learning outcomes

- [O1] State what an algebraic equation shows.
- [O2] Rearrange an algebraic equation.
- [O3] Use a graph to state whether the relationship between two variables exhibits direct proportionality.

Skills development

- Use key terms correctly.
- Use equations.
- Use conventional units.

Maths focus

- Substitute values into and rearrange algebraic equations.
- Plot graphs.
- Recognise graphs as either showing direct proportionality or not.

Resources needed apparatus as listed in Technician's notes, GCSE Physics equation sheet; Worksheets 3.14.1, 3.14.2 and 3.14.3, Practical sheet 3.14, Technician's notes 3.14

Key vocabulary algebraic equation, direct proportion, linear relationship, constant of proportionality

Teaching and learning

Engage

Issue the GCSE Physics equation sheet. Ask students to choose any one of the equations on the sheet (except advise students not to use Equations 10, 11 or 12) and to rearrange their chosen equation as many times as there are variables in the equation, to make each variable the subject of the formula. Discuss strategies for rearranging equations as necessary with the students. Students could test that they have rearranged the equation correctly by substituting in simple numbers as a check. [O1, O2]

Challenge and develop

- Take a moment to focus on the 'equals' sign in any of the chosen equations – ensure that students are clear about what it actually means. [O1]
- Introduce the sign used to show that one variable is proportional to another ($\propto$) and discuss what the term means. You could do this by referring to the equation $V = IR$, which the students have met in this unit and state that $V \propto I$. [O1]
- Ask students to collect a single set of data as outlined on Practical sheet 3.14.

Explain

- Students should then analyse their results by plotting the graph of potential difference against current, determining the gradient and using the gradient to calculate the resistance (1 ÷ gradient).

 © HarperCollins*Publishers* Limited 2016

- Use the shape of the graph to introduce the idea of direct proportionality and point out that, if variables are in direct proportion, when a graph of one variable against the other is plotted there will be a line through the origin with an unchanging gradient. [O3]

Consolidate and apply

Issue a worksheet at an appropriate level: [O1, O2, O3]

- Low demand: Worksheet 3.14.1
- Standard demand: Worksheet 3.14.2
- High demand: Worksheet 3.14.3

Extend

Ask students able to progress further to do the following:

- Design a help sheet for their peers that describes: (a) what algebraic equations are and (b) how algebraic equations can be transposed into graphical form to show the nature of the relationship between the variables in the equation. [O1, O2, O3]

Plenary suggestions

Ask me a question: Ask students to write a question about something from the lesson and then devise a mark scheme for the answer. Encourage them to come up with questions worth more than one or two marks and to try their questions out on each another. [O1, O2, O3]

Learning triangle: At the end of the lesson, ask students to draw a large triangle with a smaller inverted triangle that just fits inside it (so they have four triangles). Ask students to think back over the lesson and identify and write in the outer triangles:

- something they saw
- something they did
- something they discussed.

Ask students to write in the central triangle something they learned. [O1, O2, O3]

Answers to questions

Worksheet 3.14.1

1. The = sign shows that both sides of the equation are worth the same, and will have the same numerical value.
2. $\propto$
3. (a) $P = VI$ (b) $V = P \div I$ (c) $I = P \div V$
4. power = 2300 W (or 2.3 kW)
5. The graph will be a straight line through the origin (0,0).
6. a) Yes, because it is a straight line through the origin.

   b) Determine the gradient of the line; the resistance is equal to 1 ÷ gradient.

Worksheet 3.14.2

1. The = sign shows that both sides of the equation are worth the same, and will have the same numerical value.
2. $\propto$
3. (a) $P = VI$ (b) $V = P \div I$ (c) $I = P \div V$
4. current = 10 A
5. The graph will be a straight line through the origin (0,0).
6. a) Yes, because it is a straight line through the origin.

   b) Determine the gradient of the line; the resistance is equal to 1 ÷ gradient.

Worksheet 3.14.3

1. The = sign shows that both sides of the equation are worth the same, and will have the same numerical value.
2. $\propto$
3. (a) $P = VI$ and $V = IR$ (b) $P = I^2R$ ($P = IR \times R$, substituting IR for V) (c) $R = P \div I^2$
4. power = 150,000 W (or 150 kW)
5. The graph will be a straight line through the origin (0,0).
6. a) Yes, because it is a straight line through the origin.

   b) Determine the gradient of the line; the resistance is equal to 1 ÷ gradient.

 © HarperCollins*Publishers* Limited 2016

When and how to use these pages: Check your progress, Worked example and End of chapter test

Check your progress

Check your progress is a summary of what students should know and be able to do when they have completed the chapter. Check your progress is organised in three columns to show how ideas and skills progress in sophistication. Students aiming for top grades need to have mastered all the skills and ideas articulated in the final column (shaded pink in the Student book).

Check your progress can be used for individual or class revision using any combination of the suggestions below:

- Ask students to construct a mind map linking the points in Check your progress
- Work through Check your progress as a class and note the points that need further discussion
- Ask the students to tick the boxes on the Check your progress worksheet (Teacher Pack CD). Any points they have not been confident to tick they should revisit in the Student Book.
- Ask students to do further research on the different points listed in Check your progress
- Students work in pairs and ask each other what points they think they can do and why they think they can do those, and not others

Worked example

The worked example talks students through a series of exam-style questions. Sample student answers are provided, which are annotated to show how they could be improved.

- Give students the Worked example worksheet (Teacher Pack CD). The annotation boxes on this are blank. Ask students to discuss and write their own improvements before reviewing the annotated Worked example in the Student Book. This can be done as an individual, group or class activity.

End of chapter test

The End of chapter test gives students the opportunity to practice answering the different types of questions that they will encounter in their final exams. You can use the Marking grid provided in this Teacher Pack or on the CD Rom to analyse results. This shows the Assessment Objective for each question, so you can review trends and see individual student and class performance in answering questions for the different Assessment Objectives and to highlight areas for improvement.

- Questions could be used as a test once you have completed the chapter
- Questions could be worked through as part of a revision lesson
- Ask Students to mark each other's work and then talk through the mark scheme provided
- As a class, make a list of questions that most students did not get right. Work through these as a class.

 © HarperCollins*Publishers* Limited 2016

Marking Grid for End of Chapter 3 Test

	Getting Started [Foundation Tier]				Going Further [Foundation and Higher Tiers]					More Challenging [Higher Tier]				Most demanding [Higher Tier]				Total marks	Percentage
Student Name	Q. 1 (AO1) 1 mark	Q. 2 (AO1) 1 mark	Q. 3 (AO1) 2 marks	Q. 4 (AO2) 2 marks	Q. 5 (AO1) 1 mark	Q. 6 (AO1) 2 marks	Q. 7 (AO1) 1 mark	Q. 8 (AO1) 2 marks	Q. 9 (AO2) 2 marks	Q. 10 (AO1) 1 mark	Q. 11 (AO2) 2 marks	Q. 12 (AO2) 2 marks	Q. 13 (AO3) 2 marks	Q. 14 (AO1) 1 mark	Q. 15 (AO2) 2 marks	Q. 16 (AO2) 2 marks	Q. 17 (AO3) 2 marks		

© HarperCollinsPublishers Limited 2016

Check your progress

You should be able to:

Describe how insulating materials can become charged	Recall that there are two types of charge and that like charges repel and unlike charges attract	Explain how a person can get an electric shock and explain static electricity in terms of electric fields
Recall that an electric current is a flow of electrical charge and is measured in amperes (A)	Remember that charge is measured in coulombs (C) and recall and use the equation $Q = It$	Explain the concept that current is the rate of flow of charge. Rearrange and apply the equation $Q = It$
Recognise and use electric circuit symbols in circuit diagrams	Draw and recognise series and parallel circuits. Compare the brightness of lamps connected in series and parallel	Recall that the current in a series circuit is always the same and that the total current in a parallel circuit is the sum of the currents through each branch
Recall that the current through a component depends on the resistance of the component and the potential difference across it	Recall and apply the equation $V = IR$ and for series circuit $R_{total} = R_1 + R_2$	Explain the effect of adding more resistors to series and parallel circuits
Set up a circuit to investigate the relationship between V, I and R for a fixed resistor	Draw I–V graphs for a fixed resistor	Analyse and interpret I–V graphs for a fixed resistor
State the main properties of a diode, thermistor and light-dependent resistor (LDR)	Describe the behaviour of a thermistor and LDR in terms of changes to their resistance	Describe applications of diodes, thermistors and LDRs and explain their uses
Draw I–V graphs for a filament lamps	Explain the properties of components using I–V graphs	Use I–V graphs to determine if the characteristics of components are ohmic or non-ohmic
Recall that cells and batteries produce low-voltage direct current	Recall that domestic supply in the UK is 230 V a.c. and 50 Hz	Explain the difference between direct and alternating potential difference
Understand that everyday electrical appliances bring about energy transfer	Recall and use the equation energy transferred $E = Pt$	Recall and apply the equation energy transferred $E = QV$
Recall that power is measured in watts (W) and 1 kW = 1000 W	Recall and use the equation $P = V \times I$	Recall and apply the equation $P = I^2R$

 © HarperCollinsPublishers Limited 2016

Worked example

The diagram below shows a circuit used to investigate the resistance of a piece of thin wire.

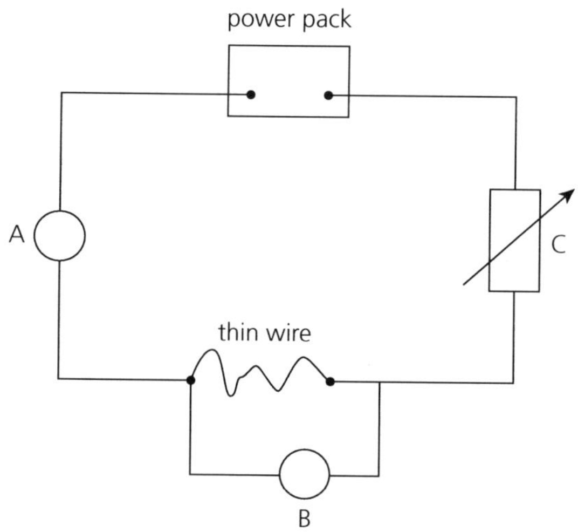

1 **Name the three components in the circuit labeled A, B and C.**

A is an ammeter, B is a voltmeter and C is a thermistor.

A and B are correct, but C is a variable resistor.

2 **Give the purpose of component C.**

To regulate the temperature.

Component C is used to change the potential difference across the test material.

3 **The student recorded the potential difference across the thin wire and the current passing through it. She plotted her results on a graph. Explain which variable she should plot on each axis.**

She should plot potential difference on the x-axis and current on the y-axis.

This is correct, but the explanation has not been given. Potential difference is plotted on the *x*-axis because it is the independent variable, and current is plotted on the *y*-axis because it is the dependent variable.

4 **The student increased the potential different to 12 V. Explain what you think would have happened.**

The wire gets very hot.

The wire gets hot because of the large current passing through it. Large currents transfer lots of energy.

5 **Explain how you would expect the graph to look if the wire had been replaced by a filament lamp.**

The same

The line would initially be straight but then curve with a shallower gradient as the temperature of the filament increases.

 © HarperCollins*Publishers* Limited 2016

Magnetism and magnetic fields : Introduction

When and how to use these pages

This unit builds upon the learning from Chapter 3: Electricity as it requires drawing and interpretation of circuit diagrams. It may also build on Key Stage 3 learning on magnetism. The unit also has a small connection with sound, as we look at the structure of speakers and microphones and link the sound produced to how they work. The unit involves the use of maths skills. There are two equations that students will need to rearrange in order to calculate certain variables.

Overview of the unit

In this unit, students will learn about the link between electricity and magnetism, the effect of a magnetic field on a wire when current is moving, and the applications of these effects. This includes studying motors, generators, speakers, microphones and transformers.

This unit offers a number of opportunities for the students to use their practical skills to make observations and to draw conclusions from them about the links between electricity, magnetism and force. It also offers opportunities for students to practise their maths skills including using equations, rearranging equations and graph interpretation.

Obstacles to learning

The students may need extra guidance with the following terms and concepts:

- Students may get confused to learn that the north pole of a compass is actually the south pole of the magnet due to the fact that opposite poles attract.
- Students may think that all metals are attracted to magnets when actually only iron, nickel and cobalt are attracted to magnets.
- Before this topic, students may think that only magnets have magnetic fields.

If you are a non-specialist physics teacher and you would like some background reading, then the following websites might be useful:

- The OCR GCSE physics specification, examination papers and other documents
- Electromagnetism practicals from the Institute of Physics
- Case study of Hans Christian Oersted from the Institute of Physics
- Dr Physics A – a very informative YouTube video series on several physics topics.

Practicals in this unit

In this unit students will do the following practical work:

- investigate which materials can be magnetic
- investigate what a magnetic field looks like
- investigate how to induce magnetism in different materials
- discover the shape of a magnetic field created by a current
- determine the shape of the magnetic field around a solenoid
- demonstrate the force on a wire carrying a current in a magnetic field
- conduct left-hand rule experiment
- investigate how moving a magnet through a coil induces a current
- investigate the factors that affect the speed and direction of a motor
- demonstrate how a microphone works

 © HarperCollins*Publishers* Limited 2016

- demonstrate what affects the motion of a loudspeaker
- investigate how a motor can be used as a dynamo
- investigate how electricity and magnetism are related.

There are no suggested practical activities for this chapter.

	Lesson title	Overarching objectives
1	Magnetism and magnetic forces	• Explain what is meant by the poles of a magnet. • Plot the magnetic field around a bar magnet. • Describe magnetic materials and induced magnetism.
2	Compasses and magnetic fields	• Describe the Earth's magnetic field. • Describe the magnetic effect of a current.
3	Magnetic effects	• Draw the magnetic field around a conducting wire and a solenoid. • Describe the force on a wire in a magnetic field. • Apply the left-hand rule to work out the direction of a magnetic field, a current or a force around a wire.
4	Calculating the force on a conductor	• Explain the meaning of magnetic flux density, *B*. • Calculate the force on a current-carrying conductor in a magnetic field.
5	The generator effect	• Describe how the current is induced in a wire when it moves in a magnetic field. • Identify apparatus needed to demonstrate induced current. • Identify the factors that affect the size and direction of the induced current or the potential difference.
6	Electric motors	• List equipment that uses motors. • Describe how motors work. • Describe how to change the speed and direction of rotation of a motor.
7	Generators and transformers	• Explain how an alternator generates alternating current. • For an alternator, draw and interpret graphs of potential difference generated in the coil against time.

 © HarperCollins*Publishers* Limited 2016

		• Understand how induced alternating electric currents are used in transformers. • Apply the transformer equation linking the numbers of turns and the potential differences induced.
8	Loudspeakers and microphones	• Describe how a moving-coil loudspeaker works. • Compare loudspeakers and headphones. • Describe how a moving-coil microphone works.
9	Key concept: The link between electricity and magnetism	• Explore how electricity and magnetism are connected. • Describe how electromagnetic induction occurs. • Describe the principle of the electric motor.
10	Maths skills: Rearranging equations	• Know how to rearrange equations. • Know how to use the transformer equation. • Know how to calculate the force on a conductor.

 © HarperCollins*Publishers* Limited 2016

Lesson 4.1: Magnetism and magnetic forces

Lesson overview

OCR Specification reference

P4.1a, b

Learning objectives

- Explain what is meant by the poles of a magnet.
- Plot the magnetic field around a bar magnet.
- Describe magnetic materials and induced magnetism.

Learning outcomes

- [O1] Students are able to plot the field of a bar magnet,
- [O2] Students are able to explain how to identify a magnet using its poles.
- [O3] Students are able to explain the difference between induced magnetism and permanent magnetism.

Skills development

- Describe how to plot the magnetic field pattern using a compass.
- Draw the magnetic field pattern of a bar magnet showing how strength and direction change from one point to another.
- Explain how the behaviour of a magnetic compass is related to evidence that the core of the Earth must be magnetic.

Resources needed bar magnets, compass, small objects made of different metals (some magnetic, some non-magnetic metals), paper, pencil, paperclips; Worksheet 4.1, Practical sheet 4.1, Technician's notes 4.1.

Digital resources Magnet and Compass PhET animation; PowerPoint presentation 4.1

Key vocabulary pole, magnetic field, magnet, magnetic material, induced magnetism

Teaching and learning

Engage

- Each student has a compass that will point north. Students discuss in pairs why the compass points north. Students discuss the poles and what they think the earth's poles are. Students link this to the idea that magnets have poles. [O1]

Challenge and develop

- Students discuss how they know whether something is a magnet. They then use a magnet on some small metal substances. [O3]
- Students use a compass to draw a magnetic field around a magnet. You can also demonstrate this experiment on the Magnet and Compass PhET animation. [O2]
- Students pick up small iron or steel objects (such as paperclips) with a magnet to demonstrate induced magnetism – that the paperclips can attract more paperclips if they are on the magnet, but they won't if they are not. [O3]

Explain

- Students explain how they can tell whether a substance is a magnet, a magnetic material or not magnetic. [O3]
- Students draw the magnetic field around a magnet, and explain the direction in which a magnetic field flows. [O2]

 © HarperCollins*Publishers* Limited 2016

Consolidate and apply

- Issue each student with a copy of Worksheet 4.1.
  - Standard demand: Questions 1–4
  - High demand: Questions 5–6 [O1, O2, O3]

Extend

Ask students able to progress further to do the following:

- Explain how the behaviour of a compass demonstrates that the Earth behaves like a giant magnet (question 5 on Worksheet 4.1). [O1]

Plenary suggestions

Learning triangle: At the end of the lesson ask students to draw a large triangle with a smaller inverted triangle that just fits inside it (so they have four triangles). Ask students to think back over the lesson and identify and write in the outer triangles:

- something they saw
- something they did
- something they discussed.

Ask students to write in the central triangle something they learned.

Using the Magnet and Compass PhET animation students can predict where the compass will point when put in a particular position. They can also do the same for the Earth.

Answers to questions

Worksheet 4.1

1. The coins with nickel in them will be attracted to the magnet.
2. (a) A compass needle would point to the north pole of a magnet.
   (b) Another magnet would be attracted to the magnet when different poles were put next to each other and it would be repelled when like poles were put next to each other.
3. The track is made from a permanent magnet as this is the only material that can repel another magnet.
4. The magnet is inducing magnetism in the paperclips so they attract other paperclips.
5. 5. A compass needle points to the north pole of a magnet. A compass needle also points to the north pole of the Earth, which means that the Earth's core is like a magnet.

Practical sheet 4.1 Evaluation

1. Iron, nickel and cobalt are attracted to a magnet. Non-magnetic materials are not attracted to a magnet. The like poles of a magnet repel. The opposite poles of a magnet attract.
2. The field should look something like this:

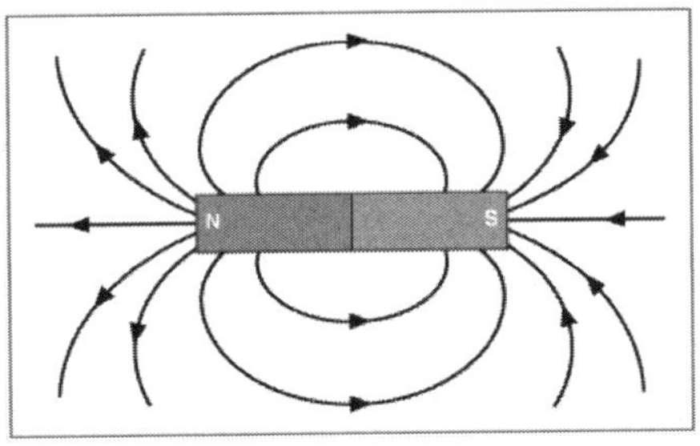

3. The paperclips that are joined to the magnet act magnetic. The paperclip joined directly to the magnet will be weaker than the magnet but will act as the strongest paperclip magnet. The paperclip joined to that will be even weaker but will act as the next strongest paperclip magnet and so on, until the magnetic field is so weak that the chain will not be able to pick up any more paperclips. These paperclips will have no magnetic field.

 © HarperCollins*Publishers* Limited 2016

Lesson 4.2: Compasses and magnetic fields

Lesson overview

OCR Specification reference

P4.1c, d, e, f

Learning objectives

- Describe the Earth's magnetic field.
- Describe the magnetic effect of a current.

Learning outcomes

- [O1] Draw a diagram of the Earth's magnetic field.
- [O2] Draw a diagram of a magnetic field around a wire.
- [O3] Use the right-hand grip rule to determine the direction of a field or a current.

Skills development

- Describe how the magnetic effect of a current can be demonstrated.
- Draw the magnetic field pattern for a straight wire carrying a current and for a solenoid (showing the direction of the field).
- Record the effects of the size of the current and the distance from the wire on the magnetic field.

Resources needed power pack, wires, a piece of cardboard with holes in (big enough for wires to fit through), iron filings, plotting compasses; Worksheet 4.2, Practical sheet 4.2, Technician's notes 4.2

Digital resources PowerPoint presentation 4.2

Key vocabulary magnetic field, pole, right-hand grip rule, north-seeking pole

Teaching and learning

Engage

- **What's in the picture?** Have a picture of a bar magnet surrounded by compasses. Students identify the picture and explain why the compasses point in the direction that they do. [O1]
- Distribute compasses to the students. Students state that the compass points in a particular direction. You can assess whether they know that since the compass points in a particular direction, the Earth must be like a magnet. [O1]

Challenge and develop

- Students make the link that since the Earth is a magnet, it should have the same magnetic field (drawn in Lesson 4.1). Students draw the magnetic field around the Earth. [O1]
- Students can plan a practical to discover the magnetic field of a wire or use Practical sheet 4.2 to do the practical. [O2]
- Students carry out the practical and draw diagrams of the position of the iron filings with the current. [O2]

Explain

- Students explain that the iron filings and compass needles move in opposite directions due to the right-hand grip rule. They state that the direction of a magnetic field around a wire depends on the direction of the current. [O3]

Consolidate and apply

- Students complete Worksheet 4.2 where they have to work out the direction of a magnetic field or the current based on the diagrams. [O3]

 © HarperCollins*Publishers* Limited 2016

Extend

Ask students able to progress further to do the following:

- Think about the poles of a compass. Is the north pole of a compass actually the north pole? Students use the idea that opposite poles attract to explain that the north pole of the compass is actually pointing to the magnetic south pole, which is the geographic north pole. What we call the north pole is just where the north pole of a magnet points. [O1]

Plenary suggestions

Learning triangle: At the end of the lesson ask students to draw a large triangle with a smaller inverted triangle that just fits inside it (so they have four triangles). Ask students to think back over the lesson and identify and write in the outer triangles:

- something they saw
- something they did
- something they discussed.

Ask students to write in the central triangle something they learned.

Answers to questions

Worksheet 4.2

1.

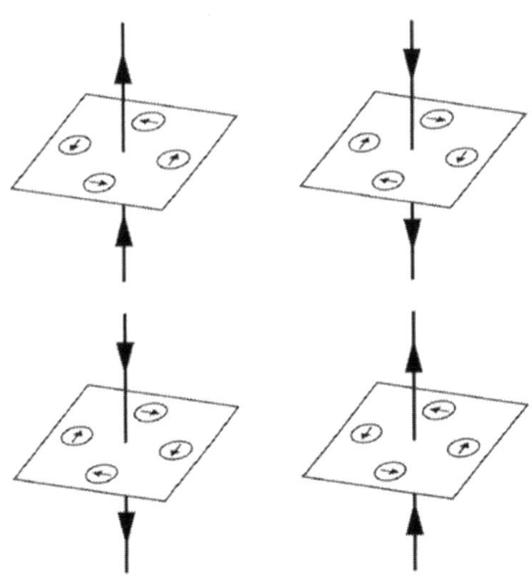

2. Like poles of a magnet repel and opposite poles attract. This means that the north pole of a magnet is attracted to the south pole of the Earth even though it is called the north pole.

 © HarperCollins*Publishers* Limited 2016

Lesson 4.3: Magnetic effects

Lesson overview

OCR Specification reference

P4.1g

Learning objectives

- Draw the magnetic field around a conducting wire and a solenoid.
- Describe the force on a wire in a magnetic field.
- Apply the left-hand rule to work out the direction of a magnetic field, a current or a force around a wire.

Learning outcomes

- [O1] Compare the magnetic field of a single wire to the magnetic field of a solenoid.
- [O2] Demonstrate the force on a wire in a magnetic field.
- [O3] Apply the left-hand rule to unfamiliar situations with magnets.

Skills development

- Draw the magnetic field pattern for a solenoid (showing the direction of the field).
- Describe the techniques and apparatus that should be used to record the magnetic field around a solenoid.
- Use Fleming's left-hand rule to predict the direction of the force, current or magnetic field of a wire.

Resources needed d.c. power supply, small cylinder, piece of cardboard with two slits, electrical wire, plotting compass, iron filings, horseshoe magnet or two bar magnets, copper wire, crocodile clips; Worksheets 4.3.1 and 4.3.2, Practical sheet 4.3, Technician's notes 4.3.1 and 4.3.2

Digital resources Generator PhET animation; PowerPoint presentation 4.3

Key vocabulary solenoid, field lines, polarity, perpendicular, Fleming's left-hand rule

Teaching and learning

Engage

- Using an overhead projector, a diagram or a video, remind the students of the shape of a magnetic field around a wire. Remind students of the right-hand rule. Using whiteboards, students can draw a picture of what they think the magnetic field around a solenoid is based on that knowledge. [O1]

Challenge and develop

- Students can attempt Practical 4.3 demonstrating the magnetic fields of a coil by sketching the magnetic fields using a plotting compass. Alternatively, use the electromagnet tab from the Generator PhET animation. [O1]
- Students can compare their predictions of the shape of a magnetic field around a solenoid with the experimental results. [O1]
- Demonstrate the kicking wire to the students. Students think about what might affect the force on the wire. Demonstrate the kicking wire with different current, different length wire and different strength of magnet. [O2]
- Demonstrate how changing the current direction changes the direction of the force on the kicking wire. Introduce Fleming's left-hand rule to work out the direction of the force. [O3]

Explain

- Students can explain why the strength of the force on a wire is affected by the length of the wire, the current passing through the wire, or the length of the wire. [O3]

 © HarperCollins*Publishers* Limited 2016

Consolidate and apply

- Students complete Worksheet 4.3.1 using the left-hand rule. [O3]

Extend

Ask students able to progress further to do the following:

- Think about the applications of solenoids and how having a magnet that you can switch on and off and change the strength of is useful. [O3]
- Complete Worksheet 4.3.2. (High demand) [O1]

Plenary suggestions

Learning triangle: At the end of the lesson ask students to draw a large triangle with a smaller inverted triangle that just fits inside it (so they have four triangles). Ask students to think back over the lesson and identify and write in the outer triangles:

- something they saw
- something they did
- something they discussed.

Ask students to write in the central triangle something they learned.

Answers to questions

Worksheet 4.3.1

1. (a) Outwards
   (b) Inwards
   (c) Inwards
   (d) Outwards

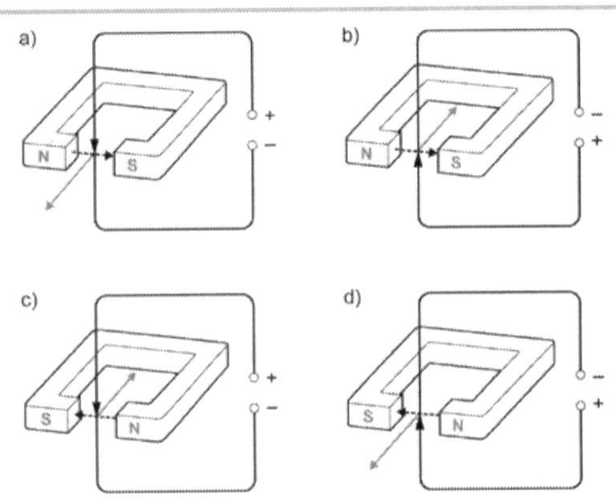

2.

Action	Effect on the size of the force	Effect on the direction of the force
Using stronger magnets	Increases the size of the force	No effect
Decreasing the current	Decreases the size of the force	No effect
Using a longer wire	Increases the size of the force	No effect
Changing the direction of the current	No effect	Changes the direction of the force
Swapping the poles of the magnet around	No effect	Changes the direction of the force

Worksheet 4.3.2

1. The image does agree with the right-hand rule.
2.

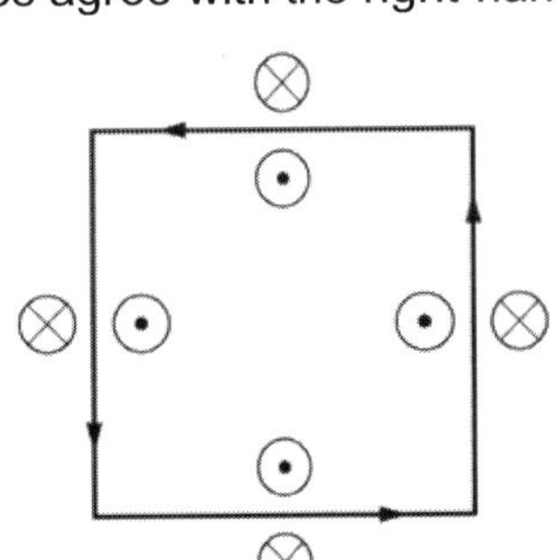

3.

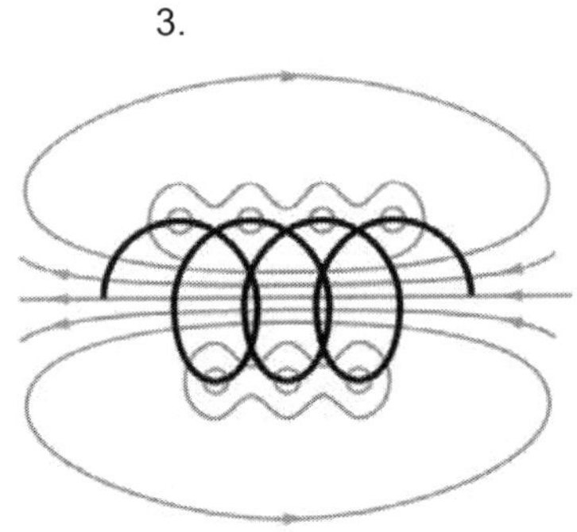

 © HarperCollinsPublishers Limited 2016

Lesson 4.4: Calculating the force on a conductor

Lesson overview

OCR Specification reference

P4.2c

Learning objectives

- Explain the meaning of magnetic flux density, *B*.
- Calculate the force on a current-carrying conductor in a magnetic field.

Learning outcomes

- [O1] State that magnetic flux density is measured in tesla, *T*.
- [O2] Explain what the size of the force on a conductor depends on.
- [O3] Use the equation $F = BIL$ to calculate the force on a conductor.

Skills development

- Substitute values into an equation.

Maths focus

- Substitute numerical values into and rearrange equations.

Resources needed kicking wire demonstration; Worksheet 4.4

Digital resources PhET simulation – Faraday's Electromagnetic Lab; PowerPoint presentation 4.4

Key vocabulary tesla (*T*), magnetic flux density, current-carrying conductor

Teaching and learning

Engage

- Show the students a range of different machines that each contain a motor and ask them what all the images have in common.
- Following on from this, ask them to recall what the motor effect is.

Challenge and develop

- Demonstrate the idea of magnetic flux density using the PhET simulation (Faraday's Electromagnetic Lab). If available, the students could use laptop computers to access this themselves, allowing them to investigate the simulation.
- From this, state that the spacing between the magnetic field lines gives an indication of the strength of the magnetic field and that the field inside the coil is much stronger than the field outside the coil. Introduce the new unit of tesla, *T*, here as the unit that measures the strength of the magnetic field, also known as the magnetic flux density. [O1]

Explain

- Having seen the kicking wire demonstration last lesson, it should be clear to the students that there is a force on the current-carrying wire.
- **Pair talk:** Ask the students to discuss what the force on the wire might depend on. If appropriate, the students could be shown the demonstration again, looking closely at the factors that could affect the size of the force on the wire. [O2]
- Take ideas from the pairs, before showing that the size of the force is calculated using $F = BIL$. The students could practise using this equation by answering questions 2 and 3 from pages 142–3 of the Student Book. [O2, O3]

 © HarperCollins*Publishers* Limited 2016

Consolidate and apply

- Issue the students with the calculating the force on a conductor worksheet (Worksheet 4.4).

Extend

- The students should answer question 7 from page 143 of the Student Book.
- Alternatively, they could design a summary map of all the key words learned so far, joining as many of them as they can with lines, writing how they are connected over the lines.

Plenary suggestions

Hot seat: Ask each student to think up a question, using material from the lesson. Select someone to put in the hot seat, ask students to pose their question and say at the end whether the answer is correct or incorrect.

Learning triangle: At the end of the lesson ask students to draw a large triangle with a smaller inverted triangle that just fits inside it (so that they have four triangles). Ask students to think back over the lesson and identify and write in the outer triangles:

- something they saw
- something they did
- something they discussed.

Ask students to write in the central triangle something they learned.

Answers to questions

Worksheet 4.4

1. Please refer to the diagram on page 142 of the Student Book.
2. The magnetic field lines would be closer together.
3. The current and the length of wire (in either order).
4. $F = BIL$
5. F – newtons; B – tesla; I – amperes / amps; L – metres
6. a) 0.06 N
   b) 0.18 N
7. 0.04 T

 © HarperCollins*Publishers* Limited 2016

Lesson 4.5: The generator effect

Lesson overview

OCR Specification reference

P4.2e

Learning objectives

- Describe how the current is induced in a wire when it moves in a magnetic field.
- Identify apparatus needed to demonstrate induced current.
- Identify the factors that affect the size and direction of the induced current or the potential difference.

Learning outcomes

- [O1] Describe the effects of moving a magnet through a solenoid or moving a wire between the poles of a magnet.
- [O2] Construct or draw an experiment that can create an induced current.
- [O3] Explain the effects of using a stronger magnet, moving the magnet faster or using a coil on the size of the induced potential difference.

Skills development

- Explain the effects of various factors on the size of induced potential difference.
- Select the materials to demonstrate that potential difference can be changed with a solenoid and a magnet.
- Suggest other changes that could be made to the apparatus to change the potential difference of the induced current.

Resources needed bar magnet (or neodymium magnet), magnets with different strengths (optional) ammeter (or multimeter), a wire (coiled up); Worksheet 4.5, Practical sheet 4.5, Technician's notes 4.5

Digital resources Faraday's Electromagnetic Lab PhET animation, video on how a generator works; PowerPoint presentation 4.5

Key vocabulary generator, slip rings, current, force, field

Teaching and learning

Engage

- Show students a short video clip of how a generator works (such as this one). You could also show students the generator in the Faraday's Electromagnetic Lab from PhET without sound and ask them to say what they think is happening. Students could either suggest the ideas or do a 'voice over' for it.

Challenge and develop

- Students compare a generator and a motor. They can write the similarities and differences in the form of a Venn diagram. [O1]
- Students use Practical sheet 4.5 to investigate what affects the strength of the generator. They can also investigate quantitatively how the number of turns in the coil affects the potential difference generated. [O2]

Explain

- Students can plot a graph of number of turns in the coil versus potential difference to determine the link between them. [O3]

 © HarperCollins*Publishers* Limited 2016

Consolidate and apply

- Students write down what affects the size of the voltage produced and what affects the direction of the current. [O3]
- Students evaluate their results and experiment and suggest improvements for it. [O3]
- Students use Worksheet 4.5 to calculate the average voltage produced, plot a graph and analyse the results. [O3]

Extend

Ask students able to progress further to do the following:

- Use a diagram of a power station to explain how electricity is generated.

Plenary suggestions

Ask me a question: Ask students to write a question about how generators work and then devise a mark scheme for the answer. Simple questions could involve recalling the parts of a generator and what they do. More difficult ones could involve long-answer questions on how generators work. The students try out their questions on one another.

Answers to questions

Worksheet 4.5

1. The student repeated their results to find anomalies and to make their results more accurate.
2. The anomalous result is 2.05 V, produced with 100 turns in the coil.
3. You do not use the anomalous result to calculate the mean.
4. The mean voltage is meant to have the same number of significant figures as the measured voltages. This is because the equipment cannot measure any more significant figures and so any more significant figures would be meaningless.
5.

Number of turns in the coil	Voltage measured (V)	Voltage measured (V)	Voltage measured (V)	Mean voltage measured (V)
25	0.40	0.42	0.42	**0.41**
50	0.83	0.82	0.85	**0.83**
75	1.25	1.23	1.24	**1.24**
100	1.60	2.05	1.61	**1.61**
125	2.02	2.06	2.03	**2.04**
150	2.42	2.41	2.40	**2.41**
175	2.83	2.84	2.84	**2.84**
200	3.20	3.20	3.20	**3.20**

6.

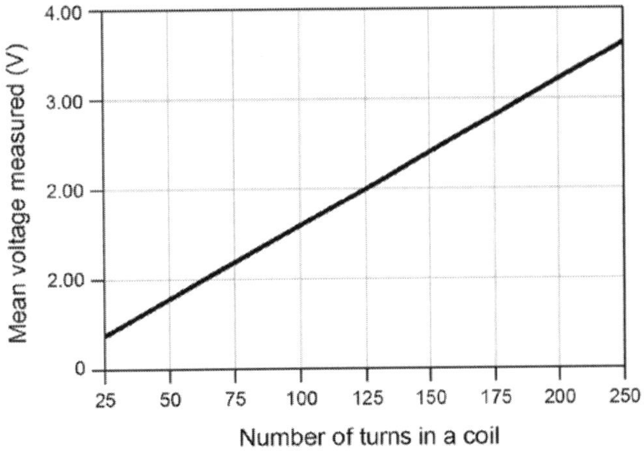

7. As the number of turns in the coil increases, the voltage increases.
8. 2.48–2.52 V
9. 215–220 turns
10. The student needed to keep the following things the same:
    - the strength of the magnet
    - the speed they put the magnet in the coil
    - the area of the coil.

 © HarperCollins*Publishers* Limited 2016

Lesson 4.6: Electric motors

Lesson overview

OCR Specification reference

P4.2d

Learning objectives

- List equipment that uses motors.
- Describe how motors work.
- Describe how to change the speed and direction of rotation of a motor.

Learning outcomes

- [O1] Describe the components of a motor and state what they do.
- [O2] Describe what the different components do to create the motor effect.
- [O3] Explain what happens to the speed and direction of the motor when the current and magnetic field are modified.

Skills development

- Plan an experiment to investigate how the current direction, current strength and magnetic field strength affect the speed and direction of rotation of the motor.
- Suggest improvements for the investigation of the factors that affect the motor.
- Interpret data linking the factors that affect a motor and the motor's speed and direction.

Resources needed d.c. power supply, magnets, armature, mild steel yoke, support base, shaft, rivets, split pins, copper wire, transparent adhesive tape, tubing, additional magnets of different strength; Worksheets 4.6.1 and 4.6.2, Practical sheet 4.6, Technician's notes 4.6

Digital resources PowerPoint presentation 4.6

Key vocabulary Fleming's left-hand rule, split-ring commutator, field, current, force, d.c. motor, a.c. motor, radial field

Teaching and learning

Engage

- Show students pictures of various items that use motors (you could use slide 2 of PowerPoint presentation 4.6). Ask the students what they have in common. Once they have established that they have motors, students can think of some more things that use motors. [O1]

Challenge and develop

- Students can build either a really simple motor or use pre-made motors and describe how the motor works. [O2]
- Students can use the pre-made motors to investigate how the motor is affected by changing the size and direction of the current, the direction of the magnetic field, and possibly the strength of the magnets. [O3]

Explain

- Show the students a curved magnet and tell them that motors use curved magnets to improve them. [O2]
- Show a picture or a video of a split ring commutator (such as this one) to demonstrate how it works. [O2]

Consolidate and apply

- Students complete the questions on Practical sheet 4.6, noting their quantitative observations on what affects the speed and direction of the motor. [O3]
- Students complete Worksheet 4.6.1. [O2]

 © HarperCollins*Publishers* Limited 2016

- Students complete Worksheet 4.6.2 without doing the extension questions. [O3]

Extend

Ask students able to progress further to do the following:

- Use the left-hand rule to explain why motors use curved magnets. This is an extension question on Worksheet 4.6.2. [O2]
- Use the left-hand rule to explain why all motors need a split ring commutator and what the motor would do if it didn't have one. This is an extension question on Worksheet 4.6.2. [O2]

Plenary suggestions

Heads and tails: Ask each student to write a question about something from the lesson on a coloured paper strip and the answer on another coloured strip. In groups of six to eight, hand out the strips so that each student gets a question and an answer. One student reads out his or her question. The student with the right answer then reads it out, followed by his or her question, and so on.

Answers to questions

Worksheet 4.6.1

1.

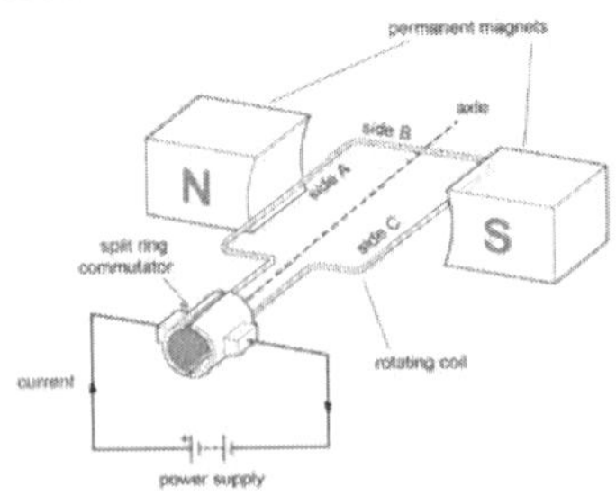

2.

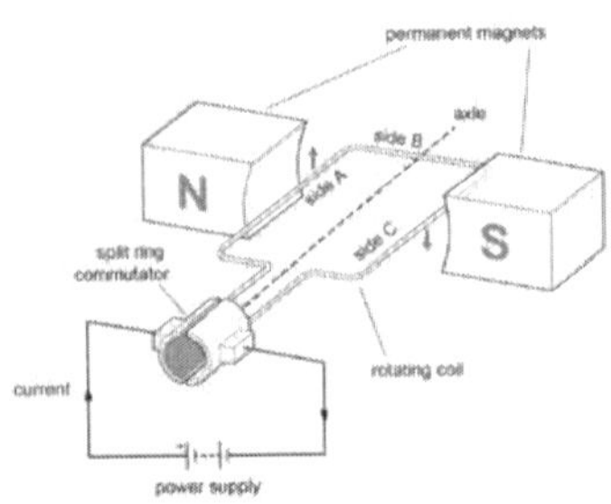

3.

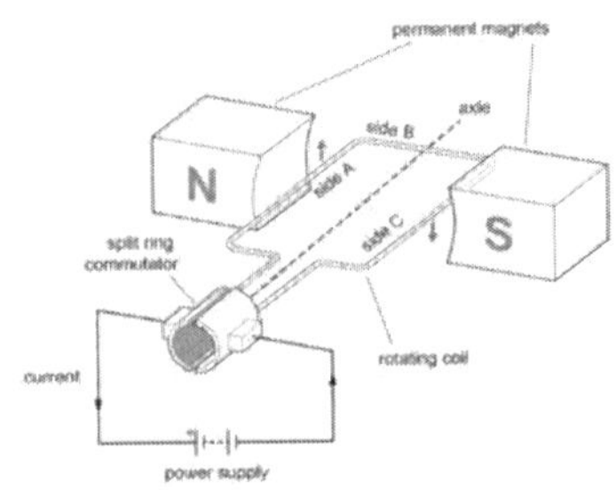

4.

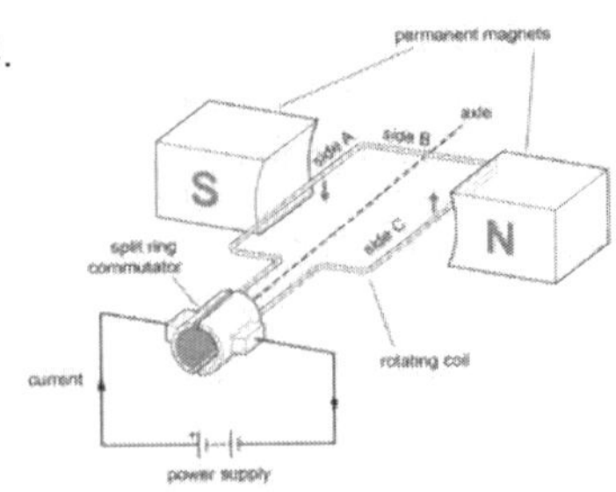

5. Increase the current/potential difference (of the coil). (Accept 'decrease resistance' and 'voltage' for potential difference.) Increase the strength of the magnetic field/electromagnet.

Worksheet 4.6.2

1. The current (in the coil) creates a magnetic field (around the coil). (Accept 'the coil is an electromagnet'.) So the magnetic field of the coil interacts with the (permanent) magnetic field of the magnets (producing a force). (Accept 'the two magnetic fields interact (producing a force)'.)
2. The current is parallel to the magnetic field. (Accept 'the current and magnetic field are in the same direction'; allow 'it/the wire is parallel to the magnetic field'.)
3. Increase the current/potential difference (of the coil). (Accept 'decrease resistance' and accept 'voltage' for potential difference.) Increase the strength of the magnetic field/electromagnet.
4. Change the direction of the current. Swap the poles of the magnet.
5. If the magnet is curved, then the magnetic field will always be perpendicular to the force. This means that the motor will always make the maximum turning force possible.
6. If there was no split ring commutator, the motor would do a half turn, but then since the current is going in the same direction, it would flip back where it came from. The motor would flip back and forth as the force cannot change direction.
7. Motors need a split ring commutator to make sure that the current flows in the same direction at each side of the ring. This ensures that the ring turns in a circle.

 © HarperCollins*Publishers* Limited 2016

Lesson 4.7: Generators and transformers

Lesson overview

OCR Specification reference

P4.2f, P4.2g

Learning objectives

- Explain how an alternator generates alternating current.
- For an alternator, draw and interpret graphs of potential difference generated in the coil against time.
- Understand how induced alternating electric currents are used in transformers.
- Apply the transformer equation linking the numbers of turns and the potential differences induced.

Learning outcomes

- [O1] Link the generator effect to its applications such as moving coil microphones.
- [O2] Explain why a dynamo and an alternator are different.
- [O3] Explain the shape of a graph of potential difference against time for a dynamo and an alternator.
- [O4] Explain how step-up and step-down transformers work.

Skills development

- Draw a graph of the potential difference in the coil against time.
- Describe the pattern in the graph of an alternator.
- Recognise the link between the position of an alternator and the pattern on the graph.

Resources needed ammeter (or galvanometer or multimeter), magnets, armature, mild steel yoke, support base, shaft, rivets, split pins, copper wire, transparent adhesive tape, tubing, magnets of different strength (optional); Worksheets 4.7.1, 4.7.2 and 4.7.3, Practical sheet 4.7, Technician's notes 4.7

Digital resources video on how power stations work (such as this one), Transformers animation link: www.youtube.com/watch?v=VucsoEhB0NA, PhET simulation link: https://phet.colorado.edu/en/simulation/faraday; PowerPoint 4.7

Key vocabulary wires, magnet, microphone, current, dynamo, alternator, potential difference, transformer, step-up transformer, step-down transformer

Teaching and learning

Engage

- Ask the students to sketch out the structure of the National Grid, labelling as many parts of it as they are able. Project an image of the National Grid and ask students to come up and label it. They should recall the function of the step-up and step-down transformers in the Grid.
- Use PowerPoint 4.7 to show students a picture of a power station (slide 2). You can assess their knowledge about this and show them a short video on how power stations work (such as this one). Ask them why they think the turning motion of the generator will make electricity. [O1]

Challenge and develop

- On the PowerPoint, show the students a diagram of a d.c. generator (a dynamo) and an a.c. generator (an alternator). Ask the students to fill in the differences and similarities. They could have a printout of the slide to stick into their books. [O2]
- Students complete Practical 4.7 where they build a generator and investigate what affects the size and direction of the current. [O2]

 © HarperCollins*Publishers* Limited 2016

Explain

- Students look at slide 4, showing a graph of voltage against time for a generator, and explain how the position of the generator links to the reading on the graph. [O2]
- Show the students the animation of a transformer twice. Allow them to watch it through once, before showing it to them again and asking them to write a script for what they think is being shown. [O4]
- This could be followed by the PhET simulation – search PhET Faraday's Electromagnetic Lab and use the Transformer tab simulation. [O4]
- If one is available, show students the set-up of a transformer and the difference between a step-up and a step-down transformer (i.e. that step-up transformers have more turns on the secondary coil than on the primary and that step-down transformers have more turns on the primary than on the secondary). If you do decide to demonstrate a transformer, ensure that the secondary voltage cannot exceed 20 V. [O4]
- Be sure to make the distinction between the 'coil' and the 'number of turns on the coil' – i.e. be clear that the number of coils remains the same (one) but the number of turns on the coil changes. [O4]
- Explain to the students that transformers only work when supplied with an alternating current to the primary coil, as it is the alternating current that produces an alternating magnetic field in the iron core. This alternating magnetic field induces a potential difference in the secondary coil, which induces an alternating current in the circuit connected to the secondary coil.
- At this stage, it might be wise for the students to draw a flow chart of the process of inducing a potential difference in the secondary coil. [O4]
- Introduce the students to the turns ratio expression ($V_p/V_s = n_p/n_s$) and state that the ratio of potential differences across the primary and secondary coils depends on the ratio of the number of turns on each coil. [O4]

Consolidate and apply

- Students should then complete one of the following worksheets appropriate to their ability:
  - Low demand: Worksheet 4.7.1 on the parts of a microphone and how it works. [O1]
  - Standard demand: Worksheet 4.7.2 on the generator. [O3]
  - High demand: Worksheet 4.7.3 on the transformer equation. [O4]

Extend

Ask students able to progress further to do the following:

- Use their knowledge of the structures of dynamos and alternators to explain why one uses a split-ring commutator and the other uses slip rings. [O2]

Plenary suggestions

Hot seat: Ask each student to think up a question, using material from the lesson. Select someone to put in the hot seat, and ask students to pose their questions and say at the end whether the answer is correct or incorrect.

 © HarperCollins*Publishers* Limited 2016

Answers to questions

Worksheet 4.7.1

1. a)

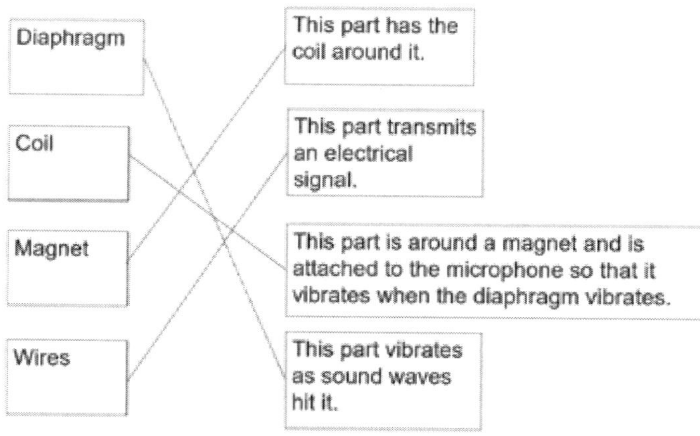

b)

Sound waves hit the **diaphragm** which vibrates → This causes the **coil** to vibrate → Since it is around a magnet, this generates a **current**

Worksheet 4.7.2

1. a) The wind causes the plastic cup to turn, which causes the magnet to turn/spin/rotate. The (magnetic) field/lines of force/flux rotate(s)/move(s)/through/in/cut(s) the coil. *(do **not** credit the idea that movement 'creates' the magnetic field)* The potential difference/p.d./voltage is induced across the coil. *(do **not** credit just 'current induced')*

   b) any **one** from:

   more powerful/stronger/lighter magnet *(do **not** credit 'a bigger magnet')*

   larger/more/bigger/lighter cups/with bigger surface area

   longer arms

   lubricate the spindle

   add more turns to the coil

   c)

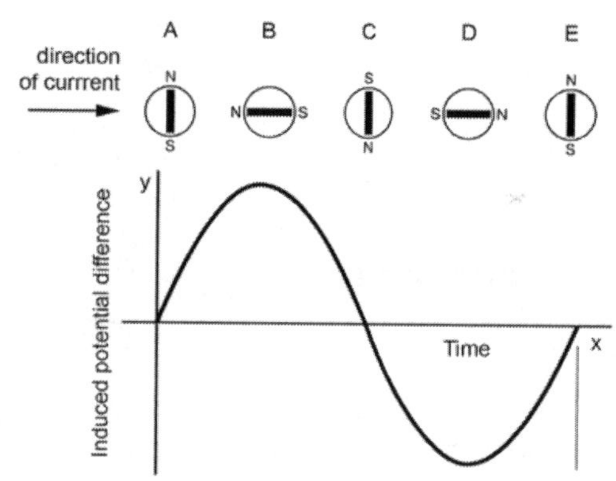

   d)

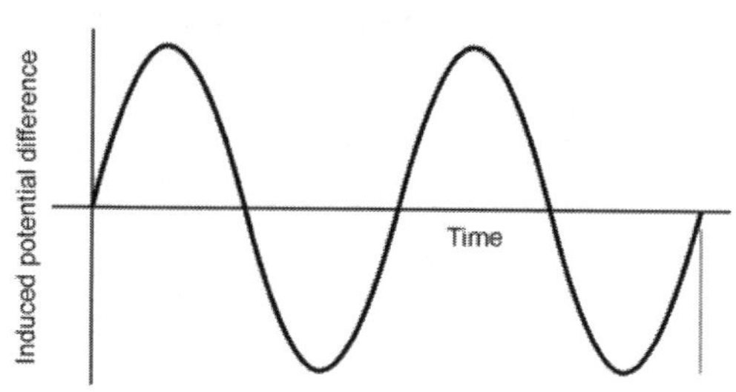

   e) There need to be twice as many waves in the same space.

Worksheet 4.7.3

1. a) 400 turns
   b) 24 V
2. a) 0.075 A
   b) 0.417 A

 © HarperCollins*Publishers* Limited 2016

Lesson 4.8: Loudspeakers and microphones

Lesson overview

OCR Specification reference

P4.2j

Learning objectives

- Describe how a moving-coil loudspeaker works.
- Compare loudspeakers and headphones.
- Describe how a moving-coil microphone works

Learning outcomes

- [O1] Label the parts of a loudspeaker and describe what they do to make it work.
- [O2] Explain how the movement of the coil affects the volume of the sound produced.
- [O3] Explain the similarities and differences between loudspeakers and microphones.

Skills development

- Students can explain how a moving coil loudspeaker and microphone work.
- Relate the effects of changing current to the effects on the sound produced by the loudspeaker.
- Evaluate the speaker and think of any improvements that could be made.

Resources needed neodymium magnets (or horseshoe magnets), iron yokes, d.c. power supply, copper wire, scissors, thread, clamp stand, card, table tennis balls or beads, oscilloscope (optional); Worksheets 4.8.1 and 4.8.2, Technician's notes 4.8

Digital resources BBC video 'How does a loudspeaker work?', 'How a Loudspeaker Works…' animation, YouTube video clip – How does a microphone work? - https://www.youtube.com/watch?v=d_crXXbuEKE; PowerPoint presentation 4.8

Key vocabulary oscillation, alternating current, radial magnetic field, pressure wave, ear drum, longitudinal wave, headphones, moving-coil loudspeaker, moving-coil microphone

Teaching and learning

Engage

- Show students a loudspeaker and play it to show it working. Ask them how they think it works. To further elaborate on this, add table tennis balls to the speaker to highlight the vibrations. [O1]
- Show students PowerPoint presentation 4.8, the picture of the wave produced by a speaker. As revision of waves, ask the students how the wave will change if the volume and pitch of the sound increase. [O2]

Challenge and develop

- You can use this video from the BBC or this animation to show how a loudspeaker works. [O1]
- Students complete Worksheet 4.8.1 where they label a diagram of a loudspeaker and state what each part does. They then put the sequence of events of how a loudspeaker works in order. [O1]

Explain

- The students should then be shown a video clip about how a microphone works. Suggested clip: https://www.youtube.com/watch?v=d_crXXbuEKE. Following the video, the students should draw a label a diagram of a moving-coil microphone. It may be a good idea to play the video clip twice: the first time the students simply watch it, the second time they take down some notes to help them draw the diagram.
- Following this, the students should revise waves. They think about the wave produced and sketch the shape of a wave if the sound has a different pitch and volume. [O2]

 © HarperCollinsPublishers Limited 2016

Consolidate and apply

- Students can use their knowledge of loudspeakers to explain the similarities and differences between a loudspeaker, a set of headphones and a microphone. They could do this in the form of a table comparing the three objects. [O3]
- Students complete Worksheet 4.8.2 where they explain how a loudspeaker works. [O2]
- Students complete Worksheet 4.8.2, Activity 2 where they explain how the wave produced changes. (Standard demand) [O2]

Extend

Ask students able to progress further to do the following:

- Explain how the sound will be different if the speaker vibrates more quickly. (High demand) [O2]

Plenary suggestions

Hot seat: Ask each student to think up a question, using material from the topic. Select someone to put in the hot seat. Ask students to pose their questions and say at the end whether the answer is correct or incorrect.

Answers to questions

Worksheet 4.8.1

1 and 2

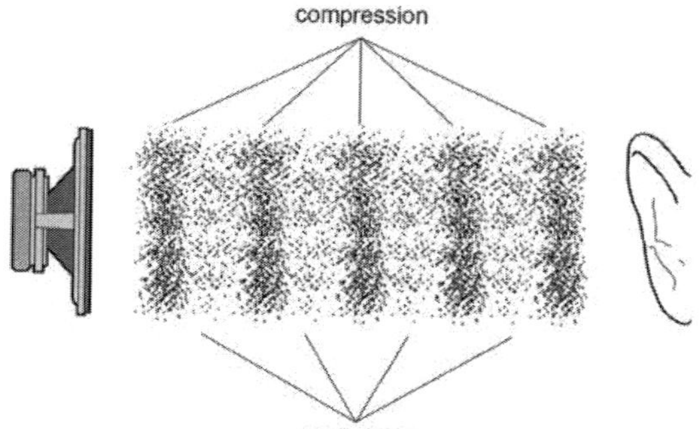

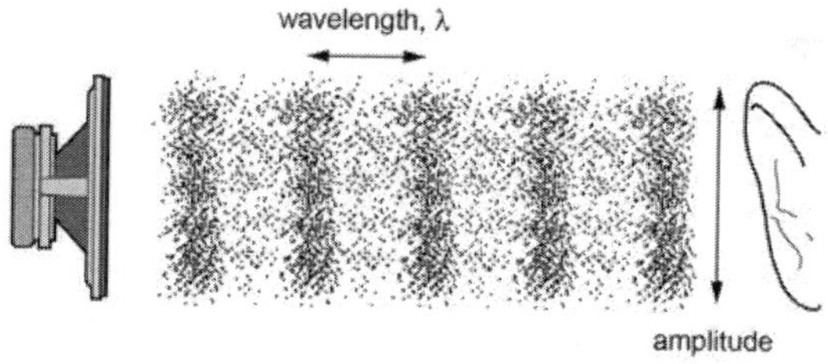

Worksheet 4.8.2

1. Indicative content:
    - The current in the electrical circuit is varying
    - The current passes through the coil
    - The coil experiences a force (inwards or outwards)
    - Reversing the current reverses the force
    - The size of the current affects the size of the force
    - The varying current causes the coil to vibrate
    - The (vibrating) coil causes the cone to vibrate
    - The vibrating cone causes the air molecules to move
    - The movement of the air molecules produces the pressure variations in the air needed for a sound wave
    - The air molecules bunch together forming compressions and spread apart forming rarefactions

Level 3: A detailed explanation linking variations in current to the pressure variations of a sound wave, with a logical sequence.

Level 2: A number of relevant points made, but not precisely. A link between the loudspeaker and a sound wave is made.

Level 1: Some relevant points but fragmented with no logical structure.

No relevant content.

 © HarperCollins*Publishers* Limited 2016

2.

a) and b)

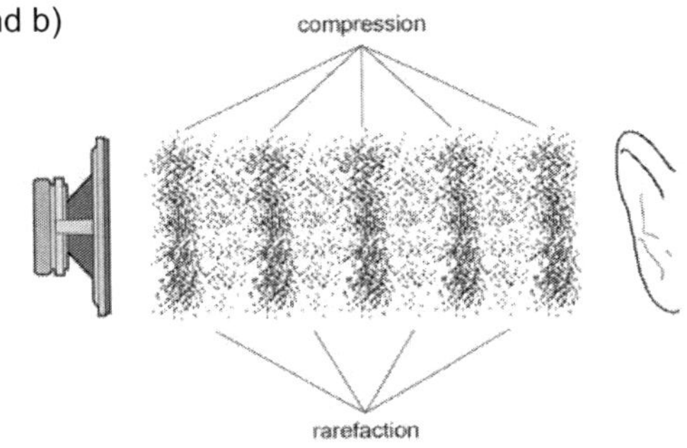

c) and d)

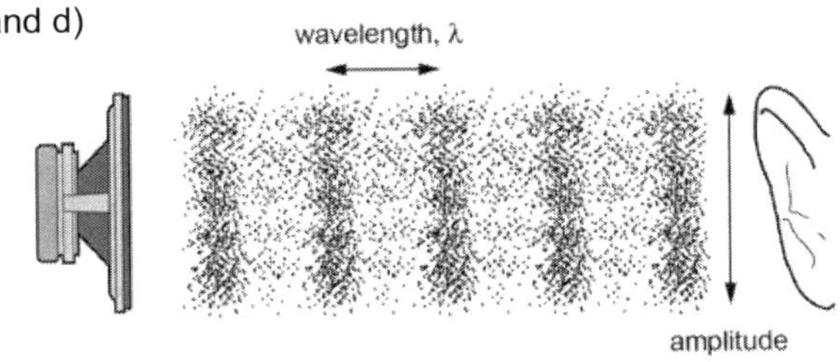

e)

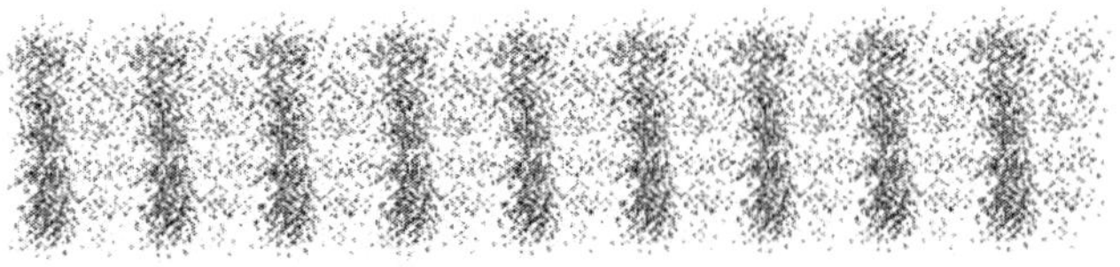

f) The frequency in the new diagram would be twice as big as the frequency in the first diagram.

g) The second sound wave would have a higher pitch.

h) The speaker with the wave that you have drawn would vibrate twice as fast.

 © HarperCollins*Publishers* Limited 2016

Key concept: Lesson 4.9: The link between electricity and magnetism

Lesson overview

OCR Specification reference

KC

Learning objectives

- Explore how electricity and magnetism are connected.
- Describe how electromagnetic induction occurs.
- Describe the principle of the electric motor.

Learning outcomes

- [O1] Identify apparatus needed to demonstrate the link between electricity and magnetism.
- [O2] Describe what causes electricity and magnetism.
- [O3] Predict the direction of a current or magnetic field on a wire, a solenoid, a U-shaped magnet and a motor.

Skills development

- Describe the apparatus that can form an electric current or a magnetic field.
- Devise experiments that can highlight how electricity and magnetism are interlinked.
- State whether the apparatus worked in a manner that was consistent with the theory; for example, Fleming's left-hand rule.

Resources needed d.c. power source, wires, plotting compass, ammeter, bar magnets, horseshoe magnet; Worksheets 4.9.1 and 4.9.2, Practical sheets 4.9.1, 4.9.2, 4.9.3 and 4.9.4, Technician's notes 4.9

Digital resources Faraday's Electromagnetic Lab PhET animation; PowerPoint presentation 4.9

Key vocabulary electricity, magnetism, conservation of energy, potential difference

Teaching and learning

Engage

- Students think about why ships that were struck by lightning had compasses that changed direction. To prepare for this lesson, they could read chapter 124 of *Moby Dick* on Worksheet 4.9.1, which describes such an occurrence, and think about why it happened. [O1]

Challenge and develop

- Students complete a carousel of practicals (4.9.1, 4.9.2, 4.9.3, 4.9.4) that demonstrate how electricity, magnetism and force are linked. [O1]
- The students could write predictions of what will happen before they do these practicals, with explanations. After the practicals, they can go back to their predictions and think about whether they agreed with their observations and why they did or did not. In the absence of equipment, the students could use the PhET animation Faraday's Electromagnetic Lab. [O3]

Explain

- Students can use the ideas of the right-hand rule and Fleming's left-hand rule to explain their observations from the practicals. [O3]

 © HarperCollinsPublishers Limited 2016

Consolidate and apply

- Students explain the similarities and differences between a dynamo and a motor. [O2]
- Students complete Worksheet 4.9.2 on the motor effect and on the difference between motors and generators. [O2]

Extend

Ask students able to progress further to do the following:

- Use their knowledge of motors and generators to explain how a hairdryer and a hydroelectric dam work. [O3]

Plenary suggestions

What do I know? Ask the students to each write down one thing that they are sure of, one thing that they are unsure of and one thing that they need to know more about. Students can work in groups of five or six to agree on group lists. Ask each group to say what they decided and agree as a class what they are confident about, what they are less sure about and what they need to know more about.

Answers to questions

Worksheet 4.9.1

1. Lightning would have caused the compass to point in a different direction because when the lightning struck, it created a magnetic field (as potential difference always has a magnetic field), which moved the compass.

Worksheet 4.9.2

1.

1. You can use the right-hand grip rule.
2. You would have to move the magnet through the coil to induce an electric current.
3. If you place the magnetic field going from north to south perpendicular to the direction of the current, then the wire will move.
4. You can use Fleming's left-hand rule.

2.

1. You need a current, a magnetic field and a force.
2. Motors and generators use similar apparatus. They both have coils in magnetic fields that spin.
3. A motor turns an electrical current into motion. A generator turns motion into an electrical current.
4. Things that use motors: cars, fans, drills, hairdryers, washing machines (there are others).
5. Things that use generators: power stations, cars (to charge the battery), bikes (to power the lights).
6. They both have coils. They both have ways of making the current stay in the direction it is going. They both have magnets.
7. d.c. motors have split-ring commutators to keep the current flowing in a particular direction. a.c. motors use slip rings to keep the current flowing in a particular direction.

 © HarperCollins*Publishers* Limited 2016

Maths skills: Lesson 4.10: Rearranging equations

Lesson overview

OCR Specification reference

P4.2h, i

Learning objectives

- Know how to rearrange equations.
- Know how to use the transformer equation.
- Know how to calculate the force on a conductor.

Learning outcomes

- [O1] Be able to work out one value from an equation if given all the other values.
- [O2] Be able to predict how values in the transformer equation change when one value changes.
- [O3] Be able to predict how values in the $F = BIL$ equation change when one value changes.

Skills development

- Use the correct terminology for the variables in the equations.
- Rearrange equations to determine the unknown variable.
- Use the appropriate number of significant figures when calculating the unknown variables in the equations.

Maths focus

- MS 3b Change the subject of an equation.
- MS 3c Substitute numerical values into algebraic equations using appropriate units for physical quantities.
- MS 3d Solve simple algebraic equations

Resources needed cards for $F = BIL$, cards for transformers; Worksheets 4.10.1 and 4.10.2

Digital resources PowerPoint presentation 4.10

Key vocabulary force, magnetic flux density, tesla, current, length, voltage, coils

Teaching and learning

Engage

- Show students the picture of the pylons on PowerPoint presentation 4.10. This may be revision of electricity. If it is, ask them to remember why pylons are used and what they are connected to (transformers). Ask them if they remember why transformers are needed. If the students need a bit more prompting, show them the picture of the plug socket and ask them what the voltage needs to do before it gets to our houses. [O1]

Challenge and develop

- Students complete the first part of Worksheet 4.10.1, including transformer calculations. [O1]
- Remind students of the $F = BIL$ equation and what each letter stands for. Students complete the first part of Worksheet 4.10.2, using the equation. [O1]

Explain

- Students complete the second part of Worksheet 4.10.1 and explain how changing one value changes the others. [O2]
- Students complete the remainder of Worksheet 4.10.2 and explain how variables in the $F = BIL$ equation change when one variable changes.

 © HarperCollins*Publishers* Limited 2016

Consolidate and apply

- Students can arrange the cards for transformers and $F = BIL$ to work out which numbers should go in the equation. They can be made progressively harder by using more sets of cards. [O2, O3]

Extend

Ask students able to progress further to do the following:

- Use the full set of cards to work out the correct values. [O2, O3]

Plenary suggestions

On the PowerPoint, show students the mistakes that people have made in their calculations. Students work out what the mistake is and what the correct answer is. [O1]

Ideas hothouse: Ask students to work in pairs to list points about what they know about a particular idea **from the lesson**. Then ask them to join together in **groups of four** and then of six to eight to discuss further and come up with an agreed list of points. Ask one person from each group to report back to the class.

Answers to questions

Worksheet 4.10.1

1.
   a) The answers are in bold in the table below (see the next answer)..
   b) The step-up transformers are in the shaded rows in the table.

Number of turns on primary coil, n_p	Potential difference on primary coil, v_p (V)	Number of turns on secondary coil, n_s	Potential difference on secondary coil, v_s (V)
1000	11 500	20	230
100	**250**	200	500
20	500	**4**	100
500	100	100	**20**
48	240	46	230
200	**400**	50	100
10	300	**2**	60
50	60	250	**300**

2.
   a) any ratio where the primary transformer has 40 times as many coils as the secondary transformer (9200 ÷ 230 = 40)
   b) You could i) increase the voltage on the primary transformer, ii) increase the number of coils on the secondary transformer, iii) decrease the number of coils on the primary transformer.
   c) i) the number of coils on the primary transformer and the number of coils on the secondary transformer
   ii) the voltage on the secondary transformer
   d) i) the voltage on the primary transformer and the number of coils on the secondary transformer
   ii) the number of coils on the primary transformer

Worksheet 4.10.2

1.

Force (N)	Current (A)	Length of wire (m)	Magnetic flux density (T)
0.006	2.0	10	3.0×10^{-4}
0.125	**1.6**	40	2.0×10^{-3}
0.060	1.5	**20**	2.0×10^{-3}
0.075	3.0	60	$\mathbf{6.3 \times 10^{-4}}$
0.150	4.0	25	1.5×10^{-3}
0.150	**2.0**	5	1.5×10^{-2}
0.080	2.0	**50**	8.0×10^{-4}
0.100	2.5	100	$\mathbf{4.0 \times 10^{-4}}$

2. a) 0.05 N
   b) i) 5 × 2 × 0.02 = 0.2 N
   ii) You could double one of the values.
   c) i) The magnetic flux density would be lower.
   ii) It will increase.
   d) He could i) quadruple the strength of the magnetic field, ii) quadruple the current, iii) quadruple the length of the wire, iv) double the strength of the magnetic field and double the current, v) double the strength of the magnetic field and double the length of the wire, vi) double the current and double the length of the wire.

 © HarperCollins*Publishers* Limited 2016

Cards for *F* = *BIL*

Force = 0.035 N

Current = 2.0 A

Magnetic flux density = 2.5×10^{-3} T

Length of wire = 7 m

Force = 0.00525 N

Current = 1.5 A

Magnetic flux density = 7.0×10^{-4} T

Length of wire = 5 m

Force = 0.04 N

Current = 4.0 A

Magnetic flux density = 1.0×10^{-2} T

Length of wire = 100 cm

Cards for transformers

Voltage on the primary coil = 275 000 V

Voltage on the secondary coil = 230 V

Number of turns on the primary coil = 55 000 turns

Number of turns on the secondary coil = 46

Voltage on the primary coil = 25 000 V

Voltage on the secondary coil = 400 000 V

Number of turns on the primary coil = 100

Number of turns on the secondary coil = 1600

Voltage on the primary coil = 230 V

Voltage on the secondary coil = 110 V

Number of turns on the primary coil = 92

Number of turns on the secondary coil = 44

What is the mistake here?

1. The primary coil of a transformer has 4000 turns and a voltage of 2000 V. The secondary coil has 250 turns. What is the voltage on the secondary coil?

   Incorrect answer: 32 000 V

   The pupil has used the equation incorrectly. They have put the primary coil voltage as the denominator.

   Correct answer: 125 V

2. The primary coil of a transformer has 7500 turns and a voltage of 25 kV volts. The secondary coil produces 11 000 V. How many turns are on the secondary coil?

   Incorrect answer: 3 300 000 turns

   They did not convert 25 kV to 25 000 V.

   The correct answer is 3300 turns.

3. The force on a wire is 0.05 N. The magnetic flux density is 2×10^{-3} T. The wire is 2.5 m long. What is the current on the wire?

   Incorrect answer: 100 A

   The student has missed by a factor of 10, possibly because they thought that 2×10^{-3} is 0.0002.

   The correct answer is 10 A.

 © HarperCollins*Publishers* Limited 2016

When and how to use these pages: Check your progress, Worked example and End of chapter test

Check your progress

Check your progress is a summary of what students should know and be able to do when they have completed the chapter. Check your progress is organised in three columns to show how ideas and skills progress in sophistication. Students aiming for top grades need to have mastered all the skills and ideas articulated in the final column (shaded pink in the Student book).

Check your progress can be used for individual or class revision using any combination of the suggestions below:

- Ask students to construct a mind map linking the points in Check your progress
- Work through Check your progress as a class and note the points that need further discussion
- Ask the students to tick the boxes on the Check your progress worksheet (Teacher Pack CD). Any points they have not been confident to tick they should revisit in the Student Book.
- Ask students to do further research on the different points listed in Check your progress
- Students work in pairs and ask each other what points they think they can do and why they think they can do those, and not others

Worked example

The worked example talks students through a series of exam-style questions. Sample student answers are provided, which are annotated to show how they could be improved.

- Give students the Worked example worksheet (Teacher Pack CD). The annotation boxes on this are blank. Ask students to discuss and write their own improvements before reviewing the annotated Worked example in the Student Book. This can be done as an individual, group or class activity.

End of chapter test

The End of chapter test gives students the opportunity to practice answering the different types of questions that they will encounter in their final exams. You can use the Marking grid provided in this Teacher Pack or on the CD Rom to analyse results. This shows the Assessment Objective for each question, so you can review trends and see individual student and class performance in answering questions for the different Assessment Objectives and to highlight areas for improvement.

- Questions could be used as a test once you have completed the chapter
- Questions could be worked through as part of a revision lesson
- Ask Students to mark each other's work and then talk through the mark scheme provided
- As a class, make a list of questions that most students did not get right. Work through these as a class.

 © HarperCollins*Publishers* Limited 2016

Marking Grid for End of Chapter 4 Test

	Getting Started [Foundation Tier]								Going Further [Foundation and Higher Tiers]					More Challenging [Higher Tier]				Most demanding [Higher Tier]		Total marks	Percentage
Student Name	Q. 1 (AO1) 1 mark	Q. 2 (AO1) 1 mark	Q. 3 (AO1) 2 marks	Q. 4 (AO1) 1 mark	Q. 5 (AO2) 2 marks	Q. 6 (AO2) 1 mark	Q. 7 (AO2) 3 marks	Q. 8 (AO1) 3 marks	Q. 9 (AO1) 1 mark	Q. 10 (AO1) 1 mark	Q. 11 (AO2) 2 marks	Q. 12 (AO2) 4 marks	Q. 13 (AO2) 2 marks	Q. 14 (AO1) 1 mark	Q. 15 (AO2) 1 mark	Q. 16 (AO2) 2 marks	Q. 17 (AO3) 6 marks	Q. 18 (AO2) 2 marks	Q. 19 (AO3) 2 marks		

 © HarperCollins*Publishers* Limited 2016

Check your progress

You should be able to:

Recall that like poles repel, unlike poles attract	Recognise that the poles of a magnet are the places where the magnetic forces are strongest	Explain that an induced magnet is a material that is only magnetic when it is placed in a magnetic field
Plot the magnetic field around a bar magnet	Recognise that the magnetic field is the region around a magnet where a force acts on another magnet or on a magnetic material	Explain how the behaviour of a magnetic compass is related to evidence that the core of the Earth must be magnetic
State how the strength of an electromagnet can be increased	Draw the magnetic field around a conducting wire and a solenoid	Explain how electromagnets are used in devices
State that a force acts on a current-carrying conductor in a magnetic field	Describe the motor effect that applies to a current-carrying conductor in a magnetic field	Explain how the direction of the force on the conductor can be identified using Fleming's left-hand rule
State that magnetic flux density is measured in tesla (T)	Explain what the size of a force on a conductor depends on	Use the equation $F = BIL$ to calculate the force on a conductor
State what the generator effect is	Recognise that the size of the induced potential difference (and so induced current) can be increased by increasing the speed of movement or by increasing the strength of the magnetic field	Explain that if the direction of motion of the conductor or the polarity of the magnetic field is reversed, the direction of the induced potential difference and any induced current is reversed
State that a dynamo generates direct current and an alternator generates alternating current	Describe that when a coil is rotated in a magnetic field an alternating current is induced in the coil Draw graphs of potential difference generated in the coil against time	Explain how the generator effect is used in a dynamo to generate d.c. with the use of commutator; and is used in an alternator to generate a.c. Describe how to draw and interpret graphs of potential difference generated in the coil against time
State what a basic transformer consists of and the difference between step-up transformer and a step-down transformer	Explain how the potential differences across the two coils depend on the number of turns on each coil and how the potential difference is induced	Use and apply the expression $\frac{V_p}{V_s} = \frac{n_p}{n_s}$

 © HarperCollinsPublishers Limited 2016

Worked example

1 How can you increase the strength of the magnetic field created by a current through a wire?

To increase the strength of the magnetic field you need to coil the wire.

Correct, but that is only part of the answer. There are two more ways in which you can strengthen the magnetic field: increase the current and wrap the coils around an iron core. Use the correct vocabulary: solenoid instead of coil.

2 Explain how continuous rotation is produced in an electric motor.

When a wire carrying a current is put into a magnetic field, one force reacts to the other and this pushes the wire down. If this is a coil then it begins to go round.

You should aim to show that you really understand exactly what is going on and where the forces are coming from.

Describe the interaction between the magnetic field in the wire and the magnetic field of the magnet.

With a complicated answer like this, it can be helpful to draw a diagram. In this case the diagram would show the coil, magnetic field and direction of movement.

3 The primary coil of a transformer has 75 turns and a potential difference of 2.5 V. Calculate the potential difference in the secondary coil when there are 225 turns in the secondary coil.

$$\frac{V_p}{V_s} = \frac{n_p}{n_s}$$

$$\text{So } V_s = \frac{V_p \times N_p}{N_s}$$

$$= \frac{2.5 \times 75}{225}$$

$$= 0.83 \text{ volts}$$

The student started with the right equation and showed their working, but the answer is incorrect. They have rearranged the equation for V_s incorrectly: n_p and n_s are the wrong way round.
Check that you can rearrange equations correctly.

 © HarperCollins*Publishers* Limited 2016

Waves: Introduction

When and how to use these pages

This unit will build on ideas students have met before, such as:

- All waves have certain things in common
- They transfer energy from one place to another
- When waves meet they can add together or cancel out
- When waves hit an object they may be absorbed by it or reflected back
- They may change direction at the point where two different materials meet
- Frequencies of waves are measured in hertz (Hz)
- Sunlight (white light) is made up of a mixture of many different colours
- Different colours are absorbed or reflected by different surfaces
- Sound waves are produced by vibrations
- Ultrasound is defined as sound with a pitch too high for humans to hear.

Overview of the unit

In this unit, students will learn about light and sound as examples of waves, as well as how other electromagnetic waves behave like light. They will discover why light and sound are different types of wave and be able to give some further examples of both transverse and longitudinal waves. They will investigate the behaviour of waves and learn about some applications of waves in medicine and other situations.

The students will learn the key terms used to describe waves and be able to use these in contexts such as describing the different parts of the electromagnetic spectrum.

The unit concludes with some work on lenses and the difference between convex and concave lenses.

This unit offers a number of opportunities for students to investigate phenomena through practical work, work collaboratively with peers and critically evaluate evidence before drawing conclusions.

Obstacles to learning

The students may need extra guidance with the following terms and concepts:

- There may be some confusion between the everyday use of the term 'wave' (in terms of a hand wave) and the definition required.
- Common misconceptions:
  - sounds cannot travel through solids and liquids
  - matter moves along waves / waves move matter (as waves in the sea move sand around)
  - big waves travel faster than small waves (in the same medium)
  - although most students are happy with the idea of sound we can't hear, they can be much less comfortable with the idea of light we can't see.

Practicals in this unit

In this unit students will do the following practical work:

- Practical: measure the speed of a water wave
- investigate reflection of light by different types of surface
- investigate refraction of light by different substances
- investigate how the amount of infrared radiation absorbed or radiated by a surface depends on the nature of that surface
- investigate the effectiveness of various substances in protecting against ultraviolet radiation
- investigate the behaviour of light as an example of an electromagnetic wave
- investigate the absorption and reflection of different colours of light
- investigate convex and concave lenses.

 © HarperCollins*Publishers* Limited 2016

	Lesson title	Overarching objectives
1	Describing waves	• Describe wave motion. • Define wavelength and frequency. • Apply the relationship between wavelength, frequency and wave velocity.
2	Transverse and longitudinal waves	• Compare the motion of transverse and longitudinal waves. • Explain why water waves are transverse waves. • Explain why sound waves are longitudinal waves.
3	Reflection and refraction of waves	• Describe reflection, transmission and absorption of waves. • Construct ray diagrams to illustrate reflection. • Construct ray diagrams to illustrate refraction.
4	Sound waves	• Describe how we hear sound and state the range of frequencies we can hear. • Explain that sound travels faster in a denser medium. • Explain about reflection, absorption and transmission of sound.
5	Exploring ultrasound	• Explain what ultrasound is. • Describe how ultrasound can be used in industry to investigate or detect hidden or buried objects. • Explain how ultrasound is used in medicine.
6	Practical: measuring the wavelength, frequency and speed of waves in a ripple tank and waves in a solid	• Develop techniques for making observations of waves. • Select suitable apparatus to measure frequency and wavelength. • Use data to answer questions.
7	Maths skills: using and rearranging equations	• Select and apply the equations: $T = \frac{1}{frequency}$ and $v = f\lambda$. • Substitute numerical values into equations

 © HarperCollins*Publishers* Limited 2016

		using appropriate units. • Change the subject of an equation.
8	Key concept: Transferring energy or information by waves	• Understand that all waves have common properties. • Understand how waves can be used to carry information. • Understand various applications of energy transfer by different types of electromagnetic waves.
9	The electromagnetic spectrum	• Recall the similarities and differences between transverse and longitudinal waves. • Recognise that electromagnetic waves are transverse waves. • Describe the main groupings and wavelength ranges of the electromagnetic spectrum.
10	Gamma rays and X-rays	• List the properties of gamma rays and X-rays. • Compare gamma rays and X-rays.
11	Ultraviolet and infrared radiation	• Describe the properties of ultraviolet and infrared radiation. • Describe some uses and hazards of ultraviolet radiation. • Describe some uses of infrared radiation.
12	Microwaves	• List some properties of microwaves. • Describe how microwaves are used for communications.
13	Radio and microwave communication	• Describe how radio waves are used for television and radio communications. • Describe how microwaves are used in satellite communications. • Describe the reflection and refraction of radio waves.
14	Reflection, refraction and wave fronts	• Explain reflection and refraction and how these may vary with wavelength. • Construct ray diagrams to illustrate refraction. • Use wave front diagrams to explain refraction in terms of the difference in velocity of

 © HarperCollins*Publishers* Limited 2016

		the waves in different substances.
15	Practical: Investigate how the amount of infrared radiation absorbed or radiated by a surface depends on the nature of that surface	• Explain reasons for the equipment used to carry out an investigation. • Explain the rationale for carrying out an investigation. • Apply ideas from an investigation to a range of practical contexts.
16	Practical: Investigating the reflection of light by different types of surface and the refraction of light by different substances	• Make and record observations of how light is reflected and transmitted at different surfaces. • Measure angles and discuss the method, apparatus and uncertainty in measurements. • Draw conclusions from experimental results.
17	Lenses	• Understand what a lens does. • Draw ray diagrams to show the formation of images by lenses. • Describe the difference between a real and a virtual image.
18	Images and magnification	• Draw ray diagrams to show the formation of real and virtual images by lenses. • Calculate the magnification of an image.
19	Colour	• Describe what happens when light of different wavelengths lands on an object. • Explain what determines the colour of an opaque object. • Explain the effect of coloured filters.

 © HarperCollins*Publishers* Limited 2016

Lesson 5.1: Describing waves

Lesson overview

OCR Specification reference

P5.1a, b, c, d

Learning objectives

- Describe wave motion.
- Define wavelength and frequency.
- Apply the relationship between wavelength, frequency and wave velocity.

Learning outcomes

- [O1] Understand that waves have velocity, wavelength and frequency.
- [O2] Use the equation $v = f \times \lambda$ to calculate wave velocity.
- [O3] Rearrange the equation $v = f \times \lambda$ to calculate the wavelength or frequency of waves.

Skills development

- Use an equation with three variables.
- Manipulate an equation with three variables.
- Use a variety of technical terms correctly.

Maths focus

- Use and rearrange an equation with three variables.

Resources needed cathode ray oscilloscope, signal generator, loudspeaker; Worksheets 5.1.1, 5.1.2, 5.1.3 and 5.1.4

Digital resources animation of cathode ray oscilloscope, signal generator and loudspeaker if equipment is not available

Key vocabulary frequency, wavelength, amplitude, time period, hertz

Teaching and learning

Engage

Arrange the students around the edge of the classroom, facing inwards and ask them to start a Mexican wave. Allow the students to watch it as it propagates around the room. Be sure to emphasise the point that the students themselves are not moving around the edge of the room, but the energy is. [O1]

Challenge and develop

- Using a signal generator, cathode ray oscilloscope and a loudspeaker, show the students an image of a sound wave on the oscilloscope screen. Use this to introduce the terms 'amplitude', 'wavelength', 'frequency' and 'time period'. [O1] (If the equipment is not available, use a suitable animation instead.)
- Change the nature of the wave on the screen and ask students to describe the new wave, using the key words of amplitude, wavelength and frequency.[O1]
- Students should be issued with Worksheets 5.1.1, 5.1.2, 5.1.3 and 5.1.4. [O1]

Explain

- Introduce the wave equation as a means of linking the wavelength and the frequency of a wave. Highlight the need for the students to be consistent with their units when they are using the wave equation. [O2]
- To get the idea across, ask students to imagine a stone thrown into a pond and the ripples moving outwards (or use a suitable video clip or animation). Suggest that the splash makes three ripples per second (i.e. the frequency) and the ripples are 6 cm apart (i.e. the wavelength). Pose the question 'how far will the first ripple

© HarperCollins*Publishers* Limited 2016

travel in the first second (or 'each second')'. Students should quickly come up with the answer of 18 cm, establishing the link summarised by the wave equation. [O2]

Consolidate and apply

- Issue the students with one Worksheet 5.1.2–5.1.4 that is appropriate to their level:
  - Low demand: Worksheet 5.1.2 [O2]
  - Standard demand: Worksheet 5.1.3 [O2]
  - High demand: Worksheet 5.1.4 [O2, O3]

Extend

Ask the students to think back to the Mexican wave set up at the start of the lesson. Ask the students to explain, in no more than 30 words, whether a Mexican wave is a real wave or not. [O1]

Plenary suggestions

- **Ideas hothouse:** Ask the students to work in pairs to list three points about what they know about waves from the lesson today. Then ask the pairs to join together into fours and then groups of six to eight to discuss this further and to come up with an agreed list of points. Ask one person from each group to report back to the class.
- **Heads and tails:** Ask each student to write a question (probably related to the wave equation) on a coloured paper strip and then answer on another coloured strip. In groups of six to eight, hand out the strips so that each student gets a question and an answer. One student reads out their question. The student with the right answer then reads it out, followed by their question, and so on.

Answers to questions

Worksheet 5.1.1

1. 1 = wavelength; 2 = amplitude; 3 = peak; 4 = trough
2. (words in this order) height, amplitude, peak, trough, wavelength, metres, frequency, number, second, hertz

Worksheet 5.1.2

1. a) 60 m/s b) 450 m/s c) 500 m/s d) 1250 m/s e) 500 m/s
2. a) 15,000 m/s b) 0.6 m/s c) 12,000 m/s d) 8000 m/s e) 1800 m/s

Worksheet 5.1.3

1. a) 60 m/s b) 450 m/s c) 500 m/s d) 1250 m/s e) 500 m/s
2. a) 21.5 Hz b) 1.1 Hz c) 480 Hz d) 0.1 Hz e) 4×10^{11} Hz

Worksheet 5.1.4

1. 2.5 m/s
2. 50 m/s
3. 16 000 m/s
4. 33Hz
5. 0.8 Hz
6. 300 Hz
7. 6 m
8. 0.42 m
9. 300 m
10. 0.67 m

 © HarperCollins*Publishers* Limited 2016

Lesson 5.2: Transverse and longitudinal waves

Lesson overview

OCR Specification reference

P5.1e, f, k

Learning objectives

- Compare the motion of transverse and longitudinal waves.
- Explain why water waves are transverse waves.
- Explain why sound waves are longitudinal waves

Learning outcomes

- [O1] Realise that waves can be transverse or longitudinal.
- [O2] Give examples of transverse and longitudinal waves.
- [O3] Explain the difference between transverse and longitudinal waves in terms of how they are produced.

Skills development

- Correct use of key terms.
- Plotting a bar chart of given data.
- Identifying patterns in data.

Maths focus

- Plotting a bar chart from given data.
- Identifying patterns in data from a bar chart.

Resources needed slinky, ripple tank (if available); Worksheets 5.2.1, 5.2.2 and 5.2.3

Digital resources video clip: https://www.youtube.com/watch?v=xPZigWFyK2o; PowerPoint presentation 5.2

Key vocabulary transverse wave, longitudinal wave, compression, rarefaction

Teaching and learning

Engage

- Watch the first 15 seconds of the Star Wars clip https://www.youtube.com/watch?v=xPZigWFyK2o and ask students: What is wrong with this clip? (The space ship shouldn't make a noise)

Challenge and develop

- Demonstrate a longitudinal wave using a slinky. Emphasise to the students where a rarefaction and a compression can be seen. Ask the students which way the waves are moving and what is happening to the individual coils in the slinky (they are oscillating backwards and forwards). Pose the question to the students: 'If the coils were not there, would the wave still move?' (This will help to introduce the idea that longitudinal waves need a medium through which to travel.) [O1, O2, O3]
- Now demonstrate a transverse wave and ask the students to describe the difference between this and the longitudinal wave they have just seen. [O1, O2, O3]
- Demonstrate, through the use of a ripple tank, some water waves. (If a ripple tank is not available, use a suitable animation instead.) Give the students time to discuss in pairs whether they think water waves are an example of longitudinal or transverse waves, giving a reason for their choice. A cork could be placed in the water to help the students decide. [O2]

 © HarperCollins*Publishers* Limited 2016

Explain

- Return to the clip from the start of the lesson. **Pair talk:** With reference to the slinky demonstrations the students have seen, ask them to decide whether sound is an example of a transverse or a longitudinal wave. [O2]

Consolidate and apply

- Issue the students with the longitudinal waves worksheet appropriate for their ability:
  - Low demand: Worksheet 5.2.1 [O1, O3]
  - Standard demand: Worksheet 5.2.2 [O1, O3]
  - High demand: Worksheet 5.2.3 [O1, O3]

Extend

Ask students able to progress further to do the following:

- Work in a pair to discuss how they could investigate the speed of longitudinal waves through different media using equipment available in the lab. They could be asked to write a plan for their investigation. [O3]

Plenary suggestions

- Show PowerPoint presentation 5.2 which gives images of different examples of waves. Ask the students to vote on which type of wave they think is depicted by holding up a green card for longitudinal and a red card for transverse. [O1, O2]
- **What do I know?** Ask students to each write down one thing about the topic that they are sure of, one thing they are unsure of and one thing they need to know more about, being specific. Ask them to work in groups of five or six to agree on group lists, ask each to say what they have decided and agree as a class what they are confident about, what they are less sure of and what things they need to know more about. [O1, O2, O3]
- **Freeze frame:** Ask students to create a freeze frame of either a transverse or a longitudinal wave. The other students are to suggest which wave it is and how they know [O1, O3]

Answers to questions

Worksheets 5.2.1–5.2.3 (Worksheet 5.2.3 also has a bar for air at 40°C)

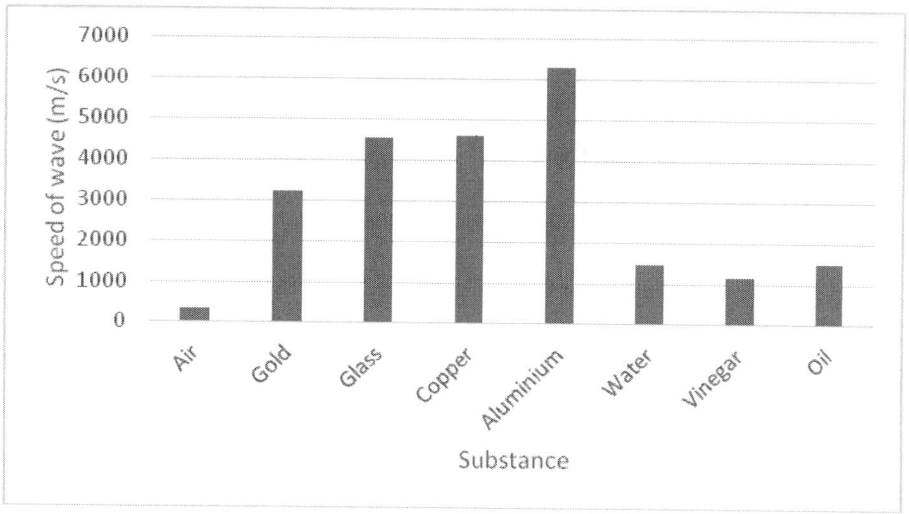

PowerPoint 5.2

Slide 1 (radiowaves) – transverse; slide 2 (X-rays) – transverse; slide 3 (tsunami) – longitudinal;

slide 4 (ultrasound) – longitudinal; slide 5 (ultraviolet) – transverse

 © HarperCollinsPublishers Limited 2016

Lesson 5.3: Reflection and refraction of waves

Lesson overview

OCR Specification reference

P5.1g

Learning objectives

- Describe reflection, transmission and absorption of waves.
- Construct ray diagrams to illustrate reflection.
- Construct ray diagrams to illustrate refraction.

Learning outcomes

- [O1] Understand that waves can be absorbed, transmitted or reflected at a surface.
- [O2] Construct a ray diagram to illustrate reflection of a wave at a boundary.
- [O3] Construct a ray diagram to illustrate refraction at a boundary.

Skills development

- Carry out fair and accurate investigations.
- Analyse data collected from experiment.
- Draw sensible conclusions from experimental data.

Maths focus

- Measure angles using a protractor.

Resources needed Worksheets 5.3.1, 5.3.2 and 5.3.3, Practical sheets 5.3.1, 5.3.2 and 5.3.3, Technician's notes 5.3.1 and 5.3.2

Key vocabulary reflection, transmission, absorption, ray diagram, normal, refraction

Teaching and learning

Engage

- **Pair talk:** Show students a picture of a fibre optic Christmas tree and ask the students to describe what they can see in the picture and ask them why the tree is only lit up at the ends of the cables, despite light only being able to travel in straight lines (i.e. ask them to think about what is actually happening to make the tree work). [O1]
- **Pair talk:** Alternatively, show students a picture of a four-eyed fish and ask them to think about why the fish has evolved in this way (note: the fish does not actually have 4 eyes, but two eyes, each divided into two parts with lenses of different thickness at the top and bottom). You could give the students the clue that these fish float at the water surface. [O1]

Challenge and develop

- The students should then carry out the practical task outlined on Practical sheet 5.3.1 to investigate reflection. [O1, O2]
- The students should then carry out the practical task outlined on Practical sheet 5.3.3 to investigate refraction. [O1, O3]

Explain

- Following the practical element to each task, the students should be asked to complete the follow-up questions related to their task which are outlined on the respective practical sheets. [O1, O2, O3]
- Students could also be asked to research an everyday use for the phenomena they have just investigated (reflection or refraction). [O1]

 © HarperCollins*Publishers* Limited 2016

Consolidate and apply

- Students should then complete the reflection and refraction worksheet appropriate to their level which will summarise the learning points from each of the practical tasks:
  - Low demand: Worksheet 5.3.1 [O1]
  - Standard demand: Worksheet 5.3.2 [O2]
  - High demand: Worksheet 5.3.3 [O3]

Extend

- Ask students able to progress further to return to the picture(s) from the start of the lesson. Can they explain what is happening in each of the pictures using the key words they have learned about in the lesson? [O1]
- Alternatively, students could be asked to use ideas about reflection to explain how concave mirrors collect light and how convex mirrors spread light out and suggest a use for each type of curved mirror. [O1]
- As a way of further investigating refraction, students could be asked to research the cause of mirages on road surfaces in summer time. [O1]

Plenary suggestions

- **Learning triangle:** At the end of the lesson ask students to draw a large triangle with a smaller inverted triangle that just fits inside (so they have four triangles). Ask them to think back over the lesson and identify and write in the outer three triangles:
  - something they saw
  - something they did
  - something they discussed.

Ask them to add to the central triangle something they learned.

- **The big ideas:** Ask students to write down three ideas they learned during the lesson. Then ask them to share their facts in groups and to compile a master list of facts, with the most important at the top. Ask for ideas to be shared and find out which other group(s) agreed with their suggestions.

Answers to questions

Worksheets 5.3.1–5.3.3

Correctly labelled diagrams

Worksheets 5.3.1 and 5.3.2

incidence, the same, angle

Refraction: glass, towards, glass, air, away from, denser, travelled slower, travelled faster

 © HarperCollins*Publishers* Limited 2016

Lesson 5.4: Sound waves

Lesson overview

OCR Specification reference

P5.1h, i

Learning objectives

- Describe how we hear sound and state the range of frequencies we can hear.
- Explain that sound travels faster in a denser medium.
- Explain about reflection, absorption and transmission of sound.

Learning outcomes

- [O1] Recognise that the speed of sound through liquids and solids is faster than in air.
- [O2] List the upper and lower limits of human hearing.
- [O3] Apply the principle of conservation of energy to the reflection, transmission and absorption of waves.

Skills development

- Take a set of data and produce a chart.
- Draw conclusions from a chart.
- Plan a valid investigation.

Maths focus

- Draw graphs (scales on axes).

Resources needed Worksheets 5.4.1 and 5.4.2

Digital resources News article about the Mosquito device.

Key vocabulary sound wave, eardum, reflection, absorption, transmission

Teaching and learning

Engage

- **The Mosquito Device.** Show the students a news article about the deployment of the Mosquito Device to deter youngsters from loitering. The article could be discussed, with the discussion including why the device is used, how it works (at the basic level), the social implications of using such a device and the discussion could end with the students voting on whether they agree with its use or not, giving a reason for their choice. [O2]

Challenge and develop

- **Sound circus:** There are two tasks for the students to complete and these should be done with half the class working on each task; they should be given a set time to work on one task before moving on to the second. The tasks address: species ranges of hearing (Worksheet 5.4.1) and a planning activity (Worksheet 5.4.2) which addresses reflection, transmission and absorption of sound waves. [O1, O2, O3]
- Worksheet 5.4.1: Students are given a set of data and are asked to plot the data on a suitable chart to show the hearing ranges of different species. [O2]
- Worksheet 5.4.2: Students are asked to design an investigation to test whether all materials reflect, absorb and transmit sound by the same amount, or whether some are better reflectors, absorbers and transmitters. An extension to this would be to investigate whether the surface of each material makes a difference to the amount of sound that is reflected. [O1, O3]

 © HarperCollins*Publishers* Limited 2016

Explain

- **Pair talk:** The students should be given the opportunity to discuss both of the tasks they have completed and to list anything that surprised them. Each student should then be given a sticky note and they should write down one thing that surprised them and two things that they learned. These sticky notes should be stuck onto the board and discussed as a group. [O1, O2, O3]

Consolidate and apply

- In listening triads, the students should explain how knowledge of how sound travels in different materials is important when designing a multiplex cinema. (In a listening triad, students work in groups of three and each student takes on the role of either talker, questioner or recorder. The talker explains their ideas. The questioner prompts and seeks clarification. The recorder makes notes and gives a report at the end of the conversation.) [O1, O3]
- Alternatively, students could be given an outline image of a multiplex cinema and asked to label the different materials that might be used in the construction of the site, giving reasons for their choices. [O1, O3]

Extend

- Students should be asked to research reverberation time in concert halls and explain their findings to the class or a small group. [O3]

Plenary suggestions

- **The big ideas:** Ask students to write down three ideas they learned during the topic. Then ask them to share their facts in groups and to compile a master list of facts, with the most important fact at the top. Ask for ideas to be shared and find out which other group(s) agreed.
- **Hot seat:** Ask each student to think of a question, using material from the topic. Select someone to put in the hot seat, ask students to pose their question and say at the end whether the answer is correct or incorrect.

Answers to questions

Worksheet 5.4.1

a) Bat, mouse, dolphin

b) Bat, rabbit, cat, dog, mouse, dolphin

c) Any frequency above 20 000 Hz and below 46 000 Hz.

 © HarperCollins*Publishers* Limited 2016

Lesson 5.5: Exploring ultrasound

Lesson overview

OCR Specification reference

P5.2i (part)

Learning objectives

- Explain what ultrasound is.
- Describe how ultrasound can be used in industry to investigate or detect hidden or buried objects.
- Explain how ultrasound is used in medicine.

Learning outcomes

- [O1] State the meaning of the term ultrasound.
- [O1] List some medical and non-medical uses of ultrasound.

Skills development

- Research a specific concept.
- Draw conclusions from research.
- Share ideas with others.

Resources needed signal generator, cathode ray oscilloscope, loudspeaker; Worksheets 5.5.1 and 5.5.2

Digital resources access to online research sources

Key vocabulary ultrasound

Teaching and learning

Engage

- Ask the students to discuss where they have come across the term 'ultrasound' before asking them to feedback and share class ideas. Now pose the question "What is ultrasound?" and take suggested answers. [O1]
- Using a signal generator, cathode ray oscilloscope and a loudspeaker, demonstrate ultrasound to the students. If a signal generator or CRO is not available, use a suitable animation instead. [O1]

Challenge and develop

- Show the students a range of images of ultrasound scans being used in the following situations and ask the students what they think is happening in each picture and if they can suggest how they might be working:
  - Foetal ultrasound scans (to view the foetus in pregnancy)
  - Opthalmic ultrasound scans (to view the structures in the eye)
  - Echocardiograms (to view the heart)
  - Abdominal ultrasound scans (to view the tissues and organs)
  - Bone sonometry (to investigate bone strength). [O2]
- Using either the text from the textbook ('How ultrasound scans work' section) or a similar explanation from a different research source, students design a story board outlining how an ultrasound scan works. [O1, O2]

Explain

- **Jigsaw:** The students are split into four groups, with students given a number between 1 and 4. All students given number one should join as a group, all number twos as another group and so on. The students should be given access to online research sources and each group will research one application of ultrasound. Once

 © HarperCollins*Publishers* Limited 2016

they have gathered information, regroup the students in groups of four so that each new group contains one student who was given number one, one student who was given number two, and so on.

- Groups should research the following applications:
  - Group 1: cleaning building, spectacles and jewellery
  - Group 2: breaking down kidney stones
  - Group 3: treating cataracts
  - Group 4: finding cracks in aircraft
- The students should then be given two minutes to present their findings to the rest of the group. [O1, O2]

Consolidate and apply

- The students should be issued with the using ultrasound worksheet appropriate to their level:
  - Standard demand: Worksheet 5.5.1 [O1]
  - High demand: Worksheet 5.5.2 [O2]

Extend

- Using ideas about speed of waves in air, frequency and wavelength, ask the students to explain why ultrasound is able to be transmitted as a beam. [O3]

Plenary suggestions

Learning triangle: Ask the students to draw a large triangle with a smaller inverted triangle that just fits inside it (so they have four triangles). Ask them to think back over the lesson and write in the outer three triangles:

- something they saw
- something they did
- something they discussed.

Ask them to add to the central triangle something they learned.

What do I know? Ask students to each write down one thing about the topic they are sure of, one thing they are unsure of and one thing they need to know more about, being specific. Ask them to work in groups of five to six to agree on group lists, ask each group to say what they decided and agree as a class what they are confident about, what they are less sure of and what things they want to know more about.

Answers to questions

Worksheet 5.5.1

Jewellery: energy, frequency, bubbles, bubbles, dislodge
Kidney stones: energy, probe, vibrate, stones, break
Eyes: cornea, probe, ultrasound, eye, lens, replaced
Aircraft: waves, cracks, scanner, path, reflected, monitor

 © HarperCollins*Publishers* Limited 2016

Lesson 5.6: Practical: Measuring the wavelength, frequency and speed of waves in a ripple tank and waves in a solid

Lesson overview

Learning objectives

- Develop techniques for making observations of waves.
- Select suitable apparatus to measure frequency and wavelength.
- Use data to answer questions.

Learning outcomes

- [O1] State how the speed of a wave can be calculated.
- [O2] Carry out a practical investigation to measure the speed of a wave.
- [O3] Evaluate practical procedures.

Skills development

- Carry out a valid investigation.
- Draw conclusions from an investigation.

Resources needed ripple tank, power pack, strobe light, shallow tray, rulers, stop clocks; Practical sheet 5.6, Technician's notes 5.6

Digital resources video clip of the ripples from a stone being thrown into a pond

Teaching and learning

Engage

- Show the students a video clip of a stone being thrown into a pond and ask the students to suggest how they could measure the speed of the waves in the water. [O1]
- Alternatively, ask students why we see lightning before we hear thunder and ask them to suggest how we could work out the speed of sound in air. [O1]

Challenge and develop

- The students should be reminded of the wave equation. As a way of reintroducing this, you could return to the stone being thrown across the pond and suggest that the splash makes three ripples per second (i.e. the frequency) and the ripples are 6 cm apart (i.e. the wavelength). Pose the question: "How far will the first ripple travel in the first second (or each second)?" Students should quickly come up with the answer of 18 cm, establishing the link summarised by the wave equation. [O1]

Explain

- Students should then carry out the Practical 5.6. [O2, O3]
- Once they have completed the practical, they should complete the analysis of their results. [O2, O3]

Consolidate and apply

- The students should now consider the questions on page 175 of the Student Book appropriate to their ability:
  - Low demand: questions 9 and 10 [O1, O2, O3]
  - Standard demand: questions 10 and 11 [O1, O2, O3]
  - High demand: questions 11 and 12 [O1, O2, O3]

 © HarperCollins*Publishers* Limited 2016

Extend

- Ask students able to progress further to design an investigation in which the speed of sound waves through water could be measured.
- Alternatively, ask students to suggest two improvements to the practical they carried out, giving reasons for their suggestions.

Plenary suggestions

Ask me a question: Ask students to write a question about something from the topic and then devise a mark scheme for the answer. Encourage them to come up with questions worth more than one or two marks and to try them out on each other.

Learning triangle: At the end of the lesson ask students to draw a large triangle with a smaller inverted triangle that just fits inside it (so they have 4 triangles). Ask them to think back over the lesson and write in the outer three:

- something they saw
- something they did
- something they discussed.

Ask them to add in the central triangle something they learned.

Heads and tails: Ask each student to write a question about something from the topic on a coloured paper strip and the answer on another coloured strip. In groups of six to eight, hand out the strips so that each student gets a question and an answer. One student reads out their question. The student with the right answer reads it out, followed by their question, and so on.

 © HarperCollins*Publishers* Limited 2016

Lesson 5.7: Maths skills: Using and rearranging equations

Lesson overview

OCR Specification reference

PM5.1i

Learning objectives

- Select and apply the equations: $T = \frac{1}{frequency}$ and $v = f\lambda$.
- Substitute numerical values into equations using appropriate units.
- Change the subject of an equation.

Learning outcomes

- [O1] Investigate the link between time period and frequency.
- [O2] Rearrange $T = \frac{1}{frequency}$ and $v = f\lambda$.
- [O3] Apply the equation $v = f\lambda$ to the electromagnetic spectrum.

Skills development

- Follow algebraic symbol conventions.

Maths focus

- Substitute values into and rearrange equations.

Resources needed mini whiteboards and whiteboard pens; Worksheets 5.7.1, 5.7.2 and 5.7.3

Key vocabulary proportional, subject of an equation, SI units, substitute

Teaching and learning

Engage

- As the students enter the room, issue them with a small piece of paper. They should write down what they understand by the terms time period, frequency and wavelength on one side of the paper, and on the other, they should write down the equation they should use in order to calculate the time period when given the frequency.
- They should use a different coloured pen to label the equation with the correct units for each quantity in the equation. [O1]

Challenge and develop

- The students should be reminded that the equation to calculate the time period is $T = 1/f$ and they should use this equation to answer questions 1 and 2 from page 176 of the Student Book. They could do this on a mini whiteboard to allow assessment of the answers given by the students.
- They should be encouraged to write out the equation first, check the units in the question and then to substitute the values into the equation. [O1]

Explain

- The students should then be encouraged to put the equation $T = 1/f$ into an equation triangle (encourage them to use the equation they know to be correct) and then to list the other rearrangement of the equation to make frequency the subject. [O2]

 © HarperCollins*Publishers* Limited 2016

- You may like to return to the example of speed = distance / time from Chapter 2 Forces in order to remind the students how to construct equation triangles. [O2]
- The students should then be asked to recall how frequency, wave speed and wavelength are linked by an equation and, again, to practise rearranging it to make each quantity the subject of the equation.
- The students should then answer questions 3–6 from page 177 of the Student Book.

Consolidate and apply

- The students should then be issued with the maths skills worksheet appropriate to their ability:
  - Low demand: Worksheet 5.7.1 [O1, O2, O3]
  - Standard demand: Worksheet 5.7.2 [O1, O2, O3]
  - High demand: Worksheet 5.7.3 [O1, O2, O3]

Extend

Ask students able to progress further to do one of the following:

- Design a single-sided summary sheet instructing their peers on how to rearrange equations with more than three quantities, and how to combine equations. [O1, O2, O3]
- Alternatively, they should answer questions 7 and 8 from page 177 of the Student Book. [O3]

Plenary suggestions

Learning triangle: At the end of the lesson, ask students to draw a large triangle with a smaller inverted triangle that just fits inside it (so they have four triangles). Ask them to think back over the lesson and write in the outer three:

- something they saw
- something they did
- something they discussed.

Ask them to add in the central triangle something they learned.

Heads and tails: Ask each student to write a question about something from the lesson on a coloured paper strip and the answer on another coloured strip. In groups of six to eight, hand out the strips so that each student gets a question and an answer. One student reads out their question. The student with the right answer then reads it out, followed by their question. This could take the form of calculations using and rearranging $Q = It$ and $V = IR$.

What do I know? Ask students to each write down one thing about the topic they are sure of, one thing they are unsure of and one thing they need to know more about, being specific. Ask them to work in groups of five or six to agree on group lists, ask each group to say what they decided and agree as a class what they are confident about, what they are less sure of and what things they want to know more about.

Answers to questions

Worksheets 5.7.1 to 5.7.3

1.

Region of spectrum	V (m/s)	F (Hz)	λ (m)
Radio waves	$\mathbf{3.0 \times 10^8}$	1×10^8	**3**
Microwaves	$\mathbf{3.0 \times 10^8}$	$\mathbf{3 \times 10^{10}}$	1×10^{-2}
Infra-red	$\mathbf{3.0 \times 10^8}$	1×10^{13}	$\mathbf{3 \times 10^{-5}}$

 © HarperCollins*Publishers* Limited 2016

Visible light	**3.0×10^{8}**	**3×10^{14}**	1×10^{-6}
Ultra violet	**3.0×10^{8}**	1×10^{16}	**3×10^{-3}**
X-rays	**3.0×10^{8}**	**3×10^{17}**	1×10^{-9}
Gamma rays	**3.0×10^{8}**	1×10^{20}	**3×10^{-12}**

2. 545.45 m
3. red = 1×10^{14} Hz; orange = 3×10^{14} Hz; yellow = 3.3×10^{14} Hz; green = 5×10^{14} Hz; blue = 6×10^{14} Hz; violet = 3×10^{15} Hz

 © HarperCollins*Publishers* Limited 2016

Lesson 5.8: Key concept: Transferring energy or information by waves

Lesson overview

OCR Specification reference

Learning objectives

- Understand that all waves have common properties.
- Understand how waves can be used to carry information.
- Understand various applications of energy transfer by different types of electromagnetic waves.

Learning outcomes

- [O1] Use the key words in the correct context.
- [O2] Research a given application.
- [O3] Present research findings clearly.

Skills development

- Develop scientific thinking, through researching the timeline of laser development.
- Analyse and evaluate through research and draw conclusions from research.
- Present findings following research.

Resources needed access to research material (probably online); Worksheet 5.8

Digital resources https://www.youtube.com/watch?v=_J8YcQETyTw

Key vocabulary energy, transfer, vibration, amplitude, absorb

Teaching and learning

Engage

- As the students enter the room, give them a small piece of paper. They should answer the following question (on their own to start with): "What does a wave do?" You could prompt those who are slow to begin by asking them to think about a water wave and what it does. This should give them enough ideas to get started. Ask the students to share their ideas with a partner, before taking feedback from a number of different pairs. The concept of waves transferring energy should be a result of the discussion.
- Alternatively, show the students a clip of Morse code (suggested clip – https://www.youtube.com/watch?v=_J8YcQETyTw) and ask the students to suggest when it might be useful.

Challenge and develop

- The students should be given a research brief that details their task:

  Using waves to transfer energy – Worksheet 5.8 [O1, O2]

Explain

- A select number of students should present their findings to the rest of the class. The rest of the class could be given an opportunity to question the presenter, before peer-marking their work against the success criteria outlined on the research brief. [O3]

Consolidate and apply

- The students could be set a challenge to design a new product that makes use of waves in a way they have investigated or researched over the course of the unit. The students should be grouped so they are working

 © HarperCollins*Publishers* Limited 2016

in teams of six to eight and each student be given a role in the team. Roles might include: project manager, researchers, sub-team manager, presenter, designer, advertising manager, etc. [O1, O2]

- Alternatively, students could be asked to write all of the waves key words randomly scattered on a sheet of A3 plain paper. They should then link the words in as many ways as they can (linking them with lines) and write above the lines what the link between the words is. For example, the students could link 'pitch' and 'frequency' with a line, and above the line write 'the higher the frequency, the higher the pitch'. They should be sure to include the new learning from the lesson, therefore including applications and uses of waves. [O1]

Extend

- The students could be asked to write two exam-style questions and the corresponding mark schemes. They should write one question which earns only one or two marks and a second question which includes sub-questions totalling at least four marks. [O1]
- Alternatively, students could be asked to consider the following question: 'If energy is being transmitted away from the Sun, then why isn't the Sun continually cooling down?'

Plenary suggestions

- **Silent animation:** Show students a short video clip without sound (perhaps a use of a laser) and ask them to say what they think is happening. They could either suggest the ideas or do a 'voice over' for it.
- **Hot seat:** Ask each student to think up a question, using material from the topic. Select someone to be put in the hot seat, ask students to pose their question and say at the end whether the answer is correct or incorrect.

 © HarperCollins*Publishers* Limited 2016

Lesson 5.9: The electromagnetic spectrum

Lesson overview

OCR Specification reference

P5.2a, b, c, d, e, f

Learning objectives

- Recall the similarities and differences between transverse and longitudinal waves.
- Recognise that electromagnetic waves are transverse waves.
- Describe the main groupings and wavelength ranges of the electromagnetic spectrum.

Learning outcomes

- [O1] Distinguish between transverse and longitudinal waves.
- [O2] Describe the main groupings in the spectrum.
- [O3] Explain the differences in wavelength and frequency of the main groups in the spectrum.

Skills development

- Use a model to represent transverse and longitudinal waves.
- Discuss the limitations of models.

Maths focus

- Convert from km/h to km/s.
- Use standard form to describe the frequencies and wavelengths of the main groups in the spectrum.

Resources needed slinky spring, ray box and prism; Worksheets 5.9.1, 5.9.2 and 5.9.3

Key vocabulary longitudinal wave, transverse wave, spectrum, electromagnetic wave, visible spectrum

Teaching and learning

Engage

- Ask students to estimate the distance from the Earth to the Sun (about 149 million kilometres) and the time it takes the light from the Sun to travel this distance (about 8 minutes).
- Ask students to calculate how fast the visible light is travelling in kilometres per hour (149 000000 km / 0.133·hour = 1.1×10^9 km/h).
- Finally, ask them to convert this speed into kilometres per second (1.1×10^9 km/h / 3600 = 300 000 km/s).

Challenge and develop

- Using a slinky spring, remind the students of the difference between transverse and longitudinal waves. [O1]
- Ask students to give an example of a transverse wave (waves in water, for example) and an example of a longitudinal wave (sound waves, for example) and ask them to identify the key difference in the way in which these waves transfer energy from one place to another. [O1]
- Ask students to list the properties that all waves have: they can be reflected, refracted and diffracted; they can interfere (both constructively and destructively).
- Remind the students of the key wave words, such as amplitude, frequency and wavelength, and ask them to give the definition of each of these during the demonstration. You may wish to have two students controlling the spring, one at each end, with other students giving directions to create waves with a bigger amplitude, longer wavelength, shorter frequency, etc. to make the definitions of these words clear. [O1]

 © HarperCollins*Publishers* Limited 2016

Explain

- Ask students why we can see light from the Sun despite there being about 149 million kilometres of empty space between the Earth and the Sun. Introduce the idea that some waves need a medium through which to travel, but a particular type of wave – electromagnetic waves – do not need a medium to travel through: they can travel through a vacuum. [O2]
- Demonstrate using a prism and a ray box that white light is a mixture of waves of different wavelengths, which we see as different colours. State that visible light is one part of the electromagnetic spectrum and that there are waves that are invisible to us on either side of the visible light spectrum. [O2]
- Show students a diagram of the electromagnetic spectrum, such as the one on page 181 of the Student Book, and be sure to emphasise that the wavelength decreases from left to right, where the frequency and energy of the waves increases from left to right. [O2, O3]
- The students should now write a mnemonic to help them remember the order of the main groups in the spectrum (help the students by pointing out that the 'infra**red** waves are next to the **red** end of the visible spectrum and the ultra**violet** waves are next to the **violet** end of the visible spectrum'). [O2, O3]

Consolidate and apply

Students should be issued with the electromagnetic spectrum worksheet appropriate for their ability:

- Low demand: Worksheet 5.9.1 [O1, O2, O3]
- Standard demand: Worksheet 5.9.2 [O1, O2, O3]
- High demand: Worksheet 5.9.3 [O1, O2, O3]

Extend

- Ask students able to progress further to give an everyday use for at least four groups of the electromagnetic spectrum.
- Alternatively, they could be asked to suggest why a shorter wavelength results in more dangerous radiation.

Plenary suggestions

Learning triangle: At the end of the lesson, ask students to draw a large triangle with a smaller inverted triangle that just fits inside it (so they have four triangles). Ask them to think back over the lesson and write in the outer three triangles:

- something they saw
- something they did
- something they discussed.

Ask them to add in the central triangle something they learned.

Ask me a question: Ask students to write a question about something from the lesson and then devise a mark scheme for the answer. Encourage them to come up with questions worth more than one or two marks and to try their questions out on one another.

Where's the answer? Give each student a card with a word written on it. Then ask a question, ask them if they think their card shows the answer, and why.

Answers to questions

Worksheets 5.9.1, 5.9.2 and 5.9.3

1. b) The first diagram should be labelled as 'transverse' and the second should be labelled as 'longitudinal'.

2. Approximate ranges are:

 © HarperCollins*Publishers* Limited 2016

Group of waves	**Frequency (Hz)**	**Wavelength (m)**
Radio waves	10^{4}–10^{9}	**10^{5}–10^{-1}**
Microwaves	**10^{10}–10^{11}**	10^{-1}–10^{-3}
Infrared	10^{12}–10^{14}	**10^{-3}–10^{-6}**
Visible	**10^{15}**	10^{-6}
Ultraviolet	**10^{15}–10^{16}**	**10^{-7}–10^{-8}**
X-rays	**10^{17}–10^{19}**	10^{-8}–10^{-10}
Gamma rays	**10^{19}–10^{20}**	**10^{-11}–10^{-12}**

 © HarperCollins*Publishers* Limited 2016

Lesson 5.10: Gamma rays and X-rays

Lesson overview

OCR Specification reference

P5.2g (part), h

Learning objectives

- List the properties of gamma rays and X-rays.
- Compare gamma rays and X-rays.

Learning outcomes

- [O1] State the properties of gamma rays and X-rays.
- [O2] List the similarities and differences of gamma rays and X-rays.
- [O3] Structure an answer clearly.

Skills development

- Read information with a purpose.
- Use information to draw conclusions.

Resources needed Student Book or other suitable textbook containing information on the nature and uses of gamma rays and X-rays, X-ray films or images of them, image of a CAT scan; Worksheets 5.10.1, 5.10.2 and 5.10.3

Digital resources pedoscope image

Key vocabulary radiation dose, X-ray, gamma ray, tracer

Teaching and learning

Engage

- Show students an image of a pedoscope from the 1930s, such as the one listed above, and tell the students that these used to be used in shoe shops to measure shoe sizes, making use of X-rays to measure the feet. However, the dangers of using X-rays were not known at the time, and a number of workers went on to develop cancer due to their repeated exposure to the X-rays. [O1]

Challenge and develop

- Show students X-ray films (if available, or images of X-ray films if they are not) – ask students to read the text on page 182 of the Student Book about the use of radiation in medicine and discuss why we are able to see the bone and not the surrounding tissue on the film. [O1]
- Also show an image of a CAT scan and tell the students that these scans also use X-rays.
- Finally, draw the students' attention of the use of gamma radiation in tracer scans, such as the use of technetium-99m. Discuss the need for a half-life in the region of hours, as opposed to minutes or years, to ensure no damage is caused to healthy tissue. [O1]
- Tell students that in the next unit they will be learning more about the properties of these rays, which allow them to be used in these medical applications. [O1]

Explain

- Explain to students that the basis of this lesson will be the development of skills in answering questions worth more than three marks.

 © HarperCollins*Publishers* Limited 2016

- By way of an introduction to developing these skills, provide students with a piece of text, perhaps from a recent newspaper article, which addresses an argument (i.e. a piece of text that has a for and against argument). An example might be the introduction of the charge of 5p for plastic bags in most shops.
- Ask students to read the article twice, once without doing anything and a second time highlighting the arguments for in one colour and against in another, thereby modelling how to read information with a purpose. [O3]
- Give students five minutes to draft a response to the task 'Outline the arguments for and against the introduction of a 5p charge for plastic bags, before deciding on a conclusion'. Tell the students that the question is worth six marks and encourage them to structure their answer clearly. [O3]
- Ask students to swap their responses with a partner and give them time to read the answer. Return the student's original answer and allow them time to improve it in the light of what they have read in their partner's work. [O3]
- Take feedback from the group before discussing how to approach answering such a question. Be sure to emphasise that a well-structured answer will detail at least two advantages, at least two disadvantages, followed by a concluding sentence. [O3]

Consolidate and apply

The students should now be issued with the gamma rays and X-rays worksheet suitable for their ability:

- Low demand: Worksheet 5.10.1 [O1, O2, O3]
- Standard demand: Worksheet 5.10.2 [O1, O2, O3]
- High demand: Worksheet 5.10.3 [O1, O2, O3]

Extend

- Ask students able to progress further to suggest how X-rays are used to detect faults in welds. [O1]
- Alternatively, they could be asked to use the ideas they have met during the lesson to return to the original image of the pedoscope and describe why some of the shoe shop workers went on to develop cancer. [O1]

Plenary suggestions

Ask me a question: Ask students to write a question about something form the topic and then devise a mark scheme for the answer. Encourage them to come up with questions worth more than one or two marks and to try their questions out on one another.

Learning triangle: At the end of the lesson ask students to draw a large triangle with a smaller inverted triangle that just fits inside (so they have four triangles). Ask them to think back over the lesson and write in In the outer three triangles:

- something they saw
- something they did
- something they discussed.

Ask them to add in the central triangle something they learned.

 © HarperCollins*Publishers* Limited 2016

Lesson 5.11: Ultraviolet and infrared radiation

Lesson overview

OCR Specification reference

P5.2g (part), h (part)

Learning objectives

- Describe the properties of ultraviolet and infrared radiation.
- Describe some uses and hazards of ultraviolet radiation.
- Describe some uses of infrared radiation.

Learning outcomes

- [O1] Describe the properties of ultraviolet radiation and explain the risks associated with their use.
- [O2] State the uses of infrared radiation.
- [O3] Explain the use of thermal imaging cameras.

Skills development

- Read information with a purpose.
- Present information clearly to peers.
- Carry out valid practical work.

Maths focus

Resources needed Worksheet 5.11, Practical sheet 5.11, Technician's notes 5.11

Key vocabulary infrared radiation, ultraviolet radiation

Teaching and learning

Engage

- Show the students a suitable video clip showing the use of a thermal imaging camera, perhaps in fighting fires or catching criminals. Police helicopter clips may be useful here. Ask them to think about what the camera is picking up and discuss their thoughts in pairs before feeding back to a class discussion.
- Tell the students that anything that is warmer than its surroundings emits energy by giving out infrared radiation. [O2]

Challenge and develop

- **Jigsaw:** Give each student a number between 1 and 3 and then ask all the 1s to sit together, all the 2s to sit together and all the 3s to sit together.
- Working in these new expert groups, the students should be given one of the following topics to research.
  - Group 1 should be asked to research the uses of UV radiation. [O1]
  - Group 2 should be issued with Practical sheet 5.11, where they will investigate the risks associated with UV radiation and how we can protect ourselves from these risks. [O1]
  - Group 3 should be asked to research the uses of infrared radiation. [O2, O3]

Explain

- The students should now be regrouped so that they are working in groups of three, each student having carried out a different task. They should be asked to present their research/practical findings to their peers.
- In order to structure this feedback, the students should complete the table on Worksheet 5.11. [O1, O2, O3]

 © HarperCollins*Publishers* Limited 2016

Consolidate and apply

- Students should answer the following questions on pages 184–5 of the Student Book:
  - Low demand: questions 1, 4 and 5 [O1, O2, O3]
  - Standard demand: questions 2, 4 and 6 [O1, O2, O3]
  - High demand: questions 3, 4 and 7 [O1, O2, O3]

Extend

Ask students able to progress further to do one of the following:

- Suggest how knowledge of infrared radiation is useful when designing security alarms. [O2, O3]
- Suggest how a thermal imaging camera can be used to help find survivors in a fire. [O2, O3]

Plenary suggestions

What do I know? Ask students to each write down one thing about the topic they are sure of, one thing they are unsure of and one thing they need to know more about, being specific. Ask them to work in groups of five or six to agree on group lists. Ask each group to say what they decided and agree as a class what they are confident about, what they are less sure of and what they want to know more about.

The big ideas: Ask students to write down three ideas they have learned during the topic so far. Then ask them to share their facts in groups and to compile a master list of facts, with the most important at the top. Ask for ideas to be shared and find out which other group(s) agreed.

Heads and tails: Ask each student to write a question about something from the topic so far on a coloured strip of paper and the answer on another coloured strip. In groups of six to eight, hand out the strips so that each student gets a question and an answer. One student reads out their question. The student with the right answer then reads it out, followed by their question, and so on.

 © HarperCollins*Publishers* Limited 2016

Lesson 5.12: Microwaves

Lesson overview

OCR Specification reference

P5.2g (part), i (part)

Learning objectives

- List some properties of microwaves.
- Describe how microwaves are used for communications.

Learning outcomes

- [O1] State uses of microwave radiation.
- [O2] State how microwaves are used for cooking.
- [O3] State how microwaves are used for communication.

Skills development

- Read information with a purpose.
- Use information to draw conclusions and make sensible suggestions.

Maths focus

Resources needed Student Book or other suitable textbook containing information on the uses of microwaves in cooking and communications, old microwave oven; Worksheet 5.12

Digital resources Brainiac science: https://www.youtube.com/watch?v=Rc-ne-nIKUY

Key vocabulary microwaves

Teaching and learning

Engage

- Show the students a video clip from Brainiac science (see link above) – can you cook an egg using microwaves from mobile phones? [O1]

Challenge and develop

- Demonstrate the wavelength of a microwave by taking the turntable out of an old microwave and putting a large bar of chocolate on a plate and turning the microwave on for 3 or 4 seconds. The chocolate should melt every 3 or 4 cm. [O1, O2]
- Alternatively, quote the wavelengths of microwaves used for cooking and the wavelengths of microwaves used for communication and ask the students to use their knowledge of waves from earlier in the chapter to state which waves have the most amount of energy associated with them and which have the least amount of energy associated with them. [O1, O2]

Explain

- The students should be issued with Worksheet 5.12, which gives a structure for the research task: finding out how microwaves are used for:
    - cooking [O1, O2] and
    - communication. [O1, O3]

Consolidate and apply

- The students should now be asked to answer the following questions on pages 186–7 of the Student Book:
    - Low demand: questions 1, 3, 4 and 5 [O1, O2, O3]

 © HarperCollins*Publishers* Limited 2016

- Standard demand: questions 1, 2, 4, 5 and 6 [O1, O2, O3]
- High demand: questions 1, 3, 4, 5, 6, and 7 [O1, O2, O3]

Extend

- Ask students able to progress further to investigate the range of Bluetooth communications between mobile phones. [O1, O3]

Plenary suggestions

Ideas hothouse: Ask students to work in pairs to list points about what they know about the electromagnetic spectrum so far. Then ask the pairs to join together into fours and then into groups of six to eight to discuss this further and to come up with an agreed list of points. Ask one student from each group to report back to the class.

Learning triangle: At the end of the lesson, ask students to draw a large triangle with a smaller inverted triangle that just fits inside (so they have four triangles). Ask them to think back over the lesson and write in the outer three triangles:

- something they saw
- something they did
- something they discussed.

Ask them to add in the central triangle something they learned.

Ask me a question: Ask students to write a question about something from the topic and then devise a mark scheme for the answer. Encourage them to come up with questions worth more than one or two marks and to try out their questions on one another.

 © HarperCollinsPublishers Limited 2016

Lesson 5.13: Radio and microwave communication

Lesson overview

OCR Specification reference

P5.2g (part), j

Learning objectives

- Describe how radio waves are used for television and radio communications.
- Describe how microwaves are used in satellite communications.
- Describe the reflection and refraction of radio waves.

Learning outcomes

- [O1] Explain how worldwide transmission of radio signals is achieved.
- [O2] State how microwaves are used in communication.
- [O3] Describe reflection and refraction of radio signals.

Skills development

- Understand how a model can help us understand a concept.
- Realise the limitations of models.

Maths focus

Resources needed plane mirror, torch, Perspex prism or toy car and sand, shallow tray and two blocks; Worksheets 5.13.1, 5.13.2 and 5.13.3

Digital resources images of a range of devices that use Bluetooth technology

Key vocabulary transmitter, receiver, radio wave, microwaves, satellite

Teaching and learning

Engage

- Show the students a range of images of devices that use Bluetooth technology. Ask students to think about what the devices have in common. [O2]

Challenge and develop

- Remind students of the general behaviours of waves: reflection, refraction and diffraction. This could be done by a series of demonstrations:
  a) Reflection – shine a single incident ray of light onto a plane mirror and ask the students to state the law of reflection. [O1, O3]
  b) Refraction – either shine a single incident ray of light through a Perspex prism and ask the students to observe the change of path of the ray or take a toy car and run it at an angle into some sand from a bench / other smooth surface. Ask the students to suggest why the direction of the car changes when it comes into contact with the sand. [O1, O3]
  c) Diffraction – part-fill a shallow tray with water to a depth of about 1 cm. Place two blocks in the water so that there is a small gap between them. Set up a water wave and ask the students to observe what happens to the water wave as it goes through the gap between the blocks. [O1, O3]
- Remind students that radio waves and microwaves are also electromagnetic waves, and so will behave in a similar way to the waves they have just observed. [O1, O2, O3]

 © HarperCollins*Publishers* Limited 2016

Explain

- **Pair talk:** Ask students to discuss how they think they are able to hear sounds from in the corridor when they are sitting in a classroom.
- **Pairs to fours:** Ask the pairs to join together into groups of four to develop their ideas, before feeding back to the class. [O1, O2]
- Through class discussion, develop the idea of the diffraction of the sound waves around objects.
- Demonstrate this effect by placing a large rock in a shallow tray with water to a depth of about 1 cm and introduce a water wave across the tray. Ask students to explain what they observe.
- Discuss with students that the signal from a radio transmitter is able to reach houses on the far side of a hill for the same reason. Be sure to emphasise to students that a shorter wavelength wave, such as a microwave, would not be able to diffract around large objects as effectively. [O1, O2]

Consolidate and apply

The students should now be issued with the radio waves worksheet appropriate to their ability:

- Low demand: Worksheet 5.13.1 [O1, O2, O3]
- Standard demand: Worksheet 5.13.2 [O1, O2, O3]
- High demand: Worksheet 5.13.3 [O1, O2, O3]

Extend

- Tell the students that radio waves can be produced by oscillations in electrical circuits. When a current flows through a wire it creates an electric field around the wire. When the current changes, the electric field changes. The changing current produces radio waves. This is how radio transmitters work. Radio waves can induce oscillations in an electrical circuit.
- Ask the students to use this information to suggest how a radio receiver works. [O1, O2, O3]

Plenary suggestions

Ask me a question: Ask students to write a question about something from the topic and then devise a mark scheme for the answer. Encourage them to come up with questions worth more than one or two marks and to try out their questions on one another.

Learning triangle: At the end of the lesson ask students to draw a large triangle with a smaller inverted triangle that just fits inside (so they have four triangles). Ask them to think back over the lesson and write in the outer three triangles:

- something they saw
- something they did
- something they discussed.

Ask them to add in the central triangle something they learned.

Answers to questions

Worksheets 5.13.1, 5.13.2 and 5.13.3

2. Radio 1: 3.09 m, Radio 2: 3.41 m, Radio 4: 2.91 m, Radio London: 3.16 m

 © HarperCollins*Publishers* Limited 2016

Lesson 5.14: Reflection, refraction and wave fronts

Lesson overview

OCR Specification reference

P5.3a (part), b

Learning objectives

- Explain reflection and refraction and how these may vary with wavelength.
- Construct ray diagrams to illustrate refraction.
- Use wave front diagrams to explain refraction in terms of the difference in velocity of the waves in different substances.

Learning outcomes

- [O1] State the laws of reflection and refraction.
- [O2] Draw ray diagrams to show reflection and refraction.
- [O3] Interpret wave-front diagrams to state whether a wave is speeding up or slowing down as it passes through different media.

Skills development

- Use a diagram to illustrate the behaviour of electromagnetic waves.

Maths focus

- Measure angles using a protractor.

Resources needed ray boxes, power packs, slit cards, Perspex blocks, plane mirrors, protractors, image of a mirage; Worksheets 5.14.1, 5.14.2 and 5.14.3, Practical sheet 5.14, Technician's notes 5.14

Key vocabulary reflection, transmission, absorption, refraction, ray diagram, wave front

Teaching and learning

Engage

- Show students an image of a mirage on a hot road or in a desert.
- **Pair talk:** Ask students to discuss what they think causes the mirage.
- **Pairs to fours:** Ask the pairs to join up into fours and to share their ideas, before preparing their best group answer, using no more than 15 words.

Challenge and develop

- Remind students that visible light is one group in the electromagnetic spectrum and that, because we can see visible light, we can study it and then apply the findings to the rest of the electromagnetic spectrum. In other words, the behaviour of light waves can tell us how the other waves in the electromagnetic spectrum will behave.
- Using Practical sheet 5.14.1, students should remind themselves of the laws of reflection and refraction. [O1, O2]
- The students could be put into pairs, where one of the pair investigates reflection and the other investigates refraction, before coming back to share their results. [O1, O2]
- Alternatively, students could carry out both practical investigations for themselves. If practical equipment is limited, the class could be split in half and each half could begin with one of the two investigations, before swapping over half way through the time. [O1, O2]

 © HarperCollins*Publishers* Limited 2016

Explain

- Using the findings from the refraction investigation, ask the students what happened to the light ray as it entered the block and as it exited again.
- Ask students to draw particle diagrams for the air and the glass and ask students whether the light would speed up or slow down as it entered the block and as it left it again.
- Use these discussions to introduce a wave front diagram, perhaps for deep and shallow water, as shown in the Student Book. Be clear that the wave fronts get closer together if the wave is moving faster and spread out if the wave is moving slower. [O3]

Consolidate and apply

The students should now be issued with the reflection and refraction worksheet suitable for their ability:

- Low demand: Worksheet 5.14.1 [O1, O2, O3]
- Standard demand: Worksheet 5.14.2 [O1, O2, O3]
- High demand: Worksheet 5.14.3 [O1, O2, O3]

Extend

- Ask students able to progress further to draw a wave front diagram for the refraction experiment they carried out. [O3]
- Alternatively, they could be asked to use the ideas they have met during the lesson to return to the original question of what causes a mirage and improve the answer they gave at the start of the lesson.

Plenary suggestions

Heads and tails: Ask each student to write a question about something from the lesson on a coloured strip of paper and the answer on another coloured strip. In groups of six to eight, hand out the strips so that each student gets a question and an answer. One student reads out their question. The student with the right answer then reads it out, followed by their question, and so on.

Hot seat: Ask each student to think up a question, using material from the lesson. Select someone to put in the hot seat, ask students to pose their question and say at the end whether the answer is correct or incorrect.

Ideas hothouse: Ask students to work in pairs to list points about what they know about electromagnetic waves so far. Then ask the pairs to join together into fours and then into groups of six to eight to discuss this further and to come up with an agreed list of points. Ask one person from each group to report back to the class.

Answers to questions

1.

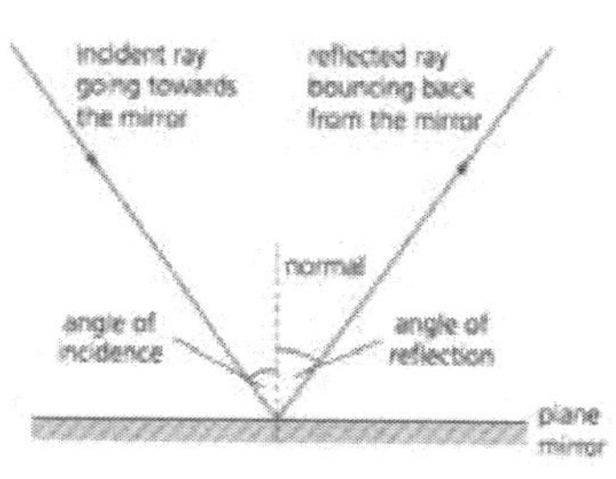

2.

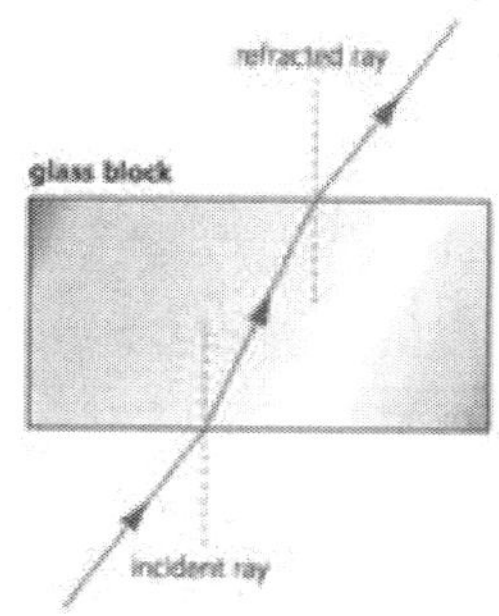

3.

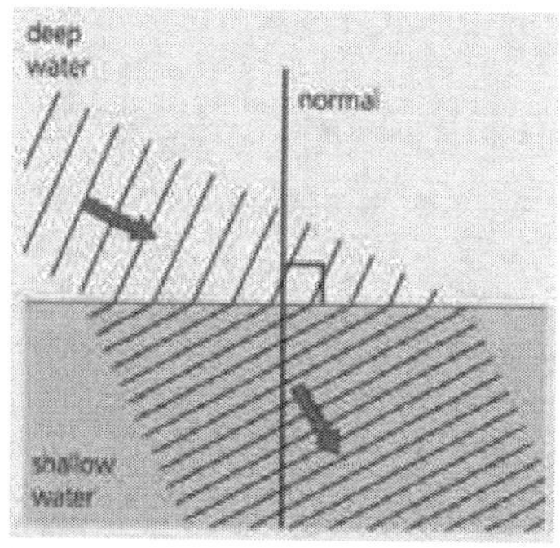

© HarperCollinsPublishers Limited 2016

Lesson 5.15: Practical: Investigate how the amount of infrared radiation absorbed or radiated by a surface depends on the nature of that surface

Lesson overview

OCR Specification reference

P5.3a (part)

Learning objectives

- Explain reasons for the equipment used to carry out an investigation.
- Explain the rationale for carrying out an investigation.
- Apply ideas from an investigation to a range of practical contexts.

Learning outcomes

- [O1] Identify independent, dependent and control variables in an investigation.
- [O2] Carry out a practical investigation to state how the nature of a surface affects the amount of infrared radiation it emits.
- [O3] Evaluate practical procedures and apply ideas learned to a different context.

Skills development

- Carry out a valid investigation.
- Draw conclusions from an investigation.

Resources needed 3 glass test tubes – one uncovered, one covered in aluminium foil and one covered in black paper, a thermometer, a stop clock, a kettle full of water; Practical 5.15, Technician's notes 5.15

Digital resources https://www.youtube.com/watch?v=hELg9A3Y8I8

Key vocabulary infrared radiation, emission, absorption, radiation

Teaching and learning

Engage

- Show the students a video clip of footage taken using an infrared camera. A suggested clip is given above.

Challenge and develop

- After the video clip, ask the students to discuss in pairs how they could investigate whether the colour of an object, as well as its temperature, would affect the amount of infrared radiation the object emitted.
- Ask the pairs to join together into fours to share their ideas. Pool ideas as a class if appropriate to your class. [O1, O2]

Explain

- Students should then carry out the practical outlined on Practical sheet 5.15. [O1, O2]
- Once they have completed the practical, they should complete the analysis and evaluation of their results. [O1, O2]

Consolidate and apply

The students should now complete the worksheet on absorption and emission of IR radiation appropriate to their ability:

- Low demand: Worksheet 5.15.1 [O1, O2, O3]
- Standard demand: Worksheet 5.15.2 [O1, O2, O3]

 © HarperCollinsPublishers Limited 2016

- High demand: Worksheet 5.15.3 [O1, O2, O3]

Extend

- Ask students able to progress further to identify another everyday application of the knowledge gained during this investigation, other than in the design of solar panels.

Plenary suggestions

Ask me a question: Ask students to write a question about something from the topic and then devise a mark scheme for the answer. Encourage them to come up with questions worth more than one or two marks and to try out their questions on one another.

Learning triangle: At the end of the lesson, ask students to draw a large triangle with a smaller inverted triangle that just fits inside it (so they have 4 triangles). Ask them to think back over the lesson and write in the outer three triangles:

- something they saw
- something they did
- something they discussed.

Ask them to add in the central triangle something they learned.

Heads and tails: Ask each student to write a question about something from the topic on a coloured paper strip and the answer on another coloured strip. In groups of six to eight, hand out the strips so that each student gets a question and an answer. One student reads out their question. The student with the right answer reads it out, followed by their question, and so on.

 © HarperCollinsPublishers Limited 2016

Lesson 5.16: Practical: Investigate the reflection of light by different types of surface and the refraction of light by different substances

Lesson overview

OCR Specification reference

P5.3c (part)

Learning objectives

- Make and record observations of how light is reflected and transmitted at different surfaces.
- Measure angles and discuss the method, apparatus and uncertainty in measurements.
- Draw conclusions from experimental results.

Learning outcomes

- [O1] Review the key words from the Waves chapter and match them with their definitions.
- [O2] Investigate how light reflects off different surfaces.
- [O3] Investigate the refraction of light by different substances.

Skills development

- Carry out fair and accurate investigations.
- Analyse data collected from experiments.
- Draw sensible conclusions from experimental data.

Maths focus

- Measure angles.

Resources Worksheets 5.16.1 and 5.16.2, Practical sheets 5.16.1, 5.16.2, 5.16.3 and 5.16.4, Technician's notes 5.16

Key vocabulary specular reflection, diffuse reflection, ray diagram, normal, refraction

Teaching and learning

Engage

- Card sort (Worksheet 5.16.1): students review the key words they have learned during the topic (and the key words that will be used in this lesson) by matching each key word with its definition. [O1]
- Alternatively, the students should complete the crossword (Worksheet 5.16.2), which recaps the key words from the waves section. [O1]

Challenge and develop

- Show the students images of children's toys which rely on the principles of the behaviour of light waves, such as a kaleidoscope, a pair of diffraction glasses, bubble mirror walls and a simple prism. Can any of the students offer an explanation for how these toys work? Encourage the use of the key words from the activity at the start of the lesson. [O1]

Explain

- The students should then carry out the practicals in four groups where they will investigate the behaviour of light when it is shone on different surfaces. Following the practicals, a spokesperson from each of the groups can share with the whole class what they did and what they found out.

 © HarperCollins*Publishers* Limited 2016

Consolidate and apply

The students should answer the questions on pages 194–5 of the Student Book appropriate to their ability:

- Low demand: questions 3, 4, 5, 8 and 9 [O2, O3]
- Standard demand: questions 3, 6, 8, 9 and 10 [O2, O3]

Extend

- Students could be asked to design a new toy for children that relies on at least one of the phenomena discovered during the investigations in the lesson. [O2, O3]

Plenary suggestions

Heads and tails: Ask each student to write a question about something from the topic on a coloured strip of paper and the answer on another coloured strip. In groups of six to eight, hand out the strips so that each student gets a question and an answer. One student reads out their question. The student with the right answer then reads it out, followed by their question.

Ask me a question: Ask students to write a question about something from the topic and then devise a mark scheme for the answer. Encourage them to come up with questions worth more than one or two marks and to try out their questions on one another.

Answers to questions

Worksheet 5.16.2

Across

4. Absorption
5. Specular reflection
6. LIght ray
7. Diffuse reflection
8. Refraction

Down

1. Reflection
2. Angle of reflection
3. Normal
4. Angle of refraction

 © HarperCollinsPublishers Limited 2016

Lesson 5.17: Lenses

Lesson overview

OCR Specification reference

P5.3c (part)

Learning objectives

- Understand what a lens does.
- Draw ray diagrams to show the formation of images by lenses.
- Describe the difference between a real and a virtual image.

Learning outcomes

- [O1] State that in a convex lens parallel rays of light are brought to a focus at the principal focus.
- [O2] Use ray diagrams to determine the nature of the image in convex lenses.
- [O3] Use ray diagrams to determine the nature of the image in concave lenses.

Skills development

- Carry out a valid investigation.
- Draw conclusions from an investigation.

Resources needed ray box and slit card, power pack, convex and concave lenses, lens holders; Worksheets 5.17.1, 5.17.2 and 5.17.3, Technician's notes 5.17

Digital resources images of several objects that use lenses

Key vocabulary convex lens, concave lens, principal focus, focal length, principal axis, real image, virtual image

Teaching and learning

Engage

- Show students images of several objects that use lenses, for example a magnifying glass, a pair of spectacles, a microscope and a telescope, and ask them to suggest what the objects have in common.
- Ask students to work in pairs to list as many other uses for lenses (in cameras, human eye, projectors, etc.) as they can.
- Ask the pairs to join together into fours and then eights to share ideas, before drawing up a list for the class.

Challenge and develop

- Issue students with a convex lens and ask them to look through it. [O2]
- Then issue them with a concave lens and ask them to describe the difference between the two lenses (convex lenses should focus the light, whereas convex lenses should cause the light to spread out). They could then investigate passing light from a ray box through each type of lens. Ideally, use the slit card in such a way as to generate three rays of light to introduce to the lenses. [O3]

Explain

- Show students the symbols for convex and concave lenses, before drawing a ray diagram showing parallel light entering a convex and a concave lens. The students should practise drawing these diagrams for themselves.
- Introduce the term focal point here, by showing the students the focal point from the light through a convex lens and by stating that light travelling parallel to the principal axis of a concave lens is refracted by the lens so that it appears to have come from the focal point. [O1]

 © HarperCollins*Publishers* Limited 2016

- Ask students to label the focal point for each type of lens on the diagrams they have drawn. [O1]
- The students should follow this by answering questions 1–6 on page 197 of the Student Book. [O1, O2, O3]

Consolidate and apply

The students should now be issued with the lenses worksheet appropriate to their ability:

- Low demand: Worksheet 5.17.1 [O1, O2, O3]
- Standard demand: Worksheet 5.17.2 [O1, O2, O3]
- High demand: Worksheet 5.17.3 [O1, O2, O3]

Extend

- Ask students able to progress further to suggest why a slide projector and a camera need different types of lens and to state what type of lens each instrument uses, adding an explanation as to why that type of lens is required.

Plenary suggestions

Ask me a question: Ask students to write a question about something from the topic and then devise a mark scheme for the answer. Encourage them to come up with questions worth more than one or two marks and to try out their questions on one another.

Heads and tails: Ask each student to write a question about something form the topic on a coloured strip of paper and then answer on another coloured strip. In groups of six to eight, hand out the strips so that each student gets a question and an answer. One student reads out their question. The student with the right answer then reads it out, followed by their question, and so on.

What's in the picture? Have a picture concealed by rectangles, each of which can be removed in turn. Ask for suggestions as to which tile should be removed and what the partially revealed graphic shows. Encourage speculation and inference; ask which tile should go next and continue until a full explanation has been produced.

 © HarperCollins*Publishers* Limited 2016

Lesson 5.18: Images and magnification

Lesson overview

OCR Specification reference

P5.3c (part)

Learning objectives

- Draw ray diagrams to show the formation of real and virtual images by lenses.
- Calculate the magnification of an image.

Learning outcomes

- [O1] State that in a convex lens parallel rays of light are brought to a focus at the principal focus.
- [O2] Use ray diagrams to determine the nature of the image formed by a lens.
- [O3]Use ray diagrams to determine the position and magnification of images.

Resources needed Worksheets 5.18.1, 5.18.2 and 5.18.3

Digital resources Short sight: http://www.myopia.org/w.optometrists.asn.au1.gif

Long sight: http://lookafteryoureyes.org/wp-content/uploads/2012/07/long-sighted.jpg

Key vocabulary ray diagram, real image, magnification, virtual image

Teaching and learning

Engage

- Ask students to discuss what they think it means to be long-sighted and short-sighted.
- Tell them that, if you are short-sighted, you are able to see close objects clearly, but that distant objects appear blurry, and the reverse if you are long-sighted.
- State that short-sightedness is caused by either the eyeball being too long or the lens being too thick and that long-sightedness is caused either by the eyeball being too short or the lens being too thin.

Challenge and develop

- Show the students images of light rays entering an eyeball that is too long (causing short-sightedness) and an eyeball that is too short (causing long-sightedness).
- Ask students to use their knowledge of lenses to suggest which type of lens should be used to correct short-sightedness (concave) and long-sightedness (convex).

Explain

- Demonstrate to students how to draw ray diagrams for both types of lenses. Begin with a convex (converging) lens. [O2]
- Choose a simple object such as an arrow or a light bulb and draw this on the board. Draw this at greater than 2× the focal length. Add in the focal length, labelled F on the principal axis. [O2]
- Introduce the first rule: draw a ray from a single point on the object (ideally the top) parallel to the principal axis which is refracted by the lens and passes through the focal point (on the other side of the lens). [O2]
- Then introduce the second rule: draw a ray from the same point on the object through the centre of the lens. The ray does not change direction. [O2]
- The point of intersection between the two rays draw gives the same point on the image that the ray was taken from on the object (usually the top). [O2]

 © HarperCollins*Publishers* Limited 2016

- Ask students to draw their own diagram, following the two rules and describe the image as either the same way up or inverted and the same size, smaller (diminished) or larger (magnified). Introduce the term real image here, as an image that can be seen on a screen or other surface. [O2]
- Using the same simple object, now redraw the principal axis and place the object inside the focal length of the convex lens. Follow the same two rules for drawing the two rays. This time, the lines representing the rays do not intersect. Extrapolate the lines back to the point at which they would intersect (on the same side of the lens as the object). Introduce the term virtual image here, as an image that cannot be seen on a screen, as the rays of light don't actually go there and no light actually reaches it. The virtual image is at the point the light appears to have come from. [O2]
- Ask students to draw this for themselves and describe the image as either the same way up or inverted and the same size, smaller (diminished) or larger (magnified). [O2]
- Show students how to construct a ray diagram for a concave lens by following the same two rules for drawing the two rays. This time, the ray coming in parallel to the principal axis is drawn as if it had come from the focal point on the same side of the lens as the object and so the light diverges from the lens. [O3]
- Students should then draw their own diagram and should describe the image as either the same way up or inverted and the same size, smaller (diminished) or larger (magnified). [O3]

Consolidate and apply

The students should now be issued with the images and magnification worksheet appropriate to their ability:

- Low demand: Worksheet 5.18.1 [O3]
- Standard demand: Worksheet 5.18.2 [O3]
- High demand: Worksheet 5.18.3 [O3]

Extend

- Ask students able to progress further to draw ray diagrams to show the effect of using a concave lens to correct short-sightedness and a convex lens to correct long-sightedness.

Plenary suggestions

Ask me a question: Ask students to write a question about something from the topic and then devise a mark scheme for the answer. Encourage them to come up with questions worth more than one or two marks and to try out their questions on one another.

Heads and tails: Ask each student to write a question about something form the topic on a coloured strip of paper and then answer on another coloured strip. In groups of six to eight, hand out the strips so that each student gets a question and an answer. One student reads out their question. The student with the right answer then reads it out, followed by their question, and so on.

What's in the picture? Have a picture concealed by rectangles, each of which can be removed in turn. Ask for suggestions as to which tile should be removed and what the partially revealed graphic shows. Encourage speculation and inference; ask which tile should go next and continue until a full explanation has been produced.

 © HarperCollins*Publishers* Limited 2016

Answers to questions

Worksheets 5.18.1 and 5.18.2

1.

Object height	Image height	Magnification
2	6	**3**
3	12	**4**
4	20 (200 mm)	5
5 (50 mm)	10	2
2 (20 mm)	**12**	6
10	**30 (3000 mm)**	3

2. The magnification factor has no units as the units for image and object height cancel each other out, leaving no units.

3. If the image height is smaller than the object height, the magnification must be less than one.

Worksheet 5.18.3

1.

Object height	Image height	Magnification
2 cm	6 cm	**3**
3 cm	12 cm	**4**
4 cm	200 mm	5
50 mm	10 cm	2
20 mm	**12 cm**	6
10 cm	**3000 mm**	3

 © HarperCollins*Publishers* Limited 2016

Lesson 5.19: Colour

Lesson overview

OCR Specification reference

P5.3e

Learning objectives

- Describe what happens when light of different wavelengths lands on an object.
- Explain what determines the colour of an opaque object,
- Explain the effect of coloured filters.

Learning outcomes

- [O1] State that each colour in the visible spectrum has its own narrow band of wavelengths.
- [O2] Describe how colour filters work by absorbing certain wavelengths and transmitting other wavelengths.
- [O3] Explain that the colour of an opaque object is determined by which wavelengths of light are more strongly reflected.

Skills development

- Carry out a valid investigation.
- Draw conclusions from an investigation.

Resources needed ray box and slit card, power pack, prism, range of opaque, translucent and transparent objects, coloured filters, range of different coloured opaque objects; Worksheets 5.19.1, 5.19.2 and 5.19.3

Key vocabulary filter, reflection, absorption, transmit, translucent, transparent, opaque

Teaching and learning

Engage

- Demonstrate to the students that white light is made up of different colours of light by shining a single ray of light into a prism.
- Ask the students to discuss why we see the split light as different colours – what is it about the different colours of the spectrum?
- Students could be asked to calculate the wavelengths of the different colours, given the frequencies of the colours and using the wave equation. [O1]

Challenge and develop

- Remind students of the difference between opaque, translucent and transparent.
- This could be done by showing the students a range of different materials and asking them to classify them as either opaque, translucent or transparent. [O1]
- Alternatively, ask students to think about objects they have in their homes, and to write three lists: opaque objects, translucent objects and transparent objects. They should try to name two for each category.

Explain

- Pose the question to the students "Why does grass appear green to us?" and allow them to discuss the answer to this question in pairs. [O2]
- **Pairs to fours:** Ask the pairs of students to join into groups of four and develop their ideas, before taking feedback from the groups. Elicit the idea that grass absorbs all wavelengths of light except green, which is reflected, and so the grass appears to be green. [O2]
- The students should answer questions 1, 2 and 3 on page 200 of the Student Book.

 © HarperCollins*Publishers* Limited 2016

- Demonstrate the effect of coloured filters on the appearance of white light and the appearance of coloured objects. [O2]
- To do this, to illustrate the effect of the appearance of white light, place the coloured filters, one at a time, in front of a ray box and ask students to record the colour of light they see transmitted. [O2]
- To illustrate the appearance of coloured opaque objects in coloured light, ask students to choose one coloured light, for example blue, and shine the blue light onto objects of different colours. Again, ask the students to record their observations, before explaining what they have found out. [O2]

Consolidate and apply

Issue the students with the colour worksheet appropriate to their ability:

- Low demand: Worksheet 5.19.1 [O1, O2, O3]
- Standard demand: Worksheet 5.19.2 [O1, O2, O3]
- High demand: Worksheet 5.19.3 [O1, O2, O3]

Extend

- Ask students able to progress further to suggest what happens when light hits a translucent object that looks white.

Plenary suggestions

Hot seat: Ask each student to think up a question, using material from the topic so far. Select someone to put in the hot seat, ask students to pose their question and say at the end whether the answer is correct or incorrect.

Where's the answer? Give each student a card with a word written on it. Then ask a question, ask them if they think they have the answer, and why.

The big ideas: Ask students to write down three ideas they have learned during the topic so far. Then ask them to share their facts in groups and to compile a master list of facts, with the most important at the top. Ask for ideas to be shared and find out which other group(s) agreed.

Answers to questions

Worksheet 5.19.3

2. red light: 620–750 nm; orange light: 590–620 nm; yellow light: 570–590 nm; green light: 495–570 nm; blue light: 450–495 nm; indigo light: 446–464 nm; violet light: 380–450 nm

Worksheets 5.19.1 to 5.19.3

3.

Colour of object	Colour of filter	Appearance of object
white	blue	**blue**
red	blue	**black**
green	green	**green**
yellow	blue	**black**

 © HarperCollins*Publishers* Limited 2016

When and how to use these pages: Check your progress, Worked example and End of chapter test

Check your progress

Check your progress is a summary of what students should know and be able to do when they have completed the chapter. Check your progress is organised in three columns to show how ideas and skills progress in sophistication. Students aiming for top grades need to have mastered all the skills and ideas articulated in the final column (shaded pink in the Student book).

Check your progress can be used for individual or class revision using any combination of the suggestions below:

- Ask students to construct a mind map linking the points in Check your progress
- Work through Check your progress as a class and note the points that need further discussion
- Ask the students to tick the boxes on the Check your progress worksheet (Teacher Pack CD). Any points they have not been confident to tick they should revisit in the Student Book.
- Ask students to do further research on the different points listed in Check your progress
- Students work in pairs and ask each other what points they think they can do and why they think they can do those, and not others

Worked example

The worked example talks students through a series of exam-style questions. Sample student answers are provided, which are annotated to show how they could be improved.

- Give students the Worked example worksheet (Teacher Pack CD). The annotation boxes on this are blank. Ask students to discuss and write their own improvements before reviewing the annotated Worked example in the Student Book. This can be done as an individual, group or class activity.

End of chapter test

The End of chapter test gives students the opportunity to practice answering the different types of questions that they will encounter in their final exams. You can use the Marking grid provided in this Teacher Pack or on the CD Rom to analyse results. This shows the Assessment Objective for each question, so you can review trends and see individual student and class performance in answering questions for the different Assessment Objectives and to highlight areas for improvement.

- Questions could be used as a test once you have completed the chapter
- Questions could be worked through as part of a revision lesson
- Ask Students to mark each other's work and then talk through the mark scheme provided
- As a class, make a list of questions that most students did not get right. Work through these as a class.

 © HarperCollins*Publishers* Limited 2016

Marking Grid for End of Chapter 5 Test

	Getting Started [Foundation Tier]								Going Further [Foundation and Higher Tiers]						More Challenging [Higher Tier]				Most demanding [Higher Tier]		Total marks	Percentage
Student Name	Q. 1 (AO1) 1 mark	Q. 2 (AO1) 1 mark	Q. 3 (AO1) 1 mark	Q. 4 (AO2) 2 marks	Q. 5 (AO1) 1 mark	Q. 6 (AO2) 1 mark	Q. 7 (AO2) 1 mark	Q. 8 (AO2) 2 marks	Q. 9 (AO1) 1 mark	Q. 10 (AO1) 1 mark	Q. 11 (AO1) 2 marks	Q. 12 (AO2) 2 marks	Q. 13 (AO2) 2 marks	Q. 14 (AO1) 2 marks	Q. 15 (AO1) 2 marks	Q. 16 (AO2) 1 mark	Q. 17 (AO2) 2 marks	Q. 18 (AO3) 6 marks	Q. 19 (AO3) 2 marks	Q. 20 (AO2) 2 marks		

 © HarperCollins*Publishers* Limited 2016

Check your progress

You should be able to:

Describe the amplitude, wavelength, frequency and period of a wave	Use the wave equation $v = \lambda \times f$ to calculate wave speed	Rearrange and apply the wave equation
Realise that waves can be transverse or longitudinal	Give examples of longitudinal and transverse waves	Explain the difference between transverse and longitudinal waves
Describe how sound waves travel through air or solids	Describe how to measure the speed of sound waves in air	Explain how to calculate the depth of water using echo sounding
Understand that waves transfer energy or information	Give examples of energy transfer by waves (including electromagnetic waves)	Describe evidence that, for e.g. ripples on a water surface, it is the wave and not the water itself that travels
Understand that waves can be absorbed, transmitted or reflected at a surface	Describe examples of reflection, transmission and absorption of waves (including electromagnetic waves) at material interfaces	Describe how different substances may absorb, transmit, refract or reflect electromagnetic waves in ways that vary with wavelength
Draw a labelled ray diagram to illustrate reflection of a wave at a boundary	Construct ray diagrams to illustrate refraction at a boundary	Use wavefront diagrams to explain refraction in terms of a change in wave velocity
Describe the range of normal human hearing Define the term ultrasound	Describe how ultrasound waves can be used for medical and industrial imaging	Explain how P and S waves can be used to deduce information about the structure of the Earth
Name the main groupings of the electromagnetic spectrum	Compare the groupings of the electromagnetic spectrum in terms of wavelength and frequency	Describe how radio waves are produced
Describe the hazardous effects of gamma rays, X-rays and ultraviolet radiation	Explain the risks associated with the use of ionising and ultraviolet radiation	Evaluate the risks and consequences of exposure to radiation
Give examples of the uses of the main groupings of the electromagnetic spectrum	Describe examples of energy transfer by electromagnetic waves	Explain why each type of electromagnetic wave is suitable for the application
State that each colour in the visible spectrum has its own narrow band of wavelength	Describe that colour filters absorb certain wavelengths and transmit other wavelengths	Explain that the colour of an opaque object depends on which wavelengths are more strongly reflected
State that in a convex lens parallel rays of light are brought to a focus at the principal focus	Use ray diagrams to determine the nature of the image formed by a lens	Use ray diagrams to determine the position and magnification of images

 © HarperCollins*Publishers* Limited 2016

Worked example

The table below shows the electromagnetic spectrum.

A	microwave	infrared	visible light	B	X-rays	gamma rays

1. **State the names of the waves labelled A and B.**

   A = radio waves   B = ultraviolet waves

   Both answers are correct. Use a mnemonic to remember the correct order.

2. **X-rays are dangerous to humans. Explain how they can also be used in medical therapy without lasting harm.**

   We can use them for X-rays to see our bones and for treating cancer by killing the cancer cells.

   This answer is a good start but is incomplete. It doesn't give a full explanation. The answer should also say that we can use them by controlling the exposure dose.

3. **Complete the ray diagram to show how the image is formed.**

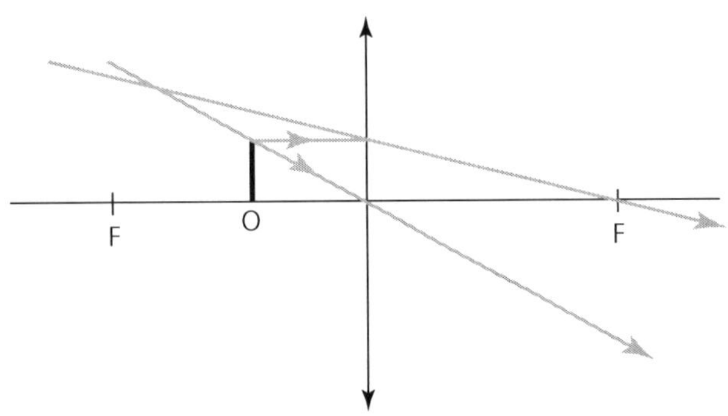

   The positions of the rays have been drawn correctly, but the virtual rays have not been drawn with dashed lines. Also the image has not been drawn in. You should always draw the image when you have completed the ray diagram.

4. **Explain why we see a red apple as red.**

   White light is a mixture of many wavelengths. We see the red apple because the red wavelengths are reflected off the apple.

   This is a good answer. Just one thing has been missed out: red wavelengths are reflected but what about the other wavelengths? They are absorbed.

5. **Complete the wave front diagram to show the refraction of light from air to water.**

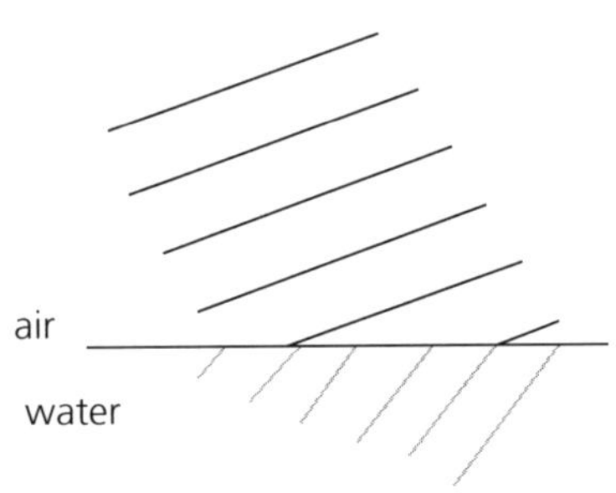

   The wave fronts being closer together is correct, but there is one crucial thing wrong. When waves go from a less dense medium to a denser medium, they are refracted **towards** the normal. The diagram shows them being refracted **away** from the normal.

    © HarperCollins*Publishers* Limited 2016

Radioactivity: Introduction

When and how to use these pages

This unit will build on ideas the students have met before, such as:

- All matter is made of atoms.
- All the atoms in an element are the same.
- Atoms cannot be created or destroyed.
- Chemical equations summarise what happens in a chemical reaction.

Overview of the unit

In this unit, students will learn the current model of the atoms, and how ideas about the structure of the atom have changed over the years. They will be introduced to the three types of ionising radiation and consider hazards related to and uses of each type of radiation.

This unit offers a number of opportunities for students to investigate changing ideas through research, work collaboratively with peers and evaluate evidence critically before drawing conclusions.

Obstacles to learning

Students may need extra guidance with the following common misconceptions:

- Students might confuse the nucleus of a cell in biology and the nucleus of an atom in physics.
- Students might not understand why the nucleus of an atom (containing positive and neutral charges) doesn't fall apart (without knowledge of the strong force).
- Some questions may arise around gamma radiation: how can it both cause and cure cancer?
- Students might struggle with the concept of half-life:
  - Students might develop a misconception that the nuclei must eventually 'die' if they have a 'half-life'.
  - Students might mistakenly think that if it takes *x* time to lose half the activity, in 2*x* time it should all be gone.

Practicals in this unit

In this unit, students will do the following practical work:

- model radioactive half-life.

 © HarperCollins*Publishers* Limited 2016

	Lesson title	Overarching objectives
1	Atomic structure	• Describe the structure of the atom. • Use symbols to represent particles. • Describe ionisation.
2	Radioactive decay	• Describe radioactive decay. • Describe the types of nuclear radiation. • Understand the processes of alpha decay and beta decay.
3	Nuclear equations	• Understand nuclear equations. • Write balanced nuclear equations.
4	Radioactive half-life	• Explain what is meant by radioactive half-life. • Calculate half-life. • Choose the best radioisotope for a task.
5	Background radiation	• Recall sources of background radiation. • Describe how different types of radiation have different ionising power.
6	Maths skills: Using ratios and proportional reasoning	• Draw a curve of best fit to calculate radioactive half-life. • Calculate the net decline.
7	Hazards and uses of radiation	• Describe radioactive contamination. • Give examples of how radioactive tracers can be used.
8	Irradiation	• Explain what is meant by irradiation. • Understand the distinction between contamination and irradiation. • Appreciate the importance of communication between scientists.
9	Uses of radiation in medicine	• Compare gamma rays and X-rays. • Describe some uses of nuclear radiation for medical diagnosis and therapy.
10	Using nuclear radiation	• Explore the risks and benefits of using nuclear radiation. • Describe how internal organs can be explored. • Understand how nuclear radiation can control or destroy unwanted tissue.
11	Nuclear fission	• Describe nuclear fission. • Explain how a chain reaction occurs. • Explain how fission is used.
12	Nuclear fusion	• Explain nuclear fusion. • Describe the conditions needed for fusion. • Describe how nuclear fusion might be an attractive energy source.

 © HarperCollins*Publishers* Limited 2016

Lesson 6.1: Atomic structure

Lesson overview

OCR Specification reference
P6.1a, b, c, h

Learning objectives
- Describe the structure of the atom.
- Use symbols to represent particles.
- Describe ionisation.

Learning outcomes
- [O1] State that the number of protons in an element is the atomic number and that the total number of protons and neutrons is the mass number.
- [O2] Understand that atoms of an element all have the same number of protons but can have different numbers of neutrons, giving different isotopes.
- [O3] Use nuclear notation to show subatomic particles in an isotope.

Skills development
- Represent atoms using accepted notation.
- Use a model to represent ionisation.
- Discuss the limitations of models.

Maths focus
- Subtraction of one number from another to give the number of neutrons.

Resources needed modelling clay (ideally three different colours), copies of the periodic table, access to emission and/or absorption spectra of different elements; Worksheets 6.1.1, 6.1.2 and 6.1.3
Digital resources access to online simulation and means of projection
Key vocabulary atomic number, energy level, mass number, nucleon, relative atomic mass, isotope, ionise, nuclear radiation

Teaching and learning

Engage
- As a reminder of the content covered in chapter 1, the students could be asked to describe, in exactly 30 words, the structure of an atom. They could choose a particular element if they like, or they could describe the general structure of all atoms. [O1]

Challenge and develop
- Issue students with a copy of the periodic table and ask students to find the element with atomic number 6. Ask them what information this gives about the element. Ask the students what else the periodic table can tell them about carbon. [O1]
- Issue pairs of students with some modelling clay (ideally three colours) and ask them to build two models: one of an atom of carbon-12 and another of carbon-14. They should use their model to define the meaning of the term 'isotope'. Ask some pairs of students to show their model to the rest of the group, explaining how they decided what their model should look like and what each part of the model shows. [O1, O2, O3]
- Alternatively, use a suitable simulation (search 'PhET model of atom') to model atoms, or use students themselves as subatomic particles and model different atoms by arranging the students in the centre of the room (if they are modelling protons or neutrons) or around the edges of the room (in 'layers') if they are modelling electrons. [O3]

Explain
- Students should now be issued with the atomic structure worksheet at an appropriate level:
  - Low demand: Worksheet 6.1.1 [O1, O2, O3]
  - Standard demand: Worksheet 6.1.2 [O1, O2, O3]
  - High demand: Worksheet 6.1.3 [O1, O2, O3]

Consolidate and apply
- Discuss with students what would happen to an atom if one of its electrons were removed. If necessary, students could return to their model of a carbon atom and count the subatomic particles once an electron had been removed. Define the term 'ionisation' as the gain or loss of one or more electrons from an atom, resulting in an ion. [O1]

 © HarperCollins*Publishers* Limited 2016

- Introduce the term 'nuclear radiation' and model this using ball bearings and a marble: model ionising radiation using small ball bearings as nucleons and electrons and a marble as incoming radiation from the nucleus of a radioactive atom – roll incoming radiation towards the electrons – the structure of the atom is changed – this models ionising radiation in that the structure of the atom becomes altered in some way. [O3]
- An alternative to ionisation is that electrons are excited and de-excited in atoms. The students should be introduced to the idea of energy being emitted from atoms as electrons are excited to higher energy levels and then 'fall back'. They could research different absorption or emission spectra and state how these can be used to identify the different elements.

Extend

Ask students able to progress further to do the following:

- Research how nuclear radiation from an isotope of americium is used in smoke alarms. [O2]

Plenary suggestions

Learning triangle: At the end of the lesson, ask students to draw a large triangle with a smaller inverted triangle that just fits inside it (so they have four triangles). Ask students to think back over the lesson and identify and write in the outer triangles:

- something they saw
- something they did
- something they discussed.

Ask students to write in the central triangle something they learned.

Ideas hothouse: Ask students to work in pairs to list points about what they know about a particular idea. Then ask the pairs to join together into groups of four and then groups of six to eight to discuss this further and to come up with an agreed list of points. Ask one person from each group to report back to the class.

Freeze frame: Ask students to create a 'freeze frame' of an idea in the topic. This involves three or four students arranging themselves as a static image and other students suggesting what it represents.

Answers to questions

Worksheets 6.1.1 and 6.1.2

1. Words in the following order: symbol; number; electrons; relative; protons; equal; isotopes

Worksheets 6.1.1 to 6.1.3

2.

Name of element	Chemical symbol	Mass number (A)	Atomic number (Z)	Number of protons	Number of neutrons	Number of electrons
Carbon	C	12	6	6	6	6
Lithium	Li	7	3	3	4	3
Magnesium	Mg	24	12	12	12	12
Uranium	U	235	92	92	143	92
Radium	Ra	226	88	88	138	88
Iron	Fe	56	26	26	30	26

Worksheets 6.1.2 and 6.1.3

3.

Formula of ion	Number of protons	Number of neutrons	Number of electrons
Na^+	11	12	10
Ca^{2+}	20	20	18
F^-	9	10	10
O^{2-}	8	8	10

 © HarperCollins*Publishers* Limited 2016

Lesson 6.2: Radioactive decay

Lesson overview

OCR Specification reference

P6.1d, e

Learning objectives

- Describe radioactive decay.
- Describe the types of nuclear radiation.
- Understand the processes of alpha decay and beta decay.

Learning outcomes

- [O1] List the three types of ionising radiation resulting from nuclear activity.
- [O2] Describe the structure of each type of ionising radiation.
- [O3] Explain the properties of each type of radiation.

Skills development

- Use the unit of activity, becquerels (Bq), correctly.
- Study the work of scientists objectively.

Maths focus

- Subtract one number from another to give the structure of the resulting atom after nuclear decay.

Resources needed samples of radioactive sources for demonstration, aluminium sheets of different thicknesses, lead sheets of different thicknesses, Geiger counter; Worksheets 6.2.1, 6.2.2 and 6.2.3

Digital resources http://science.howstuffworks.com/29296-100-greatest-discoveries-radioactivity-video.htm, https://www.youtube.com/watch?v=wsspFQn0mWM, a short video clip without sound of each type of radioactive decay

Key vocabulary radioisotope, random, alpha particle, beta particle, gamma ray, becquerel (Bq), activity, neutron radiation, nuclear radiation

Teaching and learning

Engage

- Show students a video clip of Becquerel's experiment with uranium (http://science.howstuffworks.com/29296-100-greatest-discoveries-radioactivity-video.htm). During the clip, the students could be asked to note down five key points (such as dates, names, key steps in the experiment).

Challenge and develop

- **Pair talk:** Ask students to work in pairs to suggest a few reasons why some elements are radioactive and others are not. They could be prompted to think about how the nuclei of radioactive elements might be different from stable nuclei. [O1]
- **Pairs to fours:** Pair each pair of students with another pair so that they can share their ideas. Ask one member of each group to tell these ideas to the class. Compile a list of these ideas, before discussing them with the class. From the discussion, suggest that there could be three things 'wrong' with the nuclei of unstable elements: the nuclei could be too big (causing an alpha particle to be emitted), the charge balance of the nucleus could be wrong (causing a beta particle to be emitted) or the nucleus could have too much energy (resulting in the emission of gamma rays). [O1]
- Discuss the nature of alpha particles, beta particles and gamma rays. [O1]

 © HarperCollins*Publishers* Limited 2016

Explain

- Demonstrate the differing penetration powers of alpha, beta and gamma radiation by putting different thicknesses of paper, aluminium and lead between the radioactive sources and the Geiger counter. [O2, O3]
- Alternatively, use a video clip of this demonstration. [O2, O3]
- Then issue each student with a radioactive decay worksheet at an appropriate level:
  - Low demand: Worksheet 6.2.1 [O1, O2, O3]
  - Standard demand*:* Worksheet 6.2.2 [O1, O2, O3]
  - High demand: Worksheet 6.2.3 [O1, O2, O3]

Consolidate and apply

- Students should consolidate their knowledge by splitting a plain A4-size sheet of paper into three columns, labelled 'alpha radiation', 'beta radiation' and 'gamma radiation' and giving details of each of the following:
  - The nature of each particle or ray
  - The effect the emission has on the nucleus of the parent atom
  - The products formed by the emission
  - The penetrating power of each type of radiation.

Students may need access to reference material. [O1, O2, O3]

Extend

Ask students able to progress further to do one of the following:

- Research the behaviour of each type of radiation in both electric and magnetic fields. [O3]
- Find out how a Geiger counter works.

Plenary suggestions

Hot seat: Ask each student to think of a question, using material from the topic. Select someone to put in the hot seat; ask students to ask their questions and say at the end whether the answer is correct or incorrect.

Silent animation: Show students a short video clip without sound of each type of radioactive decay. Ask the students to say which type of decay they think is taking place. They could provide a 'voice over' describing what is happening to the parent and daughter nuclei.

Where's the answer? As students enter the room, give them a card with a word written on it. At the end of the lesson, ask a question and the students should say if they think they have the answer card and explain why.

Answers to questions

Worksheets 6.2.1 and 6.2.2

1. Words in this order: nucleus; protons; electron; nucleus; energy; wave

Worksheets 6.2.1 to 6.2.3

2. (a) 150 counts (b) 3000 counts (c) 9000 counts
3. (a) 24 000 counts (b) 720 000 counts (c) 17 280 000 counts
4.

Type of radiation	Stopped by
Alpha	Thick lead or concrete
Beta	Sheet of aluminium
Gamma	Sheet of paper

(Lines drawn: Alpha – Sheet of paper; Beta – Sheet of aluminium; Gamma – Thick lead or concrete)

Worksheets 6.2.2 and 6.2.3

5. beta decay

 © HarperCollins*Publishers* Limited 2016

Lesson 6.3: Nuclear equations

Lesson overview

OCR Specification reference

P6.1f, g

Learning objectives

- Understand nuclear equations.
- Write balanced nuclear equations.

Learning outcomes

- [O1] Recognise the symbols used in a nuclear equation.
- [O2] Write nuclear equations involving alpha and beta decay.
- [O3] Write balanced nuclear equations for different types of nuclear reaction.

Skills development

- Use accepted notation (nuclear notation in this case).

Maths focus

- Balance equations.
- Use addition and subtraction.

Resources needed Worksheets 6.3.1, 6.3.2 and 6.3.3

Digital resources clip from *Harry Potter and the Philosopher's Stone* if possible, animations of alpha and beta decay

Key vocabulary nuclear equation, alpha decay, beta decay, balancing

Teaching and learning

Engage

- Read students an extract from page 77 of *Harry Potter and the Philosopher's Stone* (Bloomsbury, London, 1997), which details Professor Dumbledore and Nicolas Flamel's work on alchemy. Alternatively, show the students the clip of this scene from the film, if possible.
- **Pair talk:** Ask students what they understand by the term 'alchemy'. Share ideas with the rest of the class before stating that in this lesson they will be focusing on the aspect of turning one element into another (in alchemy, this is usually gold).

Challenge and develop

- Remind students of the processes of alpha and beta decay by showing them a suitable animation of each process. [O2, O3]
- Ask students to then draw a table with three columns: type of radiation, nature of particle and penetrating power, and fill in the details. [O2, O3]

Explain

- Using uranium-238 as an example, discuss the process of alpha decay into thorium-234 using nuclear notation. Highlight that mass and charge must both be conserved in any interaction: what can students say about the mass number and atomic number after the decay? [O2]
- The students should spot that both numbers should change, and so the element formed must be different from that with which they started.

 © HarperCollins*Publishers* Limited 2016

- **Pair talk:** Ask students if they could then suggest how to work out what element is present after the decay, given the nature of an alpha particle, previously discussed. [O2]
- Model how to write the nuclear equation for this decay: [O1, O2]

$$^{238}_{92}\mathrm{U} \xrightarrow{\alpha\text{-decay}} {}^{4}_{2}\mathrm{He} + \underset{92-2}{\overset{238-4}{{}^{234}_{90}\mathrm{Th}}}$$

- Using carbon-15 as an example, discuss the process of beta decay into nitrogen-15 using nuclear notation. Once again, highlight that the conservation of mass and charge still apply. [O3]
- Model how to write the nuclear equation for the decay: [O1, O3]

$$^{15}_{6}\mathrm{C} \xrightarrow{\beta\text{-decay}} {}^{0}_{-1}\mathrm{e} + \underset{6+1}{\overset{15-0}{{}^{15}_{7}\mathrm{N}}}$$

Consolidate and apply

- Issue the nuclear equation worksheet at an appropriate level:
  - Low demand: Worksheet 6.3.1 [O1, O2, O3]
  - Standard demand: Worksheet 6.3.2 [O1, O2, O3]
  - High demand: Worksheet 6.3.3 [O1, O2, O3]

Extend

Ask students able to progress further to do the following:

- State and explain why we do not write similar nuclear decay equations for gamma decay. Students should use mass and charge in their explanation.

Plenary suggestions

Learning triangle: At the end of the lesson, ask students to draw a large triangle with a smaller inverted triangle that just fits inside it (so they have four triangles). Ask students to think back over the lesson and identify and write in the outer triangles:

- something they saw
- something they did
- something they discussed.

Ask students to write in the central triangle something they learned.

What do I know? Ask students to each write down one thing about the topic they are sure of, one thing they are unsure of and one thing they need to know more about, being specific. Ask them to work in groups of five or six to agree on group lists; ask each group to say what they decided and agree as a class about what they are confident about, what they are less sure of and what things they want to know more about.

Silent animation: Show students a relevant short video clip without sound and ask them to say what they think is happening. They could make suggestions, or they could give a 'voice over' for the video clip.

Answers to questions

Worksheets 6.3.1–6.3.3

Completed decay chain

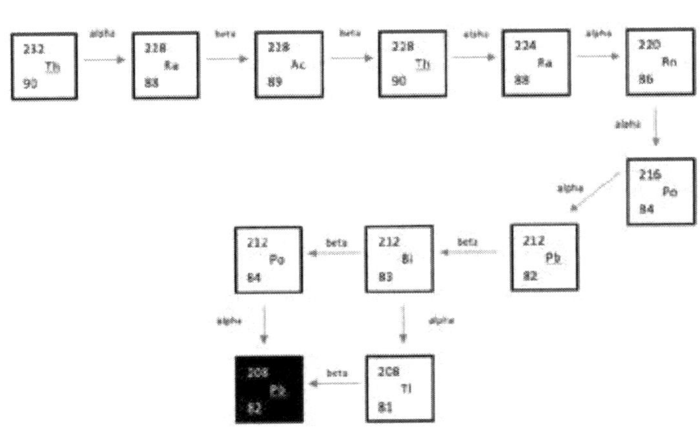

 © HarperCollins*Publishers* Limited 2016

Lesson 6.4: Radioactive half-life

Lesson overview

OCR Specification reference

P6.1j

Learning objectives

- Explain what is meant by radioactive half-life.
- Calculate half-life.
- Choose the best radioisotope for a task.

Learning outcomes

- [O1] Explain the meaning of the 'half-life' of a radioisotope.
- [O2] Determine the half-life of a radioisotope.
- [O3] Suggest a suitable radioisotope for a task using given data.

Skills development

- Develop experimental skills and strategies by modelling radioactive decay.
- Comment on the accuracy of a model.

Maths focus

- Plot a graph of data collected from the modelling experiment.
- Deduce the 'half-life' from the line graph.

Resources needed graph paper, 50 dice (or cubes with sides labelled 1–6) for each group of four students; Worksheets 6.4.1, 6.4.2 and 6.4.3, Practical sheet 6.4, Technician's notes 6.4

Digital resources video clip of a medical tracer being administered to a patient (e.g. https://www.youtube.com/watch?v=7_24jzgLCpw)

Key vocabulary radioisotope, half-life, decay, decay curve

Teaching and learning

Engage

- Show students a video clip of a medical tracer being administered to a patient (e.g. https://www.youtube.com/watch?v=7_24jzgLCpw). [O1, O2]. After the video clip, ask: "Why is it considered safe to use a radioactive element such as technetium in the body as a tracer?" Let students discuss this before taking feedback from a number of pairs of students.

Challenge and develop

- Following a discussion about the relatively safe use of radioactive elements in the body, introduce the concept of the half-life of a radioisotope. Tell the students that, at any given time, there is less than 30 g of astatine in the Earth's crust. Add that astatine decays to bismuth-206 or polonium-210. Ask the students to discuss why there could be such a small amount of astatine in the Earth's crust. [O1, O2]
- Take feedback (developing ideas where appropriate) and lead the discussion into suggesting that the element must have a very short half-life (the most stable isotope of astatine, astatine-210, has a half-life of 8.1 hours). [O1, O2]
- Define the term 'half-life' as the average time it takes for half of the nuclei present to decay, and make it clear that this is a random process: it is not possible to predict which atoms will decay and when. [O2]

Explain

- Students should now model radioactive decay and half-life using dice (Practical sheet 6.4). [O1, O2]

 © HarperCollins*Publishers* Limited 2016

- Students should then be given a half-life worksheet at an appropriate level:
  - Low demand: Worksheet 6.4.1 [O1, O2]
  - Standard demand: Worksheet 6.4.2 [O1, O2]
  - High demand: Worksheet 6.4.3 [O1, O2]

Consolidate and apply

- Display the following data of radioisotopes and their respective half-lives:

Radioisotope	**Half-life**
Polonium-212	3 ms
Protactinium-234	72 s
Technetium-99m	6 hours
Iridium-192	75 days
Cobalt-60	5.3 years
Strontium-90	28 years
Americium-241	460 years
Uranium-238	4500 million years

  - Give the following four uses: students should suggest which is the most suitable radioisotope for each use and explain why: Injecting into the body (technetium-99m – 6 hours); Destroying tumours (iridium-192 – 75 days); Smoke detectors (americium-241 – 460 years); Working out the ages of rocks (uranium-238 – 4500 million years). [O3]

Extend

Ask students able to progress further to do one of the following:

- Research one situation where having a long half-life would be useful (in smoke alarms, for example) and one other situation where a short half-life is essential for the application of the radioisotope. [O2]
- Alternatively, students could be asked to suggest why the half-life curve subtends to, but never reaches, the time axis. [O2]

Answers to questions

Worksheet 6.4.1

1. *y*-axis, count rate (counts per second); *x*-axis, time (minutes)

Worksheets 6.4.2 and 6.4.3

1. See graph

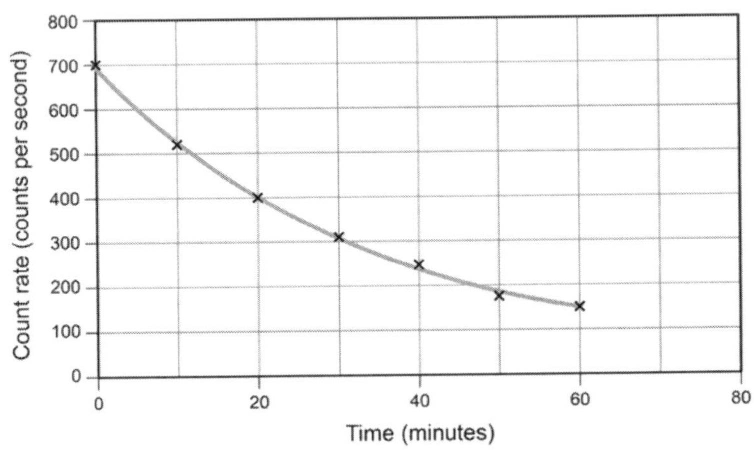

Worksheets 6.4.1 and 6.4.2

2. 25 min
3. 175 counts per second
4. 88 counts per second
5. 25 min

Worksheet 6.4.3

2. 25 min

 © HarperCollins*Publishers* Limited 2016

Lesson 6.5: Background radiation

Lesson overview

OCR Specification reference

P6.1l

Learning objectives

- Recall sources of background radiation.
- Describe how different types of radiation have different ionising power.

Learning outcomes

- [O1] Explain the meaning of background radiation.
- [O2] List different sources of background radiation.
- [O3] Explain why background radiation varies in different areas and at different times.

Skills development

- Analyse data and draw sensible conclusions from the analysis.

Maths focus

- Plot a bar chart from given data.
- Use percentages.

Resources needed graph paper, access to research sources; Worksheets 6.5.1, 6.5.2 and 6.5.3

Digital resources https://www.youtube.com/watch?v=0by4fT00yp4

Key vocabulary background radiation, natural sources, alpha, beta, gamma, penetrating power

Teaching and learning

Engage

- Show a video clip of granite kitchen work surfaces being tested for emission of radioactive radiation (e.g. https://www.youtube.com/watch?v=0by4fT00yp4) and ask students whether they think this is something we should be concerned about. [O1, O2]

Challenge and develop

- Introduce the idea of background radiation by supplying students with data (such as Figure 6.12 on page 216 in the Student Book) and giving them an opportunity to discuss the data with a partner. Ask them to calculate the percentage of background radiation that comes from natural sources. [O2, O3]
- Students should use the percentage data to plot a bar chart showing the sources of radiation.
- As a means of analysing the data, students should write two sentences: one suggesting why background radiation varies in different areas and a second suggesting why it varies with the time of day. [O2, O3]

Explain

- **Pair talk:** Ask: “Should we be concerned about background radiation?” Pair the students up to discuss the question. [O2, O3]
- **Pairs to fours:** Ask the student pairs to share their ideas and come up with their group answer. Take feedback from a number of different groups and discuss ideas as appropriate. [O2, O3]
- Remind students of the differing penetrating and ionising powers of the three types of radiation, either by showing them diagrams or suitable online simulations. Allow them to amend their ideas if appropriate.
- Take answers from each group, prompting for justifications using the key terms where appropriate.

 © HarperCollins*Publishers* Limited 2016

Consolidate and apply

- Students should then be given a background radiation worksheet at an appropriate level. It would be useful if the students had access to some research sources, either online or textbooks, when completing the task on the worksheet.
  - Low demand: Worksheet 6.5.1 [O1, O2, O3]
  - Standard demand: Worksheet 6.5.2 [O1, O2, O3]
  - High demand: Worksheet 6.5.3 [O1, O2, O3]

Extend

Ask students able to progress further to do the following:

- Research the change in levels of cosmic background radiation at increasing altitudes and use the data they collect to justify the precautions taken by pilots and cabin crew in airlines. [O1, O2, O3]

Plenary suggestions

Ask me a question: Ask students to write a question about something from the last four lessons (atomic structure, radioactive decay, nuclear equations and background radiation) and then devise a mark scheme for the answer. Encourage them to come up with questions worth more than one or two marks and to try their questions out on each other.

The big ideas: Ask students to write down three ideas they have learned during the last four lessons. Then ask them to share their facts in groups and to compile a master list of facts, with the most important at the top. Ask for ideas to be shared and find out which other group(s) agreed.

Ideas hothouse: Ask students to work in pairs to list points about what they know about radioactive decay so far. Then ask the pairs to join together into groups of four and then groups of six to eight to discuss this further and to come up with an agreed list of points. Ask one person from each group to report back to the class.

 © HarperCollins*Publishers* Limited 2016

Lesson 6.6: Maths skills: Using ratios and proportional reasoning

Lesson overview

OCR Specification reference

M1c, M3c

Learning objectives

- Draw a curve of best fit to calculate radioactive half-life.
- Calculate the net decline.

Learning outcomes

- [O1] Recall the definition of radioactive half-life.
- [O2] Use a graph of activity against time to determine the half-life of some radioisotopes.
- [O3] Explain how to draw a line of best fit.

Skills development

- Use the correct unit for time and activity.
- Draw a line graph.
- Use a line graph to determine half-life.

Maths focus

- Read scales.
- Plot graphs.
- Draw lines of best fit.
- Use a line of best fit to determine half-life.

Resources needed graph paper

Key vocabulary decay, exponential decay, line of best fit, scale, axes, ratio, net decline

Teaching and learning

Engage

- Issue students with scrap paper and ask them, in no more than 15 words, to write what they understand by the term 'half-life of a radioisotope'. [O1]
- **Pair talk:** Students should share their definitions in pairs, before being given the chance to improve their definition (still in no more than 15 words). [O1]
- Take feedback from a number of pairs, developing ideas as appropriate. [O1]

Challenge and develop

- Ask students to read the section in the Student Book on how to use a graph of activity against time to work out the half-life of a radioisotope. [O2]
- Remind the students of the units for both time and activity.
- Having read the section in the Student Book, students should answer question 1 in the Student Book, before pairing up to check their answers. Any pairs that don't agree should re-read the section and check what they have done, before asking for help. [O2]

 © HarperCollins*Publishers* Limited 2016

Explain

Ask students to discuss why the half-life curve never reaches the time axis, only ever subtending to it (they will need to use their knowledge of radioactive decay from the unit to help them explain this).

Consolidate and apply

- Low demand: Answer questions 2 and 3 from the Student Book. [O2, O3]
- Standard demand: Answer questions 2 and 4 from the Student Book. [O2, O3]
- High demand: Answer questions 2 and 5 from the Student Book. [O2, O3]

Extend

Ask students able to progress further to suggest situations when

- a short half-life of a radioisotope and
- a long half-life of a radioisotope

are necessary for the application they have been chosen for.

Plenary suggestions

Ask students to produce a set of instructions on how to draw a line of best fit and how this is useful when determining the half-life of a radioisotope. [O3]

 © HarperCollins*Publishers* Limited 2016

Lesson 6.7: Hazards and uses of radiation

Lesson overview

OCR Specification reference

P6.2a (part), b, c (part)

Learning objectives

- Describe radioactive contamination.
- Give examples of how radioactive tracers can be used.

Learning outcomes

- [O1] Define radioactive contamination.
- [O2] List the hazards of radioactive contamination and state how contaminated waste can be disposed of.
- [O3] State how radioisotopes can be used as tracers.

Skills development

- Analyse information in an article to answer questions.
- Summarise information.
- Recognise hazard symbols.

Resources needed access to research materials, ideally online; Worksheets 6.7.1, 6.7.2 and 6.7.3

Digital resources http://www.theguardian.com/world/2016/jan/16/alexander-litvinenko-murder-mystery-solved-london-scientists, http://www.bbc.co.uk/news/uk-19647226, access to research materials, ideally online

Key vocabulary radioactive contamination, irradiation, hazard, disposal, tracer

Teaching and learning

Engage

- Display the radiation hazard symbol without the word 'radiation' shown. Ask the students what the symbol is for.
- Ideally, follow this up by showing samples of alpha, beta and gamma sources with the hazard symbol shown.
- Ask the students why there has to be a radiation officer in each school and why radiation sources have to be kept in a locked cabinet clearly labelled as containing radioactive sources. [O1, O2]

Challenge and develop

- Show students a video clip of hospital workers disposing of hospital waste – why do they have to be so careful? Take feedback to develop ideas about radioactive contamination. [O2]
- Ask students why the disposal sites (such as bins) are always marked with the radiation hazard symbol. [O2]

Explain

- Issue each student with a hazards of radiation worksheet at an appropriate level. The students will research the case of Alexander Litvinenko. (Useful links: http://www.theguardian.com/world/2016/jan/16/alexander-litvinenko-murder-mystery-solved-london-scientists, http://www.bbc.co.uk/news/uk-19647226)
    - Low demand: Worksheet 6.7.1 [O1, O2]
    - Standard demand: Worksheet 6.7.2 [O1, O2]
    - High demand: Worksheet 6.7.3 [O1, O2]

Consolidate and apply

- Students should then be given an opportunity to research a particular use as a tracer of radioisotopes. [O3]

 © HarperCollins*Publishers* Limited 2016

- **Jigsaw:** In 'home' groups of four or five, students should be allocated one of the following uses of radioisotopes as tracers: detecting leaks or blockages in underground pipes; tracking the dispersal of waste; monitoring the uptake of fertilisers in plants; checking for blockages in a patient's blood vessel. They should then be regrouped into 'expert' groups, where they research together on their chosen area. They should then return to their 'home' group to report back on their area of expertise. [O3]
- The 'home' group should then be asked to complete a table with the headings 'Use of radioisotope' and 'How it works'. [O3]

Extend

Ask students able to progress further to do the following:

- Explain why people administering X-rays wear a badge containing some photographic film.

Plenary suggestions

Ideas hothouse: Ask students to work in pairs to list points about what they know about a particular idea. Then ask the pairs to join together into groups of four and then groups of six to eight to discuss this further and to come up with an agreed list of points. Ask one person from each group to report back to the class.

Ask me a question: Ask students to write a question about something from the topic and then a mark scheme for the answer. Encourage them to come up with questions worth more than one or two marks and to try their questions out on each other.

Where's the answer? As students enter the room, give them a card with a word written on it. At the end of the lesson, ask a question, and the students should say if they think they have the answer card and explain why.

Answers to questions

Worksheet 6.7.1

1. polonium (specifically, polonium-210)
2. London
3. They put the polonium in his cup of tea.
4. 1 November 2006
5. vomiting, hair loss and blistering in the mouth
6. The radioisotope was an alpha emitter and so the radiation could not penetrate the body (alpha radiation is stopped just by a layer of skin).
7. A silver disc attracted the polonium (polonium-210).
8. a nuclear weapons assembly plant at Avangard in Russia

 © HarperCollins*Publishers* Limited 2016

Lesson 6.8: Irradiation

Lesson overview

OCR Specification reference

P6.2a (part)

Learning objectives

- Explain what is meant by irradiation.
- Understand the distinction between contamination and irradiation.
- Appreciate the importance of communication between scientists.

Learning outcomes

- [O1] Define irradiation.
- [O2] List the possible effects of irradiation on body cells.
- [O3] Compare and contrast irradiation and contamination.

Skills development

- Appreciate the importance of communication between scientists.
- Plot a bar chart from given data.
- Draw conclusions from data.

Maths focus

- Plot a bar chart, using appropriate (and more complex) scales.

Resources needed graph paper, Student Book (or other suitable textbooks); Worksheets 6.8.1, 6.8.2 and 6.8.3

Digital resources video clip of food being irradiated to kill bacteria

Key vocabulary irradiation, mutation, contamination, peer review

Teaching and learning

Engage

- Ask students: "Would you eat food that had been exposed to nuclear radiation?"
- Ask students to discuss their initial thoughts before taking feedback. [O1]

Challenge and develop

- Follow the initial discussion by showing the students a video clip of food being irradiated to kill bacteria.
- Ask students to consider how else our bodies are irradiated from different sources every day by plotting a bar chart of the following data: (suggested scale of 1×10^5 on the *y*-axis to make plotting a little easier; *y*-axis: dose each hour, *x*-axis: source): [O1, O2]
  - **From the sky:** About 400 000 cosmic rays pass through us each hour.
  - **From the air:** About 30 000 atoms of radioactive gases breathed in disintegrate in our lungs each hour.
  - **From food:** About 15 million potassium-40 atoms disintegrate inside our bodies each hour.
  - **From soil and building materials:** More than 200 million gamma rays pass through us each hour.

Explain

- Students should now consider the difference between contamination (e.g. Alexander Litvinenko) and irradiation. They should work in pairs; one of the pair should research the definition of contamination while the other should find out what irradiation means. [O3]

 © HarperCollins*Publishers* Limited 2016

- They should then discuss their research findings and try to identify the distinction between the two. [O3]
- Take feedback from the group, developing ideas where necessary. A useful comparison can be made with irradiation and turning a tap on and off. When the tap is turned on, we see water flowing. When it is turned off, the water no longer appears. The same can be said for irradiation, in that the substance irradiated does not remain radioactive (note: however, cells can be permanently damaged if unrepaired). When a substance is contaminated with ionising radiation, it remains radioactive. [O3]
- Discuss with students why they think it is important for the findings of studies into the effects of radiation on human beings to be published and reviewed by the scientific community.

Consolidate and apply

Issue each student with an irradiation worksheet at an appropriate level:

- Low demand: Worksheet 6.8.1 [O1, O2, O3]
- Standard demand: Worksheet 6.8.2 [O1, O2, O3]
- High demand: Worksheet 6.8.3 [O1, O2, O3]

Extend

Ask students able to progress further to do the following:

- Explain why irradiation could have an effect on somebody's grandchildren. Link here to Marie Curie and her daughter.

Plenary suggestions

Learning triangle: At the end of the lesson, ask students to draw a large triangle with a smaller inverted triangle that just fits inside it (so they have four triangles). Ask students to think back over the lesson and identify and write in the outer triangles:

- something they saw
- something they did
- something they discussed.

Ask students to write in the central triangle something they learned.

Freeze frame: Ask students to create a 'freeze frame' of an idea in the topic. This involves three or four students arranging themselves as a static image, and other students suggesting what it represents.

What do I know? Ask students to each write down one thing about the topic they are sure of, one thing they are unsure of and one thing they need to know more about, being specific. Ask them to work in groups of five or six to agree on group lists; ask each group to say what they decided and agree as a class about what they are confident about, what they are less sure of and what things they want to know more about.

Answers to questions

Worksheets 6.8.1, 6.8.2 and 6.8.3

Key terms

Irradiation	Exposing an object to nuclear radiation
Sterilisation	The use of gamma rays to kill bacteria
Mutation	Changing of DNA
Contamination	The unwanted presence of materials containing radioactive atoms
Misrepair	A cell is repaired incorrectly in the body
Accurate repair	A cell is repaired correctly in the body

 © HarperCollins*Publishers* Limited 2016

Lesson 6.9: Uses of radiation in medicine

Lesson overview

OCR Specification reference

P6.2c

Learning objectives

- Compare gamma rays and X-rays.
- Describe some uses of nuclear radiation for medical diagnosis and therapy.

Learning outcomes

- [O1] State how radiation can be used in medicine.
- [O2] Compare the use of gamma rays and X-rays in medicine.
- [O3] Describe how radiation can be used to destroy tumours.

Skills development

- Research efficiently and record findings clearly.
- Use their research to inform further work.

Resources needed X-ray films (if possible), range of research sources (ideally books and online); Worksheets 6.9.1, 6.9.2, 6.9.3 and 6.9.4

Digital resources research sources; video clip of someone having an X-ray scan, image or video clip of someone having a CAT scan

Key vocabulary toxic, radiotherapy, tumour, X-rays, gamma rays

Teaching and learning

Engage

- Show students some X-ray films (if possible). Alternatively, show a video clip of someone having an X-ray scan. [O1]
- In addition, the students could be shown an image or video clip of someone having a CAT scan. (Be sure to make clear to students that X-rays are used in this type of scan as well as the more familiar 'X-ray'.)

Challenge and develop

- Ask students to discuss how they think the image on the film is produced (rays pass through soft tissue but are stopped by bone, which is denser, so soft tissue shows up as black (or grey), whereas bone shows up much whiter). [O1, O2]

Explain

- Students should now be given time to research the uses of both X-rays and gamma rays and complete the table on Worksheet 6.9.1. [O1, O2]
- They could do this individually, or they could work as a pair, with one person researching X-rays and the other gamma rays; students then explain their research to each other.

Consolidate and apply

Issue each student with a uses of radiation in medicine worksheet at an appropriate level. They may need further access to research sources. The guides could be produced using a word processing package.

- Low demand: Worksheet 6.9.2 [O1, O2, O3]
- Standard demand: Worksheet 6.9.3 [O1, O2, O3]
- High demand: Worksheet 6.9.4 [O1, O2, O3]

 © HarperCollins*Publishers* Limited 2016

Extend

Ask students able to progress further to do the following:

- Research proton beam therapy as an alternative to traditional radiotherapy.

Plenary suggestions

The big ideas: Ask students to write down three ideas they have learned during the last three lessons. Then ask them to share their facts in groups and to compile a master list of facts, with the most important at the top. Ask for ideas to be shared and find out which other group(s) agreed.

Heads and tails: Ask each student to write a question about something from the topic on a coloured paper strip and the answer on another coloured strip. Hand out the strips to groups of six to eight students, so that each student gets a question and an answer. One student reads out his or her question. The student with the correct answer then reads it out, followed by his or her question, and so on.

Hot seat: Ask each student to think of a question, using material from the topic. Select someone to put in the hot seat; ask students to ask their questions and say at the end whether the answer is correct or incorrect.

 © HarperCollins*Publishers* Limited 2016

Lesson 6.10: Using nuclear radiation

Lesson overview

OCR Specification reference

P6.2c

Learning objectives

- Explore the risks and benefits of using nuclear radiation.
- Describe how internal organs can be explored.
- Understand how nuclear radiation can control or destroy unwanted tissue.

Learning outcomes

- [O1] State how radiation can be used to destroy tumours.
- [O2] List some of the potential risks of using nuclear radiation.
- [O3] Evaluate the risk by considering the radiation dose.

Skills development

- Research and present findings clearly.
- Draw a conclusion, having considered benefits and risks.

Resources needed suitable image of a tumour, Student Book (or other suitable textbook); Worksheets 6.10.1, 6.10.2 and 6.10.3

Digital resources
https://www.youtube.com/watch?v=jmiaBPf0TkY, https://www.youtube.com/watch?v=JKxvs2FMCzo

Key vocabulary benefit, risk, radiation dose, radiotherapy

Teaching and learning

Engage

- Begin by showing students a suitable image of a tumour. Be mindful of students who have experience of cancer or who may be sensitive to medical images. Ask the students what they think it is an image of. Extend this by asking them how they think tumours are formed. [O1]
- Follow this by playing a video clip of how radiotherapy works, e.g. https://www.youtube.com/watch?v=jmiaBPf0TkY. [O1, O2]
- After the video, ask students to discuss why they think the beam of radiation rotates around the tumour during the radiotherapy session. Elicit the idea that the radiation will destroy both the cancerous and non-cancerous cells, and that by rotating the beam, there will be minimal exposure to healthy cells while concentrating the beam on the tumour in the centre. [O1, O2, O3]

Challenge and develop

- Play the video of treating a cancerous tumour in a horse: https://www.youtube.com/watch?v=JKxvs2FMCzo.
- During the video, students should keep a tally of the number of times the following words are mentioned: isotope, tumour, radiation, dose and tracer.
- Play the video through, stopping it as appropriate for the class to discuss the ideas that are introduced through the video (it recaps the unit so far very nicely, as well as covering the objectives here). [O1, O2, O3]

Explain

Students should then be asked to work through the questions in the Student Book at an appropriate level:

- Low demand: Questions 1, 2 and 3 [O1, O2]
- Standard demand: Questions 1, 3 and 5 [O1, O2, O3]

 © HarperCollins*Publishers* Limited 2016

- High demand: Questions 1, 5, 6 and 7 [O1, O2, O3]

Consolidate and apply

Issue each student with a using nuclear radiation worksheet at an appropriate level:

- Low demand: Worksheet 6.10.1 [O1, O2, O3]
- Standard demand: Worksheet 6.10.2 [O1, O2, O3]
- High demand: Worksheet 6.10.3 [O1, O2, O3]

Extend

Ask students able to progress further to do the following:

- Research the use of radioactive isotopes in dating rocks.

Plenary suggestions

Learning triangle: At the end of the lesson, ask students to draw a large triangle with a smaller inverted triangle that just fits inside it (so they have four triangles). Ask students to think back over the lesson and identify and write in the outer triangles:

- something they saw
- something they did
- something they discussed.

Ask students to write in the central triangle something they learned.

Ideas hothouse: Ask students to work in pairs to list points about what they know about using nuclear radiation in medicine. Then ask the pairs to join together into groups of four and then groups of six to eight to discuss this further and to come up with an agreed list of points. Ask one person from each group to report back to the class.

Hot seat: Ask each student to think of a question, using material from the topic. Select someone to put in the hot seat; ask students to ask their questions and say at the end whether the answer is correct or incorrect.

 © HarperCollins*Publishers* Limited 2016

Lesson 6.11: Nuclear fission

Lesson overview

OCR Specification reference

P6.2d

Learning objectives

- Describe nuclear fission.
- Explain how a chain reaction occurs.
- Explain how fission is used.

Learning outcomes

- [O1] Define nuclear fission.
- [O2] Describe nuclear fission using a nuclear equation.
- [O3] Describe, using a nuclear equation, how an uncontrolled nuclear fission chain reaction can result in a huge release of energy.

Skills development

- Balance nuclear equations.

Maths focus

- Appreciate the magnitude of energy provided by uranium.

Resources needed Worksheets 6.11.1, 6.11.2 and 6.11.3

Digital resources means of projection, animation of nuclear fission (e.g. https://phet.colorado.edu/en/simulation/legacy/nuclear-fission), *Chain Reaction* by Diana Ross

Key vocabulary graphite, moderator, control rods, chain reaction, enriched uranium, fuel rods, nuclear fission, neutron radiation

Teaching and learning

Engage

- As the students enter the room, play the song *Chain Reaction* by Diana Ross.
- Display a picture of 1 kg of uranium and ask students: "How much coal would we need to provide the same amount of energy as 1 kg of uranium?"
- Take ideas from the students, before telling them that we would need to burn 2 million kg of coal to get the same amount of energy. It may be a good idea to write the number as 2 000 000 kg to ensure that students fully understand the magnitude of this number. [O1]

Challenge and develop

- **Pair talk:** Students discuss how they think the energy in uranium is released (it isn't burned, like the coal).
- **Pairs to fours:** Ask pairs of students to share ideas before putting their 'best' idea forward to a class discussion. Write all ideas on the board for the students to see. [O1, O2]

Explain

- Show the students a suitable animation of nuclear fission, e.g. https://phet.colorado.edu/en/simulation/legacy/nuclear-fission.
- Be sure to distinguish between nuclear fission and radioactive decay. Ensure that students are clear that, unlike radioactive decay, nuclear fission is rarely spontaneous. The unstable nucleus of uranium must absorb an extra neutron before it splits into two daughter nuclei.

 © HarperCollins*Publishers* Limited 2016

- Be sure to emphasise how an uncontrolled nuclear fission chain reaction can result in a huge release of energy. [O1, O2, O3]
- Issue each student with a nuclear fission worksheet at an appropriate level:
  - Low demand: Worksheet 6.11.1 [O1, O2, O3]
  - Standard demand: Worksheet 6.11.2 [O1, O2, O3]
  - High demand: Worksheet 6.11.3 [O1, O2, O3]

Consolidate and apply

- Students should be given a picture of a gas-cooled nuclear reactor without any labels.
- They should use this template to create a model nuclear reactor out of coloured paper. Once they have done this, they should use a suitable textbook to label the parts of the reactor with both the names of the parts and their respective functions. They could do this using white sticky labels or make flags out of cocktail sticks and labels. [O1, O2]
- Alternatively, the students could be asked to draw up a table comparing and contrasting nuclear power stations and coal-fired power stations.

Extend

Ask students able to progress further to do one of the following:

- Suggest how the waste from the nuclear reactors is disposed of.
- Alternatively, students could be asked to consider the advantages and disadvantages of living in close proximity to a nuclear power station.

Plenary suggestions

Learning triangle: At the end of the lesson, ask students to draw a large triangle with a smaller inverted triangle that just fits inside it (so they have four triangles). Ask students to think back over the lesson and identify and write in the outer triangles:

- something they saw
- something they did
- something they discussed.

Ask students to write in the central triangle something they learned.

Ideas hothouse: Ask students to work in pairs to list points about what they know about using nuclear radiation in industry and medicine. Then ask the pairs to join together into groups of four and then groups of six to eight to discuss this further and to come up with an agreed list of points. Ask one person from each group to report back to the class.

Hot seat: Ask each student to think of a question, using material from the lesson. Select someone to put in the hot seat; ask students to ask their questions and say at the end whether the answer is correct or incorrect.

 © HarperCollins*Publishers* Limited 2016

Lesson 6.12: Nuclear fusion

Lesson overview

OCR Specification reference

P6.2e

Learning objectives

- Explain nuclear fusion.
- Describe the conditions needed for fusion.
- Describe how nuclear fusion might be an attractive energy source.

Learning outcomes

- [O1] Define nuclear fusion.
- [O2] Describe nuclear fusion using a nuclear equation.
- [O3] Explain the potential benefits of nuclear fusion.

Skills development

- Research the work of scientists and consider the importance of sharing results and attempting reproducibility.
- Consider the implications of scientific discoveries.

Maths focus

- Balance nuclear equations.

Resources needed Worksheet 6.12

Digital resources means of projection, video about the birth of a star and nuclear fusion, e.g. https://www.youtube.com/watch?v=9EnBBIx6XkM, animation for nuclear fusion, e.g. http://www.atomicarchive.com/Movies/Movie5.shtml

Key vocabulary fusion, cold fusion, hydrogen bomb

Teaching and learning

Engage

- Begin by showing students a video about the birth of a star and nuclear fusion.
  Suggested clip: https://www.youtube.com/watch?v=9EnBBIx6XkM
- After the video, ask students: "What if we could find some way of harnessing this amount of energy safely on Earth?" Ask students to discuss their thoughts. [O1]

Challenge and develop

- Tell the students that, so far, we have not been able to produce the necessary conditions for nuclear fusion on Earth, but that the International Thermonuclear Experimental Reactor (ITER) is scheduled to be operational in the late 2020s. [O1]
- **Pair talk:** Ask how this could change the way we use energy in the future. [O1, O3]
- Follow up the idea of nuclear fusion in the video and ask the students to compare and contrast the processes of fusion and fission. This would best be done using a table, with similarities and differences between the processes given. (Similarities: both processes produce large amounts of energy; both involve the nuclei of atoms. Differences: fusion is the joining of nuclei, whereas fission is the splitting of nuclei; fission requires a neutron to begin the process; fusion requires very high temperatures and pressures.) [O1, O2]

Explain

- Show the students a suitable animation of nuclear fusion, e.g. http://www.atomicarchive.com/Movies/Movie5.shtml. [O2]

 © HarperCollins*Publishers* Limited 2016

- Issue students with the nuclear fusion worksheet (Worksheet 6.12). [O1, O2, O3]

Consolidate and apply

- Low demand: Students should design a one-page summary comparing and contrasting the processes of nuclear fission and nuclear fusion. They should be encouraged to use nuclear equations as well as diagrams to represent both processes. [O1, O2]
- Standard demand: Students should answer Questions 1 to 6 from the Student Book. Alternatively, they should research and then design a one-page sheet explaining how a hydrogen bomb works, being sure to include a reference to the fission needed to provide the high temperatures for the fusion to occur. [O1, O2]
- High demand: Students should research the cold fusion work of Pons and Fleischmann in 1989 and produce a one-page summary of their work, the experimental details, the fact that the results have yet to be reproduced and the potential implications of this work. [O1, O3]

Extend

Ask students able to progress further to do the following:

- Suggest the advantages of producing energy using fusion in preference to fission in the future. [O3]

Plenary suggestions

What's in the picture? Have a picture of either fusion or fission concealed by rectangles, each of which can be removed in turn. Ask students to suggest which tile should be removed next and what the partially revealed graphic shows. Encourage speculation and inference. Continue removing one tile at a time, until the picture is revealed and an explanation of the picture can be given.

The big ideas: Ask students to write down three ideas they have learned during the topic. Then ask them to share their facts in groups and to compile a master list of facts, with the most important at the top. Ask for ideas to be shared and find out which other group(s) agreed.

What do I know? Ask students to each write down one thing about the topic they are sure of, one thing they are unsure of and one thing they need to know more about, being specific. Ask them to work in groups of five or six to agree on group lists; ask each group to say what they decided and agree as a class about what they are confident about, what they are less sure of and what things they want to know more about.

Answers to questions

Worksheet 6.12

1.

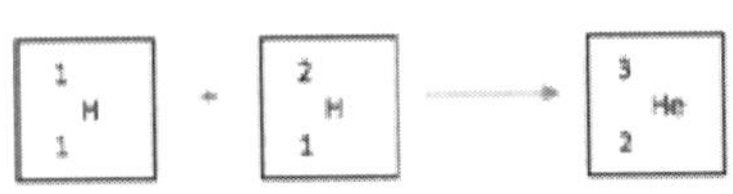

2.

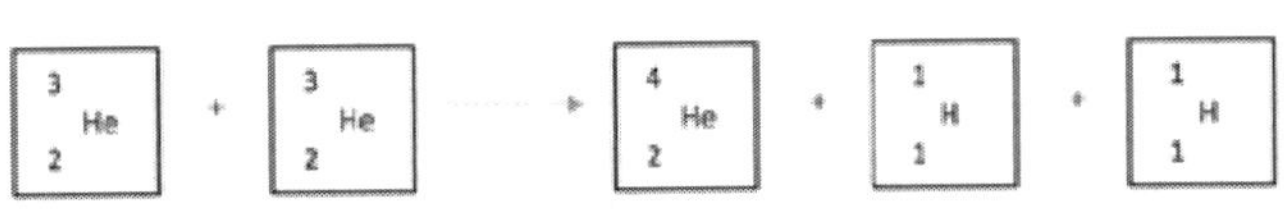

3. in the formation of stars
4. high pressure, high temperature (in either order)
5. Similarities: both produce energy, both involve nuclei
   Differences: fusion involves the joining of nuclei whereas fission involves the splitting of a nucleus; fusion needs heat and pressure to initiate the reaction whereas fission requires neutron bombardment
6. The waste products of fusion (mainly helium) are not radioactive (although it is likely that parts of the reactor would be contaminated because of the high energies used).

 © HarperCollins*Publishers* Limited 2016

When and how to use these pages: Check your progress, Worked example and End of chapter test

Check your progress

Check your progress is a summary of what students should know and be able to do when they have completed the chapter. Check your progress is organised in three columns to show how ideas and skills progress in sophistication. Students aiming for top grades need to have mastered all the skills and ideas articulated in the final column (shaded pink in the Student book).

Check your progress can be used for individual or class revision using any combination of the suggestions below:

- Ask students to construct a mind map linking the points in Check your progress
- Work through Check your progress as a class and note the points that need further discussion
- Ask the students to tick the boxes on the Check your progress worksheet (Teacher Pack CD). Any points they have not been confident to tick they should revisit in the Student Book.
- Ask students to do further research on the different points listed in Check your progress
- Students work in pairs and ask each other what points they think they can do and why they think they can do those, and not others

Worked example

The worked example talks students through a series of exam-style questions. Sample student answers are provided, which are annotated to show how they could be improved.

- Give students the Worked example worksheet (Teacher Pack CD). The annotation boxes on this are blank. Ask students to discuss and write their own improvements before reviewing the annotated Worked example in the Student Book. This can be done as an individual, group or class activity.

End of chapter test

The End of chapter test gives students the opportunity to practice answering the different types of questions that they will encounter in their final exams. You can use the Marking grid provided in this Teacher Pack or on the CD Rom to analyse results. This shows the Assessment Objective for each question, so you can review trends and see individual student and class performance in answering questions for the different Assessment Objectives and to highlight areas for improvement.

- Questions could be used as a test once you have completed the chapter
- Questions could be worked through as part of a revision lesson
- Ask Students to mark each other's work and then talk through the mark scheme provided
- As a class, make a list of questions that most students did not get right. Work through these as a class.

 © HarperCollins*Publishers* Limited 2016

Marking Grid for End of Chapter 6 Test

	Getting Started [Foundation Tier]						Going Further [Foundation and Higher Tiers]					More Challenging [Higher Tier]					Most demanding [Higher Tier]			Total marks	Percentage
Student Name	Q. 1 (AO1) 1 mark	Q. 2 (AO1) 2 marks	Q. 3 (AO1) 1 mark	Q. 4 (AO1) 2 marks	Q. 5 (AO1) 2 marks	Q. 6 (AO2) 2 marks	Q. 7 (AO1) 1 mark	Q. 8 (AO1) 1 mark	Q. 9 (AO2) 2 marks	Q. 10 (AO1) 2 marks	Q. 11 (AO2) 4 marks	Q. 12 (AO1) 1 mark	Q. 13 (AO2) 1 mark	Q. 14 (AO2) 2 marks	Q. 15 (AO2) 2 marks	Q. 16 (AO3) 4 marks	Q. 17 (AO1) 2 marks	Q. 18 (AO3) 4 marks	Q. 19 (AO2) 4 marks		

© HarperCollins*Publishers* Limited 2016

Check your progress

You should be able to:

State that the number of protons in an element is the atomic number and the total number of protons and neutrons is the mass number →	Understand that atoms of an element all have the same number of protons but can have different numbers of neutrons, giving different isotopes →	Use nuclear notation to show subatomic particles in an isotope
Recognise that some isotopes called radioisotopes are unstable and decay →	List some uses of radioisotopes in medicine →	Describe how specific radioisotopes are used medicine
Recognise that radioisotopes have a half-life →	Explain the meaning of half-life of a radioisotope →	Calculate the half-life of a radioisotope
List the three types of ionising radiation resulting from nuclear activity →	Describe the structure of each type of ionising radiation →	Explain the properties of each type of radiation
Explain the meaning of background radiation →	List different sources of background radiation →	Explain why background radiation varies in different areas and in different times
Recognise the symbols used in a nuclear equation →	Write nuclear equations involving alpha and beta decay →	Write balanced nuclear equations for different types of nuclear reaction
Define radioactive →	List the hazards of radioactive contamination →	Compare and contrast irradiation and contamination
Define nuclear fission →	Describe nuclear fission using a nuclear equation →	Describe, using a nuclear equation, how an uncontrolled nuclear fission chain reaction can result in a huge release of energy
Define nuclear fusion →	Describe nuclear fusion using a nuclear equation →	Explain the potential benefits of nuclear fusion

 © HarperCollins*Publishers* Limited 2016

Worked example

1 What is an alpha particle?

A helium atom

> Not strictly correct. It is a helium *nucleus*. It would also be helpful to give more detail of its characteristics like its charge, mass, ionising power and penetrating power.

2 Explain what you understand by the half-life of an radioisotope.

The half-life is the time it takes to lose half its size

> A good start, but it needs to be clearer and fuller in the explanation. The answer should refer to the activity of the radioisotope, not Its size. And don't forget to mention average time.

3 What is the difference between irradiation and contamination?

Contamination is where materials become radioactive. Irradiation is where an object is exposed to nuclear radiation, but it does not become radioactive itself.

> Good answer. To improve your mark you could give examples of radioactive contamination and irradiation.

4 An experiment was carried out where the activity of a radioisotope was measured over time. The results are shown on the graph.

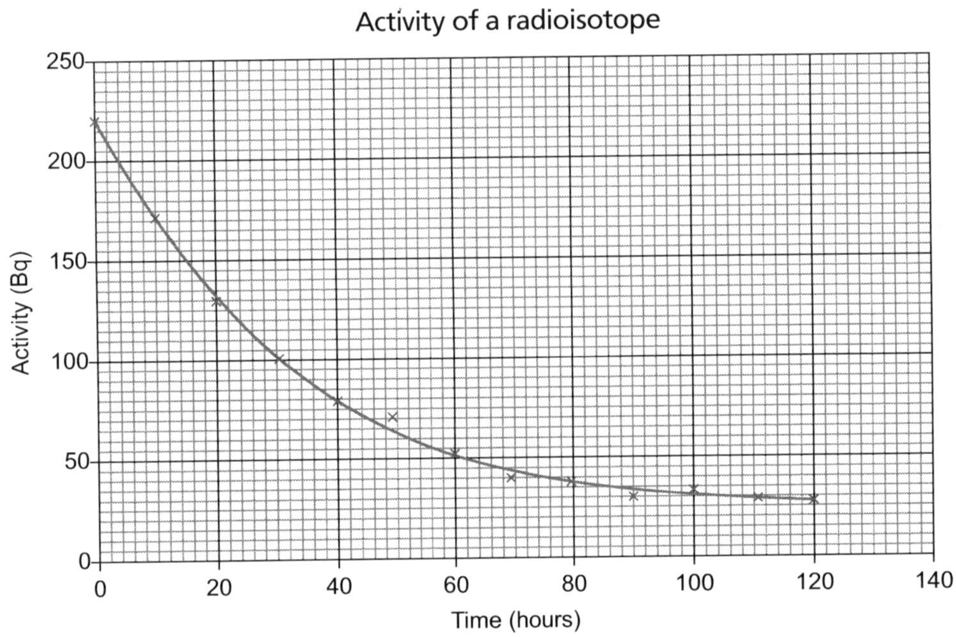

a An adjustment to the readings has not been made. Explain what it is.

The graph has levelled out above the axis. It looks like the background count has not been subtracted.

> This is a good answer.

b Explain how you would work out the half-life from the graph once the adjustment had been made

Pick a value, say 100 Bq and halve it, 50 Bq

Draw a line from 100 Bq on the vertical axis across to the graph, then down to the time axis.

Repeat for 50 Bq. Subtract the first time from the second time to get the half-life.

> This gives a good answer for calculating one value of the half-life, but you should really calculate at least one more value from another part of the graph, e.g. 150 Bq and 75 Bq and calculate the average.

 © HarperCollins*Publishers* Limited 2016

Energy: Introduction

When and how to use these pages

This unit, as the name implies, is all about energy. Nothing in the Universe can happen without a transfer of energy; gaining an understanding of energy will lead to an enhanced understanding of the Universe, how things work, why things are as they are and indeed life as we know it. This unit ingrains the idea that energy can be determined by measurements and that an appreciation of energy can help us make informed decisions about our uses of energy and their consequences. These pages are designed either for following fully or for dipping into for ideas and inspiration, including practical work, or somewhere in between.

Overview of the unit

In this unit, students will learn about the many different types of energy. They will look at transfers between different types of energy, especially elastic potential energy, gravitational potential energy and kinetic energy. Students will look at work done and power and how they are useful in different ways. Various units will be introduced, including the joule and the watt. Students will also look at temperature and how this relates to energy. Within this context, they will recognise that different substances have different specific heat capacities and the consequences of this, e.g. causing the wind between the land and the sea. Students will use the law of conservation of energy and will look at both useful and waste energies. They will learn about efficiency; not just how to calculate it but also the importance of improving the efficiency of transducers and various ways in which this can be done. Students will look at the different energy resources of Earth that are used, particularly in terms of generating electricity. They will explore the advantages and disadvantages of the different resources and how our use of such resources may change in the future, both at a local and a global scale. Students will begin to be able to make informed decisions about their own energy use in the future. Within this topic, there are two core practicals, alongside many others, that provide opportunities for students to become more familiar with the scientific process, including developing and testing a hypothesis and processing, analysing and evaluating results to reach suitable conclusions.

Obstacles to learning

Students may need extra guidance with the following terms and concepts:

- Energy itself is a tricky concept. It is often used incorrectly in everyday speech – "I have no energy today"; the use of models and experiment can be very useful in helping with this, especially in terms of being able to understand what is actually happening.
- Some of the numbers with energy can be very large or very small. Not only are these numbers difficult to understand, they can be tricky to use in calculations. It is easier to try the calculations first with easier numbers and then with trickier numbers. Indeed, this can be a good tip even when easier numbers are not provided. Sometimes students find it easier to make up easy numbers where they can easily see what to do and then apply the same logic with the more difficult numbers.
- Temperature and thermal (heat) energy are often confused by students. It may be useful to establish clearly that temperature is not an energy, although it is related to the average kinetic energy of the particles, whereas thermal (or heat) energy is the energy that is transferred when temperature changes.
- The idea that certain energy resources can lead to global warming and its implications can be alarming for some students. They need to understand that we are not past a tipping point and that we can all make positive changes for the better.
- The whole debate on energy resources is one that people can have strikingly polar views on so needs to be presented and coordinated to ensure that all views on this can be challenged but respected.

Practicals in this unit

In this unit, students will do the following practical work:

- investigate converting potential energy into kinetic energy then back into gravitational kinetic energy
- simulate meteorite strikes and determine the energy of a 'meteorite' and the force on it from the ground

 © HarperCollins*Publishers* Limited 2016

- investigate temperature changes when land and water absorb and release heat
- investigate the efficiency of different lighting appliances
- investigate the factors that affect cooling of hot water in a beaker
- Practical: investigate cooling curves for different thicknesses of materials
- use an analogy to understand the energy in the universe
- use an energy circus to describe different energy transfers

	Lesson title	**Overarching objectives**
1	Investigating kinetic energy	• Describe how the kinetic energy store of an object changes as its speed changes. • Calculate kinetic energy. • Consider how energy is transferred.
2	Work done and energy transfer	• Understand what is meant by work done. • Explain the relationship between work done and force applied. • Identify the transfers between energy stores when work is done against friction.
3	Specific heat capacity	• Understand how things heat up. • Find out about heating water. • Find out more about specific heat capacity.
4	Electrical energy and power	• Recall that energy cannot be created or destroyed, only transferred. • Describe the energy transfers in different domestic appliances. • Describe power as a rate of energy transfer. • Calculate the energy transferred.
5	Dissipation of energy	• Explain ways of reducing unwanted energy transfer. • Describe what affects the rate of cooling of a building. • Understand that energy is dissipated.
6	Energy efficiency	• Explain what is meant by energy efficiency. • Calculate the efficiency of energy transfers. • Find out about conservation of energy.
7	Practical: Investigating ways of reducing the unwanted energy transfers in a system	• Use scientific ideas to make predictions. • Analyse data to identify trends. • Evaluate an experimental procedure.
8	Key concept: Energy transfer	• Understand why energy is a key concept in science. • Use ideas about stores and transfers to explain what energy does. • Understand why accounting for energy transfers is a useful idea.
9	Maths skills: Calculations using significant figures	• Substitute numerical values into equations and use appropriate units. • Change the subject of an equation. • Give an answer using an appropriate number of significant figures.

 © HarperCollins*Publishers* Limited 2016

Lesson 7.1: Investigating kinetic energy

Lesson overview

OCR Specification reference

P7.1b (part), e (part)

Learning objectives

- Describe how the kinetic energy store of an object changes as its speed changes.
- Calculate kinetic energy.
- Consider how energy is transferred.

Learning outcomes

- [O1] Define kinetic energy.
- [O2] Calculate using the kinetic energy equation.
- [O3] Describe changes to and from kinetic energy.

Skills development

- Apply a knowledge of a range of techniques, instruments, apparatus and materials to select those appropriate to the experiment.
- Evaluate methods and suggest possible improvements and further investigations.
- Use SI units (for example, grams, metres, joules) and IUPAC chemical nomenclature unless inappropriate.

Maths focus

- Recognise and use expressions in standard form.
- Change the subject of an equation.
- Substitute numerical values into algebraic equations using appropriate units for physical quantities.
- Solve simple algebraic equations.

Resources needed magnetic linear accelerator (Gauss gun): [strong neodymium magnets, steel ball bearings, plastic channel, attached curved plastic ramp], clamp stand, boss and clamp, metre rule, mass balance; Worksheets 7.1.1, 7.1.2 and 7.1.3, Practical sheets 7.1.1 and 7.1.2, Technician's notes 7.1

Digital resources kinetic energy video of a front seat view of the Kingda Ka rollercoaster, kinetic energy and potential energy animation

Key vocabulary energy store, kinetic energy, energy transfer, gravitational potential energy

Teaching and learning

Engage

- Watch the kinetic energy video of a front seat view of the Kingda Ka rollercoaster. During the video, the students should think about, and note down their answers to the following two questions: What is kinetic energy? Where did the roller coaster have the most kinetic energy? Discuss the answers as a class.
- Following the video, pose the question "Where did the kinetic energy come from?" to the class and allow them to discuss it in pairs before taking feedback. Develop this further by discussing which two factors affect the kinetic energy of an object.
- Discuss factors that affect the kinetic energy of a body, leading to the kinetic energy equation, $E_k = ½ mv^2$. Formula triangles are trickier when there are more than three quantities, but show that it is possible with E_k at the top. Emphasise that in order to determine v, students should use the triangle to find v^2, then take the square root. Note: students will be expected to remember this equation in the examination. [O1, O2]

Challenge and develop

- Carry out practical. Demonstrate the set-up first. [O1, O2, O3]
  - Low demand: Use Practical sheet 7.1.2. Guide students through the process carefully. Note that they will not need to convert mass from grams to kilograms or height from centimetres to metres.
  - Standard demand: Use Practical sheet 7.1.1. Guide students in converting mass from grams to kilograms and height from centimetres to metres.
  - High demand: Students do Practical sheet 7.1.1 independently.

 © HarperCollins*Publishers* Limited 2016

- Discuss how the practical could have been improved, first in pairs, then groups of four, then groups of eight. Finally, someone in each group shares ideas for improvement with the class. [O1, O2, O3]
- Show the kinetic and potential energy animation to highlight the energy transfers. Link back to the practical and discuss how it relates to magnetic potential energy, kinetic energy and gravitational potential energy. [O1, O2, O3]

Explain

- Challenge students' thinking and understanding of kinetic energy and potential energy from the practical activities using the evaluation sections of the practical sheets (these sections are identical). [O1, O2, O3]

Consolidate and apply

The students should then be issued with the investigating kinetic energy worksheet appropriate to their ability:

- Low demand: Worksheet 7.1.1 [O1, O2, O3]
- Standard demand: Worksheet 7.1.2 [O1, O2, O3]
- High demand: Worksheet 7.1.3 [O1, O2, O3]

Extend

Ask students able to progress further to do one of the following:

- Each student or student pair suggests and explains other examples of changes to and from kinetic energy. [O1, O3]
- Alternatively, each student or student pair explains the changes in energy involved when a rocket is launched into space and returns. [O1, O3]

Plenary suggestions

Ask me a question: Ask the students to write a question about something from the topic and then devise a mark scheme for the answer. Encourage them to come up with questions worth more than one or two marks and to try them out on one another.

Heads and tails: Ask each student to write a question about something from the topic on a coloured paper strip and the answer on a different coloured strip. In groups of six to eight, hand out the strips so that each student gets a question and an answer. One student reads out his or her question. The student with the correct answer then reads it out, followed by his or her question, and so on.

Answers to questions

Worksheet question 1 (7.1.1 – a–f; 7.1.2 – a, d, e, f, g; 7.1.3 – a, c, f, g, h, i, j)

a) the energy of movement
b) any sensible answers; basically, anything where things have to move
c) increase mass; increase speed
d) $E_k = ½ mv^2$
e)

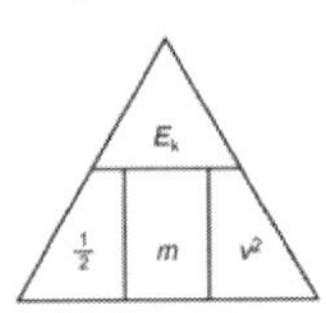

f) 6995.48 J; ≈7000 J
g) 0.058 kg
h) 341 m/s
i) 422 662.9127 kg; ≈4.23×10^5 kg
j) 960 J

Worksheet question 2 (7.1.1 – a–d; 7.1.2 – a–e; 7.1.3 – a–f)

a) 83 400 J
b) 83 400 J
c) transferred into other forms (all to kinetic energy assumed in part (d))
d) 83 400 J (assuming all transferred to kinetic energy)
e) 52.7 m/s
f) Either the motor was still on, which added to the kinetic energy, or the kinetic energy was more than zero at the top of the roller coaster (or both).

Practical sheet evaluation

1. any sensible answers, e.g. calculate change in gravitational potential energy, calculate change in kinetic energy
2. For identical ball bearings, the greater the speed, the higher the kinetic energy.
3. For ball bearings at the same speed, the greater the mass, the higher the kinetic energy.
4. to ensure almost no kinetic energy to start with (so didn't need to take away from value of starting magnetic potential energy)
5. some kinetic energy transferred into heat energy or as sound due to friction
6. any sensible answers, e.g. mechanism to hold magnets in place, determine kinetic energy of first ball bearing

 © HarperCollins*Publishers* Limited 2016

Lesson 7.2: Work done and energy transfer

Lesson overview

OCR Specification reference

P7.1b (part), c (part)

Learning objectives

- Understand what is meant by work done.
- Explain the relationship between work done and force applied.
- Identify the transfers between energy stores when work is done against friction.

Learning outcomes

- [O1] Define work done.
- [O2] Calculate, using the work done equation.
- [O3] Link kinetic energy and work done, including using calculations.

Skills development

- Use a variety of models, such as representational, spatial, descriptive, computational and mathematical to solve problems, make predictions and develop scientific explanations and understanding of familiar and unfamiliar facts.
- Make and record observations and measurements using a range of apparatus and methods.
- Recognise the importance of scientific quantities and understand how they are determined.

Maths focus

- Recognise and use expressions in standard form.
- Find arithmetic means.
- Change the subject of an equation.
- Substitute numerical values into algebraic equations using appropriate units for physical quantities.
- Solve simple algebraic equations.

Resources needed sand in a tray, metre rule, ball bearings, smoother, mass balance; Worksheets 7.2.1, 7.2.2 and 7.2.3, Practical sheets 7.2.1 and 7.2.2, Technician's notes 7.2

Key vocabulary work, energy transfer, kinetic energy, force

Teaching and learning

Engage

- As the students enter the room, give them a small piece of paper and ask them to give their best answer to the following question: what is energy?
- **Pair talk:** Ask students to share ideas with a partner, adapting their answer if they think it is appropriate.
- **Pairs to fours:** Pairs should join together to finalise their answer. Take feedback from a number of groups before developing the idea that energy is the ability to do work. Discuss what we mean by doing work after showing the work done video.
- Play the work done video showing a 'plane pull' by four-times World's Strongest Man Žydrūnas Savickas.
- After the video, ask the students the following questions: Why is the plane pull such hard work? Why couldn't many people do this? What factors would make it harder or easier? What energy transfers were there during the plane pull? [O1, O2, O3]
- Show the work done animation(s), especially, 'work' and 'work against friction'.
- Class discussion on work done leading to the definition (energy transfer when a force causes an object to move) and the equation: work done = force × distance or $W = Fs$. Put equation into the triangle. Note: students will be expected to remember this equation. [O1, O2]

 © HarperCollins*Publishers* Limited 2016

- Issue students with a piece of paper with questions related to the 'work done and kinetic energy' videos appropriate to their ability:
  - Low demand: What caused the truck to stop? Why was there sand at the end of the dragster runway?
  - Standard demand: As low demand, plus: What factors would make the vehicle take a longer or shorter distance to stop?
  - High demand: As standard demand, plus: What information do you need to know to work out the braking distance of a vehicle?
- Watch both 'Work done and kinetic energy' videos. Use the information from the videos to point out that in braking or accelerating: work done = change in kinetic energy. [O1, O2, O3]

Challenge and develop

- Watch the meteorite strike simulation. Discuss what this video shows. How are potential energy, kinetic energy and work done involved here? Note: It is estimated that about 500 meteorites hit the surface of the Earth each year (obviously not of the size that wiped out the dinosaurs!). [O1, O3]
- Carry out the practical.
  - Low demand: Use Practical sheet 7.2.2, in which it is deemed that students will find conversion between grams and kilograms as well as centimetres (and millimetres) and metres too difficult. Students need to be guided through the calculations carefully.
  - Standard demand: Use Practical sheet 7.2.1. Go through conversions between grams and kilograms as well as between centimetres (and millimetres) and metres.
  - High demand: Use Practical sheet 7.2.1 independently. [O1, O2, O3]
- Discuss how the practical could have been improved so that it more closely simulates meteorites striking the Earth – first in pairs, then groups of four, then groups of eight. Finally, someone from each group shares ideas for improvement with the class. [O1, O2, O3]

Explain

- Challenge students' thinking and understanding of work done from the practical activities using the evaluation sections of the practical sheet (these sections are identical). [O1, O2, O3]

Consolidate and apply

Distribute the worksheets, according to ability.

- Low demand: Guide students through tasks carefully.
- Standard and high demand: Work independently or in pairs or small groups. [O1, O2, O3]

Extend

Ask students able to progress further to do the following:

- Suggest why a large meteorite such as the one that killed the dinosaurs would do so much damage. [O1, O3]

Answers to questions

Worksheet question 1 (7.2.1 – a–c; 7.2.2 – a–d; 7.2.3 – a–e)

a) the greater the braking force, the shorter the braking distance
b) 75 000 J
c) 2000 N
d) 18.3 m
e) 2.1×10^{19} N

Worksheet question 2 (7.2.1 – a–c; 7.2.2 – a–c; 7.2.3 – a–d)

a) more kinetic energy; more work needs to be done to stop player in same distance; more force required
b) more kinetic energy; more work needs to be done to stop ball in same distance; more force required or a greater stopping distance
c) (i) 360 000 J (ii) 0 J (iii) 360 000 J (iv) 360 000 J (or 360 000 Nm) (v) 4800 N
d) (i) 2106 J (ii) 416 J (iii) 1690 J (iv) 1690 J (or 1690 Nm)

Practical sheet evaluation

1. Any sensible answer, including: go into ground, depth of crater depends on kinetic energy, some sand is thrown out.
2. The higher the kinetic energy of impact, the deeper the crater.
3. As it falls, gravitational potential energy is transferred into kinetic energy (or ball bearings don't reach high speeds, so little heat loss).

 © HarperCollins*Publishers* Limited 2016

4. Air resistance leads to some energy transfer to heat.
5. work done = change in kinetic energy
6. The greater the energy transfer, the greater the work done on the ball bearing.
7. transferred to heat energy and as sound, and to kinetic energy of sand
8. Any sensible suggestion, including comparing meteorites (ball bearings) of same surface area but different mass.

 © HarperCollins*Publishers* Limited 2016

Lesson 7.3: Specific heat capacity

Lesson overview

OCR Specification reference

P7.1c (part)

Learning objectives

- Understand how things heat up.
- Find out about heating water.
- Find out more about specific heat capacity.

Learning outcomes

- [O1] Describe how energy relates to changes in temperature.
- [O2] Explain why land heats up and cools down quicker than water.
- [O3] Calculate, using the specific heat capacity equation.

Skills development

- Use a variety of models, such as representational, spatial, descriptive, computational and mathematical, to solve problems, make predictions and to develop scientific explanations and understanding of familiar and unfamiliar facts.
- Use scientific theories and explanations to develop hypotheses.
- Interpret observations and other data (presented in verbal, diagrammatic, graphical, symbolic or numerical form), including identifying patterns and trends, making inferences and drawing conclusions.

Maths focus

- Change the subject of an equation.
- Substitute numerical values into algebraic equations using appropriate units for physical quantities.
- Solve simple algebraic equations.
- Translate information between graphical and numeric form.

Resources needed small pieces of metal and plastic, table lamp with 100 W filament bulb, 250 ml beakers, soil, water, thermometers, stopwatch, metre rule; Worksheets 7.3.1, 7.3.2 and 7.3.3, Practical sheet 7.3, Technician's notes 7.3

Key vocabulary specific heat capacity, temperature, mass, energy

Teaching and learning

Engage

- As the students arrive, issue them with a small piece of metal and a small piece of plastic. Ask them to hold these in their hands whilst the register is taken.
- Ask them to describe any observations that they made at the start and at the end of the time.

Challenge and develop

- Carry out Practical 7.3.
  - Low demand: Guide students through the process carefully and with the hypothesis.
  - Standard demand: Guide with the hypothesis where necessary.
  - High demand: Students work independently. [O1, O2]
- Discuss how the practical could have been improved or made fairer, first in pairs, then groups of four, then groups of eight. Finally, someone from each group shares ideas for improvement with the class. [O1, O2]

 © HarperCollins*Publishers* Limited 2016

- Lead a discussion on how the experiment has suggested that different substances will increase in temperature by different amounts when they gain the same amount of heat, or that different substances require different amounts of energy in order to increase in temperature by the same amount.
- Discuss that this is because different substances have different specific heat capacities. Use questioning to get students to come up with a definition and an equation. Some classes will need to be led far more than others. Note: students will always be given this equation in examinations, so they do not need to recall it.
- Give some typical values of specific heat capacity, for example, c_{water} = 4200 J/(kg °C), $c_{dry\ soil}$ = 800 J/(kg °C), c_{sand} = 290 J/(kg °C), c_{copper} = 385 J/(kg °C). Note: one calorie is the amount of energy that causes 1 g of water to increase in temperature from 14.5 °C to 15.5 °C. Substances actually have different specific heat capacities at different temperatures, but this is beyond the syllabus. [O1, O2, O3]

Explain

- Challenge students' thinking and understanding of heating water and soil from the practical activities using the evaluation section of Practical sheet 7.3. [O1, O2]

Consolidate and apply

- The students should then be issued with the specific heat capacity worksheet appropriate to their ability:
  - Low demand: Worksheet 7.3.1
  - Standard demand: Worksheet 7.3.2
  - High demand: Worksheet 7.3.3 [O1, O2, O3]

Extend

- Each student or student pair suggests how life would be different if water and land had similar specific heat capacities. [O1, O2]

Plenary suggestions

Ask me a question: Ask the students to write a question about something from the lesson and then devise a mark scheme for the answer. Encourage them to come up with questions worth more than one or two marks and to try out their questions on one another.

Where's the answer? As students enter the room, give them a card with a word written on it. At the end of the lesson, ask a question related to the lesson. Ask them if they think their card contains the answer and why. [O1, O2, O3]

Answers to questions

Worksheets 7.3.1–7.3.3 (7.3.1 – a, d, e, f and g; 7.3.2 – a–h; 7.3.3 – a–i)

a) The specific heat capacity of a substance is the amount of energy required to raise the temperature of one kilogram of the substance by one degree Celsius.

b) J/(kg °C)

c) $\Delta E = mc\Delta\theta$

d)

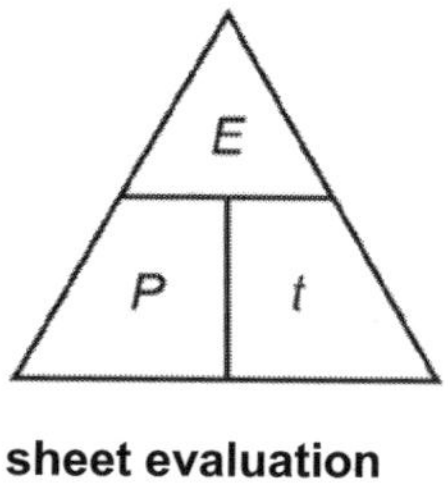

e) Specific heat capacity of water is much greater than that of land. Therefore it is much more difficult to alter the temperature of water, compared with that of sand (increase and decrease).

f) 790 J/(kg °C)

g) 6 °C

h) 50 kg

i) 87 000 J

Practical sheet evaluation

1. Any sensible answer that correctly connects the hypothesis to the findings, looking at lines on graph
2. The data for both soil and water were on the same graph so that they could be compared easily.
3. In the day, land heats up faster than water.
4. At night, land cools down faster than water.

 © HarperCollins*Publishers* Limited 2016

Lesson 7.4: Electrical energy and power

Lesson overview

OCR specification reference

P7.2a, b, c

Learning objectives

- Recall that energy cannot be created or destroyed, only transferred.
- Describe the energy transfers in different domestic appliances.
- Describe power as a rate of energy transfer.
- Calculate the energy transferred.

Learning outcomes

- [O1] Understand that everyday electrical appliances bring about energy transfers.
- [O2] Recall and use the equation energy transferred, $E = Pt$.
- [O3] Recall and apply the equation energy transferred $E = QV$.

Maths focus Use and rearrange an equation with three quantities.

Resources needed range of domestic appliances, power meters, stop clocks; Worksheets 7.4.1, 7.4.2 and 7.4.3, Practical sheet 7.4, Technician's notes 7.4

Key vocabulary power, energy transfer, joules, watts, power rating

Teaching and learning

Engage

- Show the students a range of domestic appliances and ask them to suggest the energy transfers occurring when each appliance is turned on. Suggested appliances include: a hair dryer, an electric drill, a kettle, a desk lamp, a radio and a microwave.
- If necessary, the students could be reminded of the energy stores work carried out at KS3. [O1]

Challenge and develop

- The students should then use a power meter to collect data from each appliance. Details for the investigation can be found on Practical sheet 7.4.1. [O1]

Explain

- The students should use the data gathered to calculate the electrical energy transferred by each device, by applying the equation electrical energy transferred = power × time. [O1, O2]
- They should then rank the appliances from highest to lowest in terms of the amount of energy they transfer. [O1]

Consolidate and apply

The students should then be issued with the power and energy transfers worksheet appropriate to their ability:

- Low demand: Worksheet 7.4.1 [O1, O2, O3]
- Standard demand: Worksheet 7.4.2 [O1, O2, O3]
- High demand: Worksheet 7.4.3 [O1, O2, O3]

Extend

Ask students able to progress further to do the following:

 © HarperCollins*Publishers* Limited 2016

- Use the equations $E = QV$ and $I = Q/t$ to show that $E = VIt$. They could follow this by using the equations $P = VI$ and $E = Pt$ to confirm that $E = VIt$.

Plenary suggestions

The big ideas: Ask students to write down three ideas they have learned during the topic so far. Then ask them to share their facts in groups and to compile a master list of facts, with the most important at the top. Ask for ideas to be shared and find out which other group(s) agreed.

Heads and tails: Ask each student to write a question about something from the topic on a coloured paper strip and the answer on a different coloured strip. One student reads out their question. The student with the correct answer then reads it out, followed by their question.

Hot seat: Ask each student to think up a question, using material form the lesson. Select someone to put in the hot seat, ask students to pose their questions and say at the end whether the answer is correct or incorrect

Answers to questions

Worksheets 7.4.1–7.4.3

1. a) 600 seconds

   b) 900,000 J / 900 kJ

2. a) 2500 W

   b) 2700 seconds

   c) 6,750,000 J / 6750 kJ

3. 6900 J

4. 50 C

Worksheet 7.4.3

5. 40 V

 © HarperCollins*Publishers* Limited 2016

Lesson 7.5: Dissipation of energy

Lesson overview

OCR Specification reference

P7.2a, g

Learning objectives

- Explain ways of reducing unwanted energy transfer.
- Describe what affects the rate of cooling of a building.
- Understand that energy is dissipated.

Learning outcomes

- [O1] Describe energy transfer using the law of conservation of energy.
- [O2] Describe the process of conduction in terms of energy.
- [O3] Describe the use of lubrication and thermal insulation in reducing unwanted energy transfers.

Skills development

- Appreciate the power and limitations of science and consider any ethical issues that may arise.
- Explain everyday and technological applications of science; evaluate associated personal, social, economic and environmental implications; make decisions based on the evaluation of evidence and arguments.
- Make and record observations and measurements using a range of apparatus and methods.

Maths focus

- Recognise and use expressions in decimal form.

Resources needed beakers, thermometers, insulation, stopwatches, scissors, elastic bands, kettles, graphene, ice cubes, paper towels; Worksheets 7.5.1, 7.5.2 and 7.5.3, Practical sheets 7.5.1, and 7.5.2, Technician's notes 7.5

Digital resources Honda cog advert, energy water video, thermal conductivity video, uses of high thermal conductivity video

Key vocabulary energy dissipation, conduction, radiation, thermal conductivity

Teaching and learning

Engage

- Watch the Honda Cog advert. During the clip, students should tally the number of energy transfers they see. Ask them if it would be possible to recreate this advert in the classroom. Introduce the idea of energy loss through dissipation.
- Have students draw a table in their books:

How energy is wasted	How to reduce this energy waste

- Watch the energy waste video. Students fill in the table while watching the video. After the video, students share what they learned, adding details to their own answers by drawing on other students' observations.

Challenge and develop

- Carry out Practical 7.5.1 or 7.5.2.
  - Low demand: Use Practical sheet 7.5.2. Guide students carefully through the task.
  - Standard demand: Students choose Practical sheet 7.5.1 or 7.5.2. Guide them through the process.
  - High demand: Use Practical sheet 7.5.1 independently or in small groups. [O1, O2, O3]
- Discuss how the practical could have been improved, first in pairs, then groups of four, then groups of eight. Finally, someone from each group shares ideas for improvement with the class. [O1, O2]

 © HarperCollins*Publishers* Limited 2016

- Carry out Practical 7.5.3. This can be a demonstration or students can complete it in small groups. [O3]
- Discuss Practical 7.5.3. What does this tell us about graphene? What other uses might graphene have? [O3]
- Show videos on thermal conductivity and uses of thermal conductivity. Discuss the video with class. [O3]

Explain

- Challenge students' thinking and understanding of energy transfer from the practical activities using the evaluation sections of the practical sheets. [O1, O2]

Consolidate and apply

- Issue the students with the dissipation of energy worksheet appropriate to their ability:
  - Low demand: Worksheet 7.5.1
  - Standard demand: Worksheet 7.5.2
  - High demand: Worksheet 7.5.3 [O1, O2, O3]

Extend

Ask students who are able to do the following:

- Suggest possible technological applications of graphene based on what they have seen in the lesson. [O3]

Plenary suggestions

The big ideas: Ask students to write down three ideas they have learned during the topic so far. Then ask them to share their facts in groups and to compile a master list of facts, with the most important at the top. Ask for ideas to be shared and find out which other group(s) agreed.

Hot seat: Ask each student to think up a question, using material from the lesson. Select someone to put in the hot seat. Ask students to pose their questions and say at the end whether the answer is correct or incorrect.

Answers to questions

Worksheets 1.1–1.3

1.
   a) Dissipation of energy is where some energy is not made use of but is transferred into a form that is no longer useful.
   b) Any sensible answer, including: light bulbs, engines, televisions.
   c) Any sensible answer, including: where as much heat needs to be released as possible, for example, back of a fridge.

2.
   a) (i) thermal, sound (ii) lubrication, ball bearings (iii) oil
   b) Any sensible answer, including: any machine with moving parts.
   c) Any sensible answer, including: a machine where braking or changes in direction need to take place.

Practicals 7.5.1 and 7.5.2 Analysis and evaluation

1. The beaker of water that cooled down the least has the highest temperature after 10 min; the one that cooled down the most has the lowest temperature after 10 min.
2. Most probable correct answer: On average, three layers leads to the least cooling.
3. Whichever is correct according to the results: the one with the lowest temperature after 10 min.
4. Whichever is correct according to the results: the one with the highest temperature after 10 min.
5. The higher the starting temperature, the more the temperature would change by (or the higher the final temperature would be); therefore, it is necessary for all the groups to start at the same temperature so that the results can be combined.
6. The longer the time, the more heat would have been lost and the lower the temperature. Therefore, it is necessary for all the groups to use the same time so that the results can be combined.
7. The greater the volume of water, the lower the rate of cooling. Therefore, it is necessary for the groups to use the same volume so that the results can be combined.
8. 'Volume of water' is more precise: 'amount' may mean 'volume' or 'mass'.
9. So all of water is at the same temperature.

 © HarperCollins*Publishers* Limited 2016

Practical 7.5.3 Evaluation

1. The graphene easily cut through the ice cube.
2. Heat from your hand transferred through the graphene to heat the ice cube and melt the ice.
3. Graphene is a very good conductor; it has a very high thermal conductivity.

 © HarperCollins*Publishers* Limited 2016

Lesson 7.6: Energy efficiency

Lesson overview

OCR Specification reference

P7.1a

Learning objectives

- Explain what is meant by energy efficiency.
- Calculate the efficiency of energy transfers.
- Find out about conservation of energy.

Learning outcomes

- [O1] Explain the term 'energy efficiency'.
- [O2] Calculate, using the energy efficiency equation.
- [O3] Compare and evaluate light bulbs with different efficiencies.

Skills development

- Explain everyday and technological applications of science; evaluate associated personal, social, economic and environmental implications; make decisions based on the evaluation of evidence and arguments.
- Interpret observations and other data (presented in verbal, diagrammatic, graphical, symbolic or numerical form), including identifying patterns and trends, making inferences and drawing conclusions.
- Recognise the importance of scientific quantities and understand how they are determined.

Maths focus

- Change the subject of an equation.
- Substitute numerical values into algebraic equations using appropriate units for physical quantities.
- Solve simple algebraic equations.

Resources needed large spherical object, high-up pivot point, strong cord, hand generator, connecting leads, light meter (if possible), infrared detector (if possible), mounted 1.5 V incandescent light bulb, mounted 6 V incandescent light bulb, mounted white LED, 30 cm ruler; Worksheets 7.6.1 and 7.6.2, one of Practical sheets 7.6.1–7.6.4 (depending on equipment available).

Digital resources law of conservation of energy video

Key vocabulary energy efficiency, conservation of energy

Teaching and learning

Engage

- As the students enter the room, issue them with a small piece of paper. Working independently to begin with, ask them what they understand by the term 'efficient' and ask them to give an example of a sentence where the word can be used.
- **Pair talk:** Ask the students to join with a partner to share their ideas.
- **Pairs to fours:** Pairs of students should then join up to share their ideas further, before feedback is taken from a range of students.
- Watch the law of conservation of energy video.
- Challenge students' thinking and understanding of the experiment by discussion. Suggested questions to pose to the students would be: What type of energy did the object have at the start? At the point when the object returns, does the object have more, less or the same of this type of energy? What other main energy transfers were involved?

 © HarperCollinsPublishers Limited 2016

- Introduce the idea of the law of conservation of energy. Students discuss, then share with the class what they believe is meant by this, including examples. Clarify the exact definition.

Challenge and develop

- Set up a tungsten filament lamp and an energy-saving light bulb. If available, set up a temperature probe next to each bulb and note the temperature rise of the air near each of the different bulbs. [O1]
- Carry out Practical 7.6.1 (or 7.6.2 or 7.6.3 or 7.6.4, depending on availability of equipment). [O1, O3]
- Class discussion: exactly what is efficiency? Lead the class to a full definition, including the equation, in terms of both energy and power. Make it clear that the efficiency may need to be determined as a decimal fraction or as a percentage. Students will be expected to remember these equations. [O1, O2, O3]

Explain

- Challenge students' thinking and understanding of efficiency from the practical activities using the evaluation section of Practical sheet 7.6.1 (or 7.6.2 or 7.6.3 or 7.6.4 – these sections are identical). [O1, O2]

Consolidate and apply

Students should complete each section of Worksheet 7.6.1 as appropriate:

- Low demand: Guide students carefully through the tasks.
- Standard and high demand: Students work independently or in pairs or small groups. [O1, O2, O3]

Alternatively, the students should complete the task outlined on Worksheet 7.6.2. [O1, O2, O3]

Extend

- Each student or student pair suggests how the efficiency of a machine could be improved. [O1, O2]

Plenary suggestions

Freeze frame: Ask students to create a 'freeze frame' of an idea in the topic. This involves three to four students arranging themselves as a static image and other students suggesting what it represents.

What do I know? Ask students to each write down one thing about the topic they are sure of, one thing they are unsure of and one thing they need to know more about, being specific. Ask them to work in groups of five or six to agree on group lists. Ask each group to say what they decided and agree as a class about what they are confident about, what they are less sure of and what things they want to know more about.

Answers to questions

Worksheet 7.6.1

1. a) Energy cannot be created or destroyed. It can only be converted from one form into other forms.
   b) 100 J
   c) transferred to thermal energy, 96 J

2. a) Efficiency is the percentage (or proportion) of energy that is transferred into useful energy.
   b) high – less energy wasted, lower cost, less environmental impact
   c) $\text{efficiency} = \dfrac{\text{useful output energy}}{\text{total input energy}}$
   d) $\text{efficiency} = \dfrac{\text{useful power output}}{\text{total power input}}$
   e) 0.65 = 65%
   f) 81 600
   g) 1364 J

3. a) light and thermal energy
   b) 0.083 = 8.3%
   c) 11.1 W
   d) 3.6 W

Practical sheets 7.6.1–7.6.4 Evaluation

1. whichever fits the results, most probably 6 V incandescent bulb
2. whichever fits the results, most probably 6 V incandescent bulb
3. whichever fits their results, most probably: The white LED was the most efficient.
4. A different brightness would affect the temperature and the number of turns required so it would be difficult to compare results.

 © HarperCollins*Publishers* Limited 2016

Lesson 7.7: Practical: Investigating ways of reducing the unwanted energy transfers in a system

Lesson overview

OCR Specification reference

P7.2f

Learning objectives

- Use scientific ideas to make predictions.
- Analyse data to identify trends.
- Evaluate an experimental procedure.

Learning outcomes

- [O1] Use ideas on thermal conduction and insulation to predict results of an experiment.
- [O2] Plot a graph of the results and use it to compare the rate of cooling.
- [O3] Evaluate the method and the conclusion to the experiment.

Skills development

- Explain everyday and technological applications of science; evaluate associated personal, social, economic and environmental implications; make decisions based on the evaluation of evidence and arguments.
- Use scientific theories and explanations to develop hypotheses.
- Evaluate methods and suggest possible improvements and further investigations.
- Present observations and other data using appropriate methods.
- Present reasoned explanations, including relating data to hypotheses.

Maths focus

- Translate information between graphical and numeric form.
- Draw and use the slope of a tangent to a curve as a measure of rate of change.

Resources needed mini whiteboards, marker pens, whiteboard rubbers, beaker (for example, 250 ml), thermometer, kettle, cardboard, scissors, stopwatch, insulating material, rubber bands; Worksheet 7.7, Practical sheet 7.7, Technician's notes 7.7

Digital resources tangent lines video

Key vocabulary insulation, energy transfer, conduction, thermal radiation, insulation

Teaching and learning

Engage

- Challenge students' thinking and understanding of how to determine the rate at which a substance cools down by asking: When do substances cool down? What happens when a substance cools down? How would you find out how quickly a substance cools down by experiment? What is the rate of cooling? How could you find the rate of cooling at a certain temperature?
- Explain that the class will be conducting an experiment to investigate cooling curves for different thicknesses of material.
- **Pair talk**, then share suggestions for the hypothesis for this experiment. [O1]

Challenge and develop

- Carry out Practical 7.7. [O1, O2]
    - Low demand: All students are guided through the process carefully and pool their results.

 © HarperCollins*Publishers* Limited 2016

- Standard demand: Depending on resources, either guide students in using several beakers at once or let them work independently using just one beaker.
- High demand: Carry out the practical independently, using several beakers.

- Discuss how the practical could have been improved, first in pairs, then groups of four, then groups of eight. Finally, someone in each group shares ideas with the class. [O3]

Explain

- Challenge students' thinking and understanding of cooling curves from the practical activities, using the evaluation section of Practical sheet 7.7. [O1, O2]
- Show the tangent line video. Have each student draw a cooling curve on a mini whiteboard and draw a tangent at some point on the curve. Rub out then draw tangents again at different places on the cooling curve. [O1]

Consolidate and apply

Distribute Worksheet 7.7 and have students complete it, according to ability:

- Low demand: Guide students through tasks carefully.
- Standard and high demand: Work independently or in pairs or small groups. [O1, O2, O3]

Extend

- Each student or student pair suggests practical uses for the findings of the experiment.

Plenary suggestions

Heads and tails: Ask each student to write a question about something from the topic on a coloured strip of paper and the answer on a different coloured strip. In groups of six to eight, hand out the strips so that each student gets a question and an answer. One student reads out his or her question. The student with the right answer then reads it out, followed by his or her question, and so on.

Hot seat: Ask each student to think up a question, using material from the topic. Select someone to put in the hot seat. Ask students to pose their question and say at the end whether the answer is correct or incorrect.

Answers to questions

Worksheet 7.7

a) graph with both axes labelled, with units, scale leads to spread of points at least half page, sensible scale (not 3, 6, 7, 9), points plotted, curve of best fit

b) tangent drawn, ideally as long as possible, ending at easy-to-read numbers

c, d) horizontal dotted line drawn with correct change in time written down

c, e) vertical dotted line drawn with correct change in temperature written down

f) gradient determined correctly; expect ≈0.28 °C/s

g) The rate of cooling was greater at 80 °C than at 50 °C (steeper line).

h) The rate of cooling was less at 30 °C than at 50 °C (shallower line).

Practical sheet 7.7 evaluation

1. Any that matches results, for example: The thicker the insulation, the lower the rate of cooling. This is because with more insulation, heat takes a longer time to travel through the insulation so heat is lost at a lower rate.
2. The higher the number of layers, the shallower the cooling curve.
3. The steeper the cooling curve, the worse the water is insulated.
4. The curve shows clearly how temperature varies with time.
5. When the curves are on the same graph it is easy to compare them.
6. Any sensible answer, for example, different colours or labelling.

 © HarperCollins*Publishers* Limited 2016

Lesson 7.8: Key concept: Energy transfer

Lesson overview

OCR Specification reference

KC

Learning objectives

- Understand why energy is a key concept in science.
- Use ideas about stores and transfers to explain what energy does.
- Understand why accounting for energy transfers is a useful idea.

Learning outcomes

- [O1] Write a full list for the different types of energy.
- [O2] Explain the importance of energy in our lives.
- [O3] Use analogy to describe the conservation of energy.

Skills development

- Use a variety of models such as representational, spatial, descriptive, computational and mathematical to solve problems, make predictions and to develop scientific explanations and understanding of familiar and unfamiliar facts.
- Use scientific vocabulary, terminology and definitions.

Maths focus

- Make estimates of the results of simple calculations.

Resources needed wooden blocks, energy circus (various transducers); Worksheet 7.8, Practical sheets 7.8.1 and 7.8.2, Technician's notes 7.8.1 and 7.8.2

Digital resources energy song, for example, Buzzcocks – Energy, types of energy video, heat death of the universe video

Key vocabulary energy, energy transfer, conservation of energy

Teaching and learning

Engage

- Play the 'energy song' to set the tone for the lesson as students arrive.
- Challenge the students' thinking and understanding of energy by discussion. **Pair talk**, then share.
  - Low demand: When do you have more or less energy? When do objects have more or less energy?
  - Standard demand: As low demand, plus: What are the nine types of energy (10 if you include magnetic)?
  - High demand: As standard demand, plus: What is energy? [O1, O2]

Challenge and develop

- Carry out Practical 7.8.1 with students, according to ability:
  - Low demand: Guide students through the analogy carefully.
  - Standard and high demand: Get the students to describe the analogy themselves. [O3]
- Challenge the students' thinking and understanding of the law of conservation of energy from the analogy by using the evaluation section of Practical sheet 7.8.1. This can be done in small groups or as a class. [O3]
- Watch the types of energy video. Students then give examples of each type of energy. Lead a discussion on the law of conservation of energy with energy transfers. [O1, O2]
- Carry out Practical 7.8.2.

 © HarperCollins*Publishers* Limited 2016

- Low demand: Students only fill their results in the table when discussing as a class afterwards.
- Standard and high demand: Students work independently in small groups. [O1, O2]

- **Snowball** on how the energy circus could have been improved. [O1, O2]

Explain

- Challenge the students' thinking and understanding of energy transfers from the practical activities by using the evaluation sections of the practical sheets (these sections are identical). [O1, O2]

Consolidate and apply

- Students complete each section of Worksheet 7.8.
  - Low demand: Guide students through the tasks carefully.
  - Standard and high demand: Students do the tasks independently, in pairs or small groups. [O1, O2, O3]

Extend

- Each student, student pair or small group discuss and share what they believe are the most important energy transfers that affect their lives. [O1, O2]

Plenary suggestions

Watch the heat death of the universe video. Discuss the following as a class: Why can nothing happen in the heat death of the universe? Do we need to be worried about it? [O2]

Answers to questions

Worksheet 7.8

1.
   (a) kinetic, sound, light, thermal/heat, electrical (and magnetic), gravitational potential energy, elastic potential energy, chemical potential energy, nuclear potential energy
   (b) stored energy
   (c) gravitational, elastic, chemical and nuclear (electrical can be)
   (d) energy of movement
   (e) kinetic, sound, light, thermal (electrical can be)
   (f) techincally all of them, although nuclear energy is not transferred at all so this could be left out

2.
   (a) Electrical
   (b) 100 J
   (c) law of conservation of energy
   (d) light
   (e) 90 J
   (f) thermal/heat
   (g) 0.10 = 10%

3. Blocks represent energy, room represents the universe (or system), total number of blocks (and therefore energy) stays the same. Some of the blocks can be difficult to find (some of the energy can be difficult to measure), but they are still there.

Practical sheet 7.8.1 Evaluation

1. energy
2. the universe
3. stayed the same
4. It always stays the same.
5. Some energy is difficult to locate – it spreads out.
6. Energy cannot be created or destroyed; it can only be converted between types of energy.

Practical sheet 7.8.2 Evaluation

1. Energy can be transferred from one form into other forms.
2. Any sensible answer, for example: being able to take measurements of the total energy before and after.

 © HarperCollins*Publishers* Limited 2016

Lesson 7.9: Maths skills: Calculations using significant figures

Lesson overview

OCR Specification reference

PM7.2i

Learning objectives

- Substitute numerical values into equations and use appropriate units.
- Change the subject of an equation.
- Give an answer using an appropriate number of significant figures.

Learning outcomes

- [O1] Revise the idea of energy transfers.
- [O2] Apply and rearrange an equation with three or more variables.
- [O3] Give an answer using an appropriate number of significant figures.

Skills development

- Use an equation with at least three variables.
- Manipulate an equation with three variables.

Maths focus

- Use and rearrange an equation with at least three variables.
- Give an answer using an appropriate number of significant figures.

Resources needed clamp stand, boss and clamp, pendulum bob on a length of string, tennis ball, metre ruler, balance; Worksheets 7.9.1, 7.9.2 and 7.9.3, Technician's notes 7.9

Key vocabulary substitute, rearrange an equation, subject of an equation, significant figures

Teaching and learning

Engage

- Demonstrate a swinging pendulum to the students.
- Ask students to discuss, in pairs, why the pendulum loses height and eventually ceases to swing (that is, why it does not exhibit perpetual motion). [O1]

Challenge and develop

- Use the demonstration of the pendulum to remind the students of the concept of energy transfers. Ask them to suggest what energy transfers are causing the pendulum to slow and eventually stop. [O1]
- Show the students a second demonstration of raising a tennis ball and then allowing it to fall to the floor. Discuss the energy transfer that occurs (potential energy to kinetic energy). Remind the students of the equations to calculate (a) gravitational potential energy and (b) kinetic energy. Tell them that if we assume 100% energy transfer then we are able to equate the two values for the energy. [O1]

Explain

- Calculate with the students the amount of potential energy that the tennis ball has at its maximum height using $E_p = mgh$. [O2]
- Release the tennis ball from the maximum height and use the equation $E_k = 0.5\ mv^2$ to calculate the velocity of the ball when it hits the floor. [O1, O2]

 © HarperCollins*Publishers* Limited 2016

Consolidate and apply

Issue the Maths skills worksheets at an appropriate level. These worksheets include the consideration of significant figures. [O2, O3]

- Low demand: Worksheet 7.9.1
- Standard demand: Worksheet 7.9.2
- High demand: Worksheet 7.9.3

Extend

- Having calculated the velocity of the tennis ball through the method at the start of the lesson, challenge the students to think about whether they would get the same value if they simply used speed = distance ÷ time. If time allows, allow them to carry this out experimentally. Ask them to explain the difference.

Plenary suggestions

Ideas hothouse: Ask students to work in pairs to list points about what they have learned about rearranging equations and significant figures. Then ask the pairs to join together into groups of four and then groups of six to eight to discuss this further and to come up with an agreed list of points. Ask one person from each group to report back to the class.

Heads and tails: Ask each student to write a question (probably related to an equation) on a coloured paper strip and the answer on another coloured strip. Hand out the strips to groups of six to eight students, so that each student gets a question and an answer. One student reads out his or her question. The student with the correct answer then reads it out, followed by his or her question, and so on.

Answers to questions

Worksheets 7.9.1, 7.9.2 and 7.9.3

1. a) 58.8 J
   b) 24.2 m/s
2. a) 2352 J
   b) 6.86 m/s
3. a) 7.0
   b) 3.1
   c) 6.0
   d) 1100
   e) 4.8

 © HarperCollins*Publishers* Limited 2016

When and how to use these pages: Check your progress, Worked example and End of chapter test

Check your progress

Check your progress is a summary of what students should know and be able to do when they have completed the chapter. Check your progress is organised in three columns to show how ideas and skills progress in sophistication. Students aiming for top grades need to have mastered all the skills and ideas articulated in the final column (shaded pink in the Student book).

Check your progress can be used for individual or class revision using any combination of the suggestions below:

- Ask students to construct a mind map linking the points in Check your progress
- Work through Check your progress as a class and note the points that need further discussion
- Ask the students to tick the boxes on the Check your progress worksheet (Teacher Pack CD). Any points they have not been confident to tick they should revisit in the Student Book.
- Ask students to do further research on the different points listed in Check your progress
- Students work in pairs and ask each other what points they think they can do and why they think they can do those, and not others

Worked example

The worked example talks students through a series of exam-style questions. Sample student answers are provided, which are annotated to show how they could be improved.

- Give students the Worked example worksheet (Teacher Pack CD). The annotation boxes on this are blank. Ask students to discuss and write their own improvements before reviewing the annotated Worked example in the Student Book. This can be done as an individual, group or class activity.

End of chapter test

The End of chapter test gives students the opportunity to practice answering the different types of questions that they will encounter in their final exams. You can use the Marking grid provided in this Teacher Pack or on the CD Rom to analyse results. This shows the Assessment Objective for each question, so you can review trends and see individual student and class performance in answering questions for the different Assessment Objectives and to highlight areas for improvement.

- Questions could be used as a test once you have completed the chapter
- Questions could be worked through as part of a revision lesson
- Ask Students to mark each other's work and then talk through the mark scheme provided
- As a class, make a list of questions that most students did not get right. Work through these as a class.

 © HarperCollins*Publishers* Limited 2016

Marking Grid for End of Chapter 7 Test

	Getting Started [Foundation Tier]							Going Further [Foundation and Higher Tiers]			More Challenging [Higher Tier]					Most demanding [Higher Tier]			Total marks	Percentage
Student Name	Q. 1 (AO1) 1 mark	Q. 2 (AO1) 1 mark	Q. 3 (AO1) 1 mark	Q. 4 (AO1) 1 mark	Q. 5 (AO2) 2 marks	Q. 6 (AO2) 1 mark	Q. 7 (AO1) 1 mark	Q. 8 (AO1) 1 mark	Q. 9 (AO1) 2 marks	Q. 10 (AO2) 2 marks	Q. 11 (AO1) 2 marks	Q. 12 (AO2) 1 mark	Q. 13 (AO1) 1 mark	Q. 14 (AO2) 2 marks	Q. 15 (AO3) 6 marks	Q. 16 (AO1) 2 marks	Q. 17 (AO2) 2 marks	Q. 18 (AO3) 6 marks		

 © HarperCollins*Publishers* Limited 2016

Check your progress

You should be able to:

☐ describe how energy can be stored by raising an object up or by stretching or compressing it →	☐ use the equations for gravitational potential energy and elastic potential energy →	☐ apply the equations for gravitational potential energy and elastic potential energy in a variety of contexts, and change the subject of these equations
☐ describe how a moving object has kinetic energy →	☐ know that kinetic energy is related to mass and velocity squared and use the formula to calculate it →	☐ use the equation for kinetic energy to solve problems, including changing the subject of the equation
☐ state that some materials require more energy than others to increase a certain mass by a certain temperature rise →	☐ describe what is meant by the specific heat capacity of a material and use the equation for specific heat capacity ☐ plan an experiment to measure the specific heat capacity of a material →	☐ calculate temperature changes, masses or specific heat capacities given the other values ☐ evaluate an experiment to measure the specific heat capacity of a material
☐ recognise that some energy transfers are unwanted →	☐ describe how lubrication and insulation can be used to reduce unwanted energy transfers ☐ calculate energy efficiency ☐ describe how some energy transfers are more useful than others →	☐ explain how thermal conductivity affects the rate of energy transfer across a material and affects the rate of cooling of a building ☐ recognise that in a closed system there may be energy transfers that change the way energy is stored, but there is no net change to the total energy

 © HarperCollins*Publishers* Limited 2016

Worked example

Jo's group is investigating the energy changes that take place when a toy car runs down a ramp.

1 The mass of the car is 500 g, *g* is 10 N/kg and the vertical height of the top of the ramp is 20 cm. Calculate the GPE of the car at the top of the ramp.

$GPE = mgh = 500 \times 10 \times 20 = 100000J$

This answer correctly uses the formula to multiply the variables together but hasn't converted the mass or the height into standard units. It should be 0.5 kg × 10 N/kg × 0.2 m = 1 J

2 The pupils use a speed detector to measure the speed of the car when it gets to the bottom of the ramp. They find out it is travelling at 0.8 m/s. Calculate its kinetic energy at the bottom of the ramp.

$KE = \frac{1}{2}mv^2 = \frac{1}{2} \times 0.5 \times 0.8 \times 0.8 = 0.16$

This answer correctly uses the formula and has converted g to kg. The speed (only) is squared but there is no indication of the correct unit at the end.

3 a How could they calculate the percentage efficiency of the system at converting GPE into KE?

b Explain why it will be less than 100%.

Efficiency could be found by dividing output by input, so $\frac{KE}{GPE}$.

This is correct in that it is the output divided by the input but it is important to use the **useful** output. Furthermore, the question asked for the efficiency as a percentage, so the answer has to be multiplied by 100.

Nothing is perfect and efficiency is always less than 100%

This is true but is not an explanation. The reason it is not 100% is because not all of the GPE store has been transferred to the kinetic energy store of the car. Some energy is dissipated to the surroundings, raising the temperature of the car body and the surrounding air.

4 Jo says that if the ramp was 100% efficient then doubling the height of the top of the ramp would double the speed at the bottom. Referring to the relevant formulae, suggest whether she is correct.

$GPE = mgh$ and $KE = 1/2 mv^2$. Increasing *h* will increase GPE and this will increase KE which will increase v so Jo is correct.

This quotes the correct formulae and identifies that an increased h will mean an increased GPE. In fact, doubling *h* will double the GPE which will double the KE (if it's 100% efficient). However, KE = $1/2\,mv^2$ and the squaring means that although the speed will increase, it won't double (it will actually go up by a factor of about 1.4). If the question includes a quantitative approach (it refers to doubling) it's not enough to give a qualitative response (say **how much** it increases).

 © HarperCollins*Publishers* Limited 2016

Global challenges: Introduction

When and how to use these pages

This unit will build on ideas the students have met before, such as:

- The use of different energy sources to generate electricity for homes
- The environmental impact of different energy resources
- Our Sun is a star and it is at the centre of our Solar System
- Planets, including Earth, orbit the Sun
- The position and movement of the Earth affects our days and seasons
- In astronomy, we measure distances in light years
- The force of gravity is important to the Earth
- The force of gravity keeps the Moon in orbit around the Earth and the Earth in orbit around the Sun

Overview of the unit

In this unit, students will learn about the different energy resources such as fossil fuels and renewable alternatives, and the ways they are used to generate electricity for homes and elsewhere.

Students will also learn about the origin and life cycles of stars. They will discover how we are able to state that the Universe is expanding, and at an increasing rate. They will investigate phenomena such as red-shift and the Big Bang.

The students will learn about the role of gravity in space physics, linking this to the life cycle of stars. They will discover how and why our ideas about the Universe have developed over time.

This unit offers a number of opportunities for the students to investigate phenomena through practical work, work collaboratively with peers and critically evaluate evidence before drawing conclusions.

Obstacles to learning

The students may need extra guidance with the following terms and concepts:

- Some students will find the enormity of space difficult to comprehend.
- Some may find measuring distances in space in light years confusing, and think they are quoting a time.
- Imagining stars as gas (which is largely invisible) and dust can be challenging, as can the fact that these lumps of dust and gas can be seen despite being such a long way away (and older than most students can imagine).

Practicals in this unit

In this unit students will do the following practical work:

- research different energy types
- debate various solutions to an imaginary energy crisis
- observe visible light using a spectrometer
- make a model of how gas compresses in the Sun
- investigate the emission of radiation from two different types of surface
- make models of the Solar System to gauge distances

 © HarperCollins*Publishers* Limited 2016

- investigate the orbits of the planets using a model
- make a model to demonstrate gravity on planets orbiting the Sun
- measure the speed of water waves

	Lesson title	**Overarching objectives**
1	Keeping safe on the road	• Explain the factors that affect stopping distance. • Explain the dangers caused by large deceleration. • Estimate the forces involved in the deceleration of a road vehicle (higher tier). • Apply the idea of rate of change of momentum to explain safety features (higher tier).
2	Transmitting electricity	• Describe how electricity is transmitted using the National Grid. • Explain why electrical power is transmitted at high potential differences. • Understand the role of transformers.
3	Using energy resources	• Describe the main energy resources available for use on earth. • Distinguish between renewable and non-renewable resources. • Explain the ways in which the energy resources are used.
4	Global energy supplies	• Analyse global trends in energy use. • Understand what the issues are when using energy resources.
5	Electricity in the home	• Recall that the domestic supply in the UK is a.c. at 50 Hz and about 230 V. • Describe the main features of live, neutral and earth wires.

 © HarperCollins*Publishers* Limited 2016

6	Red-shift	• Describe red-shift. • Describe evidence for the expanding Universe.
7	The Sun and other stars	• Describe how the Sun and other stars formed. • Describe the nuclear fusion reactions in the Sun.
8	Emission and absorption of infrared radiation	• Realise that all bodies emit and absorb infrared radiation • Compare emission and absorption of radiation from different surfaces • Define a perfect black body • Explain that the intensity and distribution of wavelengths of any emission depends on the temperature of the body.
9	The Solar System	• Describe the orbits of planets and moons in the Solar System. • Distinguish between planets, dwarf planets and moons.
10	Orbits of planets, moons and artificial satellites	• Compare the orbital motion of moons, artificial satellites and planets in the Solar System. • Describe what keeps bodies in orbit around planets and stars. • Explain how, for circular orbits, an object can have a changing velocity but unchanged speed. • Explain why bodies must move at a particular speed to stay in orbit at a particular distance.
11	Key concept: Gravity: the force that binds the Universe	• Understand that gravity provides the force that keeps planets and satellites in orbits. • Understand that gravity is necessary at the start of a star's life cycle and to maintain equilibrium in a stable star. • Describe how the weight of an object depends on the gravitational field strength. • Recognise that there is still much about the universe that is not understood, e.g. dark mass and dark energy.

 © HarperCollins*Publishers* Limited 2016

12	Temperature of the Earth	• Describe how the atmosphere absorbs radiation in a way that varies with wavelength. • List the factors affecting the temperature of the Earth. • Explain how the temperature of an object is related to the radiation absorbed and radiation emitted.
13	Measuring wave speeds	• Explain how the speed of sound in air can be measured. • Explain how the speed of water ripples can be measured. • Describe the use of echo sounding.
14	Seismic waves	• Describe how earthquakes are detected. • Describe the properties of P waves and S waves. • Explain how the properties of seismic waves allow us to investigate the inside of the Earth.
15	Maths skills: Using scale and standard form	• Understand the scale of objects in the Universe. • Use standard form.

 © HarperCollins*Publishers* Limited 2016

Lesson 8.1: Keeping safe on the road

Lesson overview

OCR Specification reference

P8.1d, e, f, g, h, i

Learning objectives

- Explain the factors that affect stopping distance.
- Explain the dangers caused by large deceleration.
- Estimate the forces involved in the deceleration of a road vehicle (higher tier).
- Apply the idea of rate of change of momentum to explain safety features (higher tier).

Learning outcomes

- [O1] Define the terms 'thinking time', 'thinking distance', 'braking time', 'braking distance', 'stopping time' and 'stopping distance'.
- [O2] Categorise factors as affecting either thinking distance or braking distance.
- [O3] Research the effects of large deceleration to explain car safety features such as airbags.

Skills development

- Relate a scientific concept to car design.

Maths focus

- Rearrange equations.

Resources needed Worksheets 8.13.1, 8.13.2 and 8.13.3

Digital resources a road safety video about the dangers of speeding, video cameras, smart phones or similar, if available, access to online research material

Key vocabulary thinking distance, stopping distance, thinking time, reaction time, braking distance

Teaching and learning

Engage

- Show students a road safety video about the dangers of speeding. Discuss why there is a speed limit of 30 mph in built-up areas.

Challenge and develop

- **Pair talk:** Ask: "What must happen before a car will stop?" Take ideas and introduce the terms 'thinking time', 'thinking distance', 'braking time', 'braking distance', 'stopping time' and 'stopping distance'. [O1]
- Ask students to use the words in the terms to provide a definition for each of them (this should stop them defining the thinking distance as a time). [O1]

Explain

- Issue each student with two cards of different colours. Students should use one colour to vote 'thinking distance' and the other colour to vote 'braking distance'.
- Read out the following factors and ask the students to vote using their coloured cards: tiredness of the driver, wet road surface, poor brakes on the car, alcohol or drugs in the driver's system, speed of the car, distraction to the driver. [O2]
- Divide the students into small groups and assign each group one of these factors. Groups should design a new road safety campaign highlighting the danger of their particular factor. They could record an advert if camera equipment is available. [O2]

 © HarperCollins*Publishers* Limited 2016

Consolidate and apply

Issue students a worksheet at an appropriate level to guide their research into the effects of large deceleration:

- Low demand: Worksheet 8.13.1 (researching airbags) [O3]
- Standard demand: Worksheet 8.13.2 (researching seat belts) [O3]
- High demand: Worksheet 8.13.3 (researching surfaces on the ground in children's playgrounds – if resources are available, these students might like to investigate the effects of different surfaces on the rate of deceleration – students will use the equations $v^2 = u^2 + 2as$ and $F = ma$ to explain the surface design) [O3]

Extend

Students could be asked to consider how they could show that thinking distance and speed are proportional. How could they investigate it? They should think about the key variables in their investigation.

Plenary suggestions

Hot seat: Ask each student to think of a question, using material from the lesson. Select someone to put in the hot seat; ask students to pose their questions and say at the end whether the answer is correct or incorrect.

Learning triangle: At the end of the lesson, ask students to draw a large triangle with a smaller inverted triangle that just fits inside it (so they have four triangles). Ask students to think back over the lesson and identify and write in the outer triangles:

- something they saw
- something they did
- something they discussed.

Ask students to write in the central triangle something they learned.

 © HarperCollins*Publishers* Limited 2016

Lesson 8.2: Transmitting electricity

Lesson overview

OCR Specification reference

P8.2c, d, e

Learning objectives

- Describe how electricity is transmitted using the National Grid.
- Explain why electrical power is transmitted at high potential differences.
- Understand the role of transformers.

Learning outcomes

- [O1] Recall that the National Grid is a system of cables and transformers linking power stations to consumers.
- [O2] Describe how step-up and step-down transformers change the potential difference in the National Grid.
- [O3] Explain why electrical power is transmitted at high voltages in the National Grid.

Maths focus

- Appreciate the scale of numbers involved in power transmission.

Resources needed Worksheet 8.2

Digital resources a video of a demonstration to show why high voltages are used to minimise power loss during electricity transmission (suggested YouTube search: 'STEM power lines demonstration').

Key vocabulary National Grid, transformer, potential difference, step-up, step-down

Teaching and learning

Engage

- Ask students: "How does a light bulb work?" and give them some time to note down their ideas.
- The students should then share (and develop) their ideas with a partner.
- **Pairs to fours:** the pairs of students should join into groups of four to develop their ideas. Take ideas from a number of different groups, developing ideas where appropriate. [O1]

Challenge and develop

- Issue Worksheet 8.2, which requires students to label a diagram of the National Grid. [O1]

Explain

- Show the students a video of a demonstration to show why high voltages are used to minimise power loss during electricity transmission (suggested YouTube search: 'STEM power lines demonstration'). [O1, O2, O3]
- After the video, the demonstration should be linked to the National Grid and the students should write a sentence under their diagram stating why high voltages are used to transmit electricity over large distances. [O3]

Consolidate and apply

- Ask students to return to the original question from the start of the lesson, 'How does a light bulb work?' and let them add to their answer, knowing what they do now. [O1, O2, O3]
- They should then answer questions from the Student Book appropriate to their ability:
  - Low demand: questions 1, 2 and 3 [O1, O2]
  - Standard demand: questions 3, 4 and 5 [O1, O2, O3]

 © HarperCollins*Publishers* Limited 2016

- High demand: questions 5, 6 and 7 [O1, O2, O3]
- Alternatively, the students should be given one of the following tasks appropriate to their ability:
  - Low demand: Describe the structure of the National Grid. [O1]
  - Standard demand: Explain the structure of the National Grid. [O1, O2, O3]
  - High demand: Explain the structure of the National Grid and explain why transformers are very efficient. [O1, O2, O3]

Extend

Ask students able to progress further to do the following:

- Explain why birds do not get electrocuted when they perch on power lines, despite the high potential difference. [O3]

Plenary suggestions

The big ideas: Ask students to write down three ideas they have learned during the topic so far. Then ask them to share their facts in groups and to compile a master list of facts, with the most important at the top. Ask for ideas to be shared and find out which other group(s) agreed.

Heads and tails: Ask each student to write a question about something from the lesson on a coloured paper strip and the answer on another coloured strip. Hand out the strips to groups of six to eight students, so that each student gets a question and an answer. One student reads out his or her question. The student with the correct answer then reads it out, followed by his or her question, and so on.

Hot seat: Ask each student to think of a question, using material from the topic. Select someone to put in the hot seat. Ask students to pose their questions and say at the end whether the answer is correct or incorrect.

 © HarperCollins*Publishers* Limited 2016

Lesson 8.3: Using energy resources

Lesson overview

OCR Specification reference

P8.2b

Learning objectives

- Describe the main energy resources available for use on Earth.
- Distinguish between renewable and non-renewable resources.
- Explain the ways in which the energy resources are used.

Learning outcomes

- [O1] Describe the main energy resources available for use on Earth.
- [O2] Split the energy resources into renewable and non-renewable energy resources.
- [O3] Describe the uses of the different energy resources.

Skills development

- Appreciate the power and limitations of science and consider any ethical issues that may arise.
- Explain everyday and technological applications of science; evaluate associated personal, social, economic and environmental implications; make decisions based on the evaluation of evidence and arguments.
- Evaluate risks both in practical science and the wider societal context, including perception of risk in relation to data and consequences.

Resources needed access to internet or library; Worksheet 8.3, Practical sheet 8.3, Technician's notes 8.3

Digital resources renewable energy resources video, non-renewable energy resources video, energy resources in Costa Rica video

Key vocabulary renewable resource, non-renewable resource

Teaching and learning

Engage

- Challenge students' thinking and understanding of energy resources by asking: Name some energy resources. Why are energy resources needed? What is the difference between a renewable and a non-renewable energy resource? Describe environmental issues with energy resources.
- Alternatively, show the students images of a wind turbine, solar panel, fossil fuel, nuclear fuel and biofuel and ask the students what all the pictures have in common.

Challenge and develop

- Watch videos on renewable and non-renewable energy resources. Define a renewable energy resource as one that is being (or can be) replenished as it is used. [O2]
- The students should then carry out the research part of the task outlined on Practical sheet 8.3. The class will need to be divided up into nine different groups. [O1, O2]
  - Low demand: All students are guided through the process carefully. Give them a clear structure to follow.
  - Standard and high demand: Students work in groups or independently.

 © HarperCollinsPublishers Limited 2016

Explain

- The students should then come back together as a class to give their presentations. They should make notes on each of the different energy resources as the presentations are given.

Consolidate and apply

- Watch the video on energy resources in Costa Rica. Discuss why Costa Rica produces a far higher percentage of its electricity from renewable sources than Britain. [O1, O2, O3]
- Alternatively, have students complete each section of Worksheet 8.3. [O1, O2, O3]
    - Low demand: Guide students carefully through tasks.
    - Standard and high demand: Work independently or in pairs or small groups.

Extend

- Many people assume that solar energy is only worthwhile in very hot countries yet last year, in one day in the summer, Germany produced over 50% of its electricity using solar energy. Describe possible uses of solar energy in Britain. [O3]

Plenary suggestions

Ideas hothouse: Ask students to work in pairs to list points about what they know about a particular idea from the lesson. Then ask the pairs to join together into fours and then groups of six to eight to discuss this further and to come up with an agreed list of points. Ask one person from each group to report back to the class.

Ask me a question: Ask students to write a question about something from the topic and then devise a mark scheme for the answer. Encourage them to come up with questions worth more than 1 or 2 marks and to try them out on one another.

Answers to questions

Worksheet 8.3

1. Any sensible answers, including:
   hydroelectricity – mountains and river
   solar farm – lots of land
   tidal or wave power – next to the sea
   wind – wind indicated
   fossil fuels – brought in by motorway or rail
   nuclear fuels – brought in by motorway or rail

2.
   a) Any sensible answer, including still small percentage of total power of fossil fuel companies, lack of promotion by government.
   b) renewable energy resource, little chemical pollution, no fuel costs
   c) visual pollution – people don't like the way they look
   d) less reliable than non-renewables, depends on weather, expensive set-up costs
   e) Any sensible answer, with explanation.
   f) Any sensible answer, for example, make use of roof space, no fuel costs, fewer fossil fuels used, makes or saves money for householder.
   g) Any sensible answer, for example, high set-up costs if little or no subsidy available – some people don't like the way they look.

3.
   a) mountainous with high rainfall
   b) hot rocks close to surface, for example, volcanic region
   c) lots of land to plant trees or crops
   d) lot of wind, often more near coasts
   e) sunny climate

 © HarperCollins*Publishers* Limited 2016

Lesson 8.4: Global energy supplies

Lesson overview

OCR Specification reference

P8.2b

Learning objectives

- Analyse global trends in energy use.
- Understand what the issues are when using energy resources.

Learning outcomes

- [O1] Describe the environmental impact arising from the use of different energy resources.
- [O2] Use ethical, social and economic considerations to discuss the use of different energy resources.
- [O3] Explain the importance of scientists evaluating the different energy resources.

Skills development

- Appreciate the power and limitations of science and consider any ethical issues that may arise.
- Explain everyday and technological applications of science; evaluate associated personal, social, economic and environmental implications; make decisions based on the evaluation of evidence and arguments.
- Evaluate risks both in practical science and the wider societal context, including perception of risk in relation to data and consequences.

Maths focus

- Translate information between graphical and numerical form.

Resources needed laminated cards for Groups A–G and laminated set of instructions for Group H; Worksheets 8.4.1, 8.4.2 and 8.4.3, Practical sheet 8.4, Technician's notes 8.4

Digital resources global song, for example, *Earth Song* by Michael Jackson, global energy consumption video, global energy consumption diagram

Key vocabulary efficiency

Teaching and learning

Engage

- Play the 'global song' to set the tone for the lesson as students arrive.
- Show the global energy consumption video and then display and discuss the global energy consumption diagram.
- Challenge students' thinking and understanding of global energy consumption by asking: What has happened to the demand for energy over the last couple of hundred years? How can you tell? Explain why this has happened. Suggest solutions to this problem. Or is it not a problem?

Challenge and develop

- Hold a debate as described in Practical sheet 8.4. [O1, O2, O3]
  - Low demand: Guide students through the process carefully. Advise students to organise their cards into those that would be useful for their debate and those that aren't as useful. It may be best to take H yourself.
  - Standard demand: Students work in groups, but speak to Group H to ensure they are using suitable timings, etc.
  - High demand: Students work in groups.

 © HarperCollins*Publishers* Limited 2016

Explain

- Challenge students' thinking and understanding of the energy crisis using the Conclusion section of Practical sheet 8.4. [O1, O2, O3]

Consolidate and apply

Issue students with the global energy supplies worksheet appropriate to their ability. [O1, O2, O3]

- Low demand: Worksheet 8.4.1
- Standard demand: Worksheet 8.4.2
- High demand: Worksheet 8.4.3

Extend

- Each student or student pair suggests, with explanations, what might have happened to our use of energy resources in 100 years' time. [O1, O2, O3]

Plenary suggestions

The big ideas: Ask students to write down three ideas they have learned during the topic. Then ask them to share their facts in groups and to compile a master list of facts, with the most important at the top. Ask for ideas to be shared and find out which other group(s) agreed.

What do I know? Ask students to each write down one thing about the topic they are sure of, one thing they are unsure of and one thing they need to know more about, being specific. Ask them to work in groups of five or six to agree on group lists. Ask each group to say what they decided and agree as a class what they are confident about, what they are less sure of and what things they want to know more about.

Answers to questions

Worksheet 8.4.1

1. a) 5000 million tons of oil equivalent (2×10^{20} J)
   b) 10 000 million tons of oil equivalent (4×10^{20} J)
   c) It has doubled.
   d) Any sensible answer, for example: 2036
2. increased extinctions, changes in weather patterns, melting ice leading to rising sea levels, etc.

Worksheets 8.4.2 and 8.4.3

1. **a)** 5000 million tons of oil equivalent (2×10^{20} J)
   b) 10 000 million tons of oil equivalent (4×10^{20} J)
   c) It has doubled.
   d) Any sensible answer, for example: increased population, increased travel, increased use of electronic technology.
   e) Any sensible answer, for example: 2036

Worksheet 8.4.2

2. **a)** increased extinctions, changes in weather patterns, melting ice leading to rising sea levels, etc.
   b) Any sensible answer, for example: denial, not wanting to change their lifestyles, bias due to links to fossil fuel industry, not wanting to accept findings of scientists.

Worksheet 8.4.3

2. a) increased extinctions, changes in weather patterns, melting ice leading to rising sea levels, etc.
   b) Any sensible answer, for example: denial, not wanting to change their lifestyles, bias due to links to fossil fuel industry, not wanting to accept findings of scientists.
   c) Any sensible answer, for example: ease of use, cost, reliability, power of fossil fuel industry.

Worksheets 8.4.1, 8.4.2 and 8.4.3

3. Scientists conduct research on each of the resources. Their findings are peer reviewed.

Practical sheet 8.4 Conclusion

Any sensible answer with explanation.

 © HarperCollins*Publishers* Limited 2016

Lesson 8.5: Electricity in the home

Lesson overview

OCR Specification reference

P8.2g, h, i, j

Learning objectives

- Recall that the domestic supply in the UK is a.c. at 50 Hz and about 230 V.
- Describe the main features of live, neutral and earth wires.

Learning outcomes

- [O1] Recall that domestic supply in the UK is 230 V a.c. and 50 Hz.
- [O2] Identify live, neutral and earth wires by their colour-coded insulation.
- [O3] Explain the dangers of providing any connection between the live wire and earth or our bodies.

Skills development

- Understand labelling conventions, such as those used for the colour-coding of insulation in household appliance plugs.

Resources needed a correctly wired three-pin plug and a range of incorrectly wired three-pin plugs, research materials on the three-pin plug (ideally both textbooks and access to online resources); Worksheets 8.5.1, 8.5.2 and 8.5.3

Digital resources (if the three-pin plugs are not available) images of a properly wired three-pin plug (Figure 8.12 on page 270 of the Student Book) and of a number of incorrectly wired plugs, research materials on the three-pin plug (ideally both textbooks and access to online resources)

Key vocabulary earth, live, neutral, fuse

Teaching and learning

Engage

- Ask the students to discuss what they know about the differences between the energy provided from a cell or a battery and the energy provided by the current in the mains electricity supply. [O1]
- Alternatively, ask the students what they understand by the terms 'direct current' and 'alternating current', encouraging them to focus on the words themselves in their explanations. [O1]
- Inform students that mains electricity has a current that alternates (at a frequency of 50 Hz) and that it has a much higher potential difference (about 230 V) than that of a cell. [O1]

Challenge and develop

- Issue the students with a properly wired three-pin plug. Allow them to look at it for a few moments, before giving them a range of incorrectly wired plugs and asking them to spot the mistakes. [O2]
- Alternatively, if no plugs are available, show the students an image of a properly wired three-pin plug (for example, Figure 8.12 on page 270 of the Student Book), before showing them further images of incorrectly wired plugs and asking them to spot the mistake in each image. [O2]

Explain

The students should now be split into three groups, ideally based on ability. Issue each student with an electricity in the home worksheet at an appropriate level:

- Low demand: Worksheet 8.5.1. The students will research the function of the neutral wire in a three-pin plug. [O2]
- Standard demand: Worksheet 8.5.2. The students will research the function of the earth wire in the three-pin plug. [O2, O3

 © HarperCollins*Publishers* Limited 2016

- High demand: Worksheet 8.5.3. The students will research the function of the live wire and the fuse in the three-pin plug. [O3]

Once they have done this, the students should be regrouped so that there is someone from each 'home' group in each group of three. Students should be given the chance to explain their findings to their peers. The table on the worksheet will help students to structure this feedback. [O2, O3]

Consolidate and apply

The students should use their research findings and their knowledge of the nature of mains electricity to design a leaflet for homeowners on the safe use of electricity in the home. [O1, O2, O3]

Extend

Ask students able to progress further to do the following:

- Explain what it means if an appliance is double-insulated.

Plenary suggestions

Ask me a question: Ask students to write a question about something from the lesson and then devise a mark scheme for the answer. Encourage them to come up with questions worth more than one or two marks and to try them out on each other.

Learning triangle: At the end of the lesson, ask students to draw a large triangle with a smaller inverted triangle that just fits inside it (so they have four triangles). Ask students to think back over the lesson and identify and write in the outer triangles:

- something they saw
- something they did
- something they discussed.

Ask students to write in the central triangle something they learned.

 © HarperCollins*Publishers* Limited 2016

Lesson 8.6: Red-shift

Lesson overview

OCR Specification reference

P8.3a, b

Learning objectives

- Describe red-shift.
- Describe evidence for the expanding Universe.

Learning outcomes

- By looking at spectra from stars, determine when red-shift has taken place. [O1]
- Describe Hubble's observations that led to the idea that the Universe is expanding. [O2]

Skills development

- Understand how scientific methods and theories develop over time.
- Use a variety of models to solve problems.
- Explain and evaluate everyday and technological applications of science.

Maths focus

- Visualise and represent 2D and 3D forms including 2D representations of 3D objects.

Resources needed equipment listed in Technician's notes; Worksheet 8.6, Practical sheet 8.6, Technician's notes 8.6

Digital resources

Space song video, e.g. Jamiroquai's 'Space Cowboy' https://www.youtube.com/watch?v=OPkjnRldQXQ
Red-shift diagram, e.g. http://scienceblogs.com/startswithabang/files/2013/09/redshift.jpeg
Adam Hart Davies' 'Hubble's Expanding Universe' video: www.youtube.com/watch?v=Z_TkZM4u5rQ
Flame colours demonstration:
http://www.rsc.org/learn-chemistry/resource/res00000760/flame-colours-a-demonstration
PowerPoint presentation 8.6

Key vocabulary wavelength, red-shift, Big Bang

Teaching and learning

Engage

- As students arrive, play a space song (e.g. Jamiroquai's 'Space Cowboy' https://www.youtube.com/watch?v=OPkjnRldQXQ) to set the tone for the lesson.
- Ask the students to put the following colours in order of increasing wavelength (to remind them that different parts of the visible spectrum have different wavelengths): blue, green, red, yellow and orange.

Challenge and develop

- Introduce the idea of red-shift by stating that if an object that emits light (or any other electromagnetic wave) is travelling towards you very quickly, it will have the effect of compressing the waves in front of it. This would decrease the wavelength and increase the frequency. Similarly, if the object is moving away from you, the waves would 'stretch out', increasing in wavelength and decreasing in frequency, and this is called red-shift.
- Show a diagram linking red-shift to colours. Ensure that the students recognise that when objects such as a galaxy move away and there is an increase in wavelength, the colour shifts towards the red end of the visible light spectrum, hence the term 'red-shift'.

 © HarperCollins*Publishers* Limited 2016

- Use Practical sheet 8.6 for spectra demonstration. Alternatively, carry out the flame colours demonstration given on the Learn Chemistry website (http://www.rsc.org/learn-chemistry/resource/res00000760/flame-colours-a-demonstration). [O2]
- So that the students get the idea of red-shift, ask them to listen out as an ambulance or police car (with sirens on!) passes them. They should notice that, as the vehicle approaches, the pitch of the sound increases, and as it travels away from them, the pitch lowers.
- **Pair talk:** Show the red-shift diagram in Presentation 8.6. Pair students up to discuss what they think the diagram shows.

Explain

- Challenge students' thinking and understanding of red-shift from the practical activities by using the evaluation sections of Practical sheet 8.6. [O1, O2]
- Alternatively, part or all of the Hubble's Expanding Universe video could be shown at this stage.

Consolidate and apply

Students should then complete a copy of Worksheet 8.6 appropriate to their ability:

- Low demand: Guide the students through the tasks carefully [O1, O2, O3]
- Standard and high demand: Let the students complete the tasks independently, in pairs or in small groups. [O1, O2, O3]

Extend

Explain that Hubble's evidence indicated that the rate of expansion of the Universe was decreasing with time. Group students into pairs or fours and ask them to discuss i) why it would decrease and ii) what it would suggest if evidence was found that the rate of expansion of the Universe at some point had started to increase. [O3]

Plenary suggestions

Ideas hothouse: Ask students to work in pairs to list points about what they know about a particular idea. Then ask the pairs to join together into fours and then into groups of six to eight to discuss this further and to come up with an agreed list of points. Ask one student from each group to report back to the class.

Learning triangle: At the end of the lesson, ask students to draw a large triangle with a smaller inverted triangle that just fits inside it (so they have four triangles). Ask students to think back over the lesson and identify and write in the outer triangles:

- something they saw
- something they did
- something they discussed.

Ask students to write in the central triangle something they learned.

Answers to questions

Worksheet 8.6

1. To left of diagram: moving away, increase in wavelength, red-shift; To right of diagram: moving towards, decrease in wavelength, blue-shift
2. a) absorption of these wavelengths/colours by the hydrogen
   b) Galaxy C moving away from us
   c) red-shift (increased shift in wavelength); moving away faster
   d) moving away faster, further away; greater red-shift (increased shift in wavelength)
3. Light from almost all galaxies red-shifted, i.e. an increase in wavelength of the absorption lines, because most galaxies moving away. Suggests that Universe is expanding.

Practical sheet 8.6 Evaluation

1. white
2. full spectrum of colours
3. red, orange, yellow, green, blue, indigo, violet
4. Any sensible answer, including Richard Of York Gave Battle In Vain *or* Roy G. Biv.
5. Red has the **longest** wavelength of all the colours.
6. a few different bright colours with gaps in between
7. light not emitted at these colours

 © HarperCollins*Publishers* Limited 2016

Lesson 8.7: The Sun and other stars

Lesson overview

OCR Specification reference

P8.7c

Learning objectives

- Describe how the Sun and other stars formed.
- Describe the nuclear fusion reactions in the Sun.

Learning outcomes

- [O1] Be able to describe the origin of the Sun.
- [O2] Draw a labelled diagram to explain fusion in the Sun.

Skills development

- Use a variety of models to solve problems.
- Carry out experiments.
- Interpret observations and other data.

Maths focus

- Visualise and represent 2D and 3D forms including 2D representations of 3D objects.

Resources needed Hyman fire piston, dry organic material (e.g. cotton, charcoal or tissue paper); Worksheet 8.7, Practical sheet 8.7, Technician's notes 8.7

Digital resources

Sun song video, e.g. 'Why does the Sun shine?' by They Might be Giants https://www.youtube.com/watch?v=uLpu2UP3rGI
Hyman fire piston video, e.g. https://www.youtube.com/watch?v=jYTvaF_W8rg
Fusion and temperature animation, e.g. third down on http://www.kcvs.ca/martin/astro/au/unit3/72/chp7_2.html
Detailed fusion reaction in Sun http://astro.unl.edu/classaction/animations/sunsolarenergy/fusion01.html
Thermal and radiation pressure video, e.g. https://www.youtube.com/watch?v=2al8EFZq1pU
Magnetic fields to solar flares video, e.g. https://www.youtube.com/watch?v=7CjMgVGYHUY

Key vocabulary galaxy, Milky Way, nebula, nuclear fusion, protostar

Teaching and learning

Engage

- As students arrive, play a Sun song (e.g. 'Why does the Sun shine?' by They Might be Giants https://www.youtube.com/watch?v=uLpu2UP3rGI) to set the tone for the lesson.
- **Pair talk:** Challenge students' thinking and understanding of the Sun. Let students discuss their knowledge in pairs, then join up into fours to share their thoughts. Discussions might include:
  - Low demand: What is the Sun? Why is it important to us? [O1, O2]
  - Standard demand: As low demand, plus: Where did the Sun come from? [O1, O2]
  - High demand: As standard demand, plus: What is fuelling the Sun? [O1, O2]
- Demonstrate a Hyman fire piston. (Note: if this is not possible, then show a video, e.g. https://www.youtube.com/watch?v=jYTvaF_W8rg). Get students to explain what is happening: as volume of gas decreases (and pressure of gas increases), the temperature of the gas increases to above ignition temperature for the fuel (e.g. cotton) and so it ignites. Ask how this links to the formation of the Sun. Lead to the idea that the Sun formed from gas and dust (nebula) that compressed (due to gravity), its temperature increased so much as to allow fusion to take place (rather than the fuel ignite). Ensure that the students know the difference between degrees Celsius and Kelvin and why it doesn't matter at these temperatures. [O1, O2]

 © HarperCollins*Publishers* Limited 2016

Challenge and develop

- Recap what students know about from Chapter 1. Share ideas in a class discussion. Then show an animation of fusion and temperature (e.g. http://www.kcvs.ca/martin/astro/au/unit3/72/chp7_2.html). Why does temperature affect whether fusion can take place and the rate of fusion? Link this to the point at which the Sun switched from being a protostar to being a star. [O2]
- **Pair talk:** How could the animation have been made better? Let students discuss their ideas in pairs, then join up into fours to share their thoughts. [O2]
- Show a video on magnetic fields in the Sun and solar flares (e.g. https://www.youtube.com/watch?v=7CjMgVGYHUY). Have a class discussion on what causes the magnetic fields and on the possible consequences of large solar flares.

Explain

- Challenge students' thinking and understanding of the formation of the Sun and forces inside it by using the evaluation section of Practical sheet 8.7. [O1]

Consolidate and apply

Students should then complete a copy of Worksheet 8.7 at an appropriate level to consolidate this work:

- Low demand: Guide students through the tasks carefully. [O1, O2]
- Standard and high demand: Complete the tasks independently, in pairs or in small groups. [O1, O2]

Extend

Pairs to fours: Ask student pairs to describe what they think happens when a star larger than the Sun forms – include all aspects of today's lesson. The pairs then join up into fours to discuss their ideas. [O1, O2]

Plenary suggestions

Ask me a question: Ask students to write a question about something from the topic and then devise a mark scheme for the answer. Encourage them to come up with questions worth more than one or two marks and to try them out on each other.

Learning triangle: At the end of the lesson, ask students to draw a large triangle with a smaller inverted triangle that just fits inside it (so they have four triangles). Ask students to think back over the lesson and identify and write in the outer triangles:

- something they saw
- something they did
- something they discussed.

Ask students to write in the central triangle something they learned.

Answers to questions

Worksheet 8.7

1. Correct order: b, f, a, d, e, c

b) i) 2 ii) 2 iii) 5 iv) 5

2. a) A = tritium, $^{3}_{1}H$, B = deuterium, $^{2}_{1}H$, C = helium, $^{4}_{2}He$, D = neutron, $^{1}_{0}n$

c) A = positive, +1, B = positive, +1, C = positive, +2, D = neutral, 0

d) A and B are both positive so repel each other. They need a high energy to overcome these repelling forces and join together. OWTTE

e) Fusion will take place at a higher rate.

Practical sheet 8.7 Evaluation

1. It was set alight.
2. As the volume of air decreased, its temperature increased.
3. As the volume of air decreased, its temperature increased so much that it reached the ignition temperature of the material and so it was set alight.
4. Any sensible answer including: Similarities: force inwards larger than force outwards, decrease in volume, led to increase in temperature, fuel used above certain temperature, etc. Differences: temperature in star much higher, fusion rather than ignition in star, amount of fuel, type of fuel, natural as opposed to caused by man, etc.
5. Gravity

 © HarperCollins*Publishers* Limited 2016

Lesson 8.8: Emission and absorption of infrared radiation

Lesson overview

OCR Specification reference

P8.3d

Learning objectives

- Realise that all bodies emit and absorb infrared radiation.
- Compare emission and absorption of radiation from different surfaces.
- Define a perfect black body.
- Explain that the intensity and distribution of wavelengths of any emission depend on the temperature of the body.

Learning outcomes

- [O1] State how the surface of a body affects the amount of radiation emitted and absorbed.
- [O2] Describe that a black body does not reflect or transmit any radiation.
- [O3] Explain that when the temperature is increased the intensity of every wavelength emitted increases but shorter wavelengths increase more rapidly making an object appear whiter.

Skills development

- Carry out a valid investigation.
- Draw conclusions from an investigation.

Maths focus

Resources needed copper cans covered in aluminium foil and matt black paper, thermometers, Leslie's cube with temperature probes linked to a monitor, filament bulb; Worksheets 8.8.1, 8.8.2 and 8.8.3, Practical sheet 8.8, Technician notes 8.8

Digital resources video clip of forging iron, e.g. https://www.youtube.com/watch?v=Z8cTtWD5YaA

Key vocabulary absorb, emit, infrared radiation, black body

Teaching and learning

Engage

- Show the students a video clip of forging iron (see suggested clip above).
- During the clip, the students will see the iron glowing red, orange, yellow and white in different places. Pose the questions to the students: "What is causing the difference in observed colour? What happens to the observed colour as the iron cools and is then reheated?" [O3]
- You could follow this up by asking students where they have heard the phrase white hot and what do they think it means. [O3]
- The students should be able to link observed colour and temperature. [O3]

Challenge and develop

- Students investigate the emission of radiation by different coloured surfaces by carrying out Practical 8.8. In this practical they will investigate the emission of radiation by dull, matt surfaces and shiny, light surfaces. [O1]
- Once they have collected their results, they should plot a graph of time against temperature for each of the cans. [O1]

 © HarperCollins*Publishers* Limited 2016

Explain

- Demonstrate the difference in emission of radiation by different types of surface using a Leslie's cube with four temperature probes linked up to take the temperature of each side of the cube. One side of the cube should be matt black, one matt white, one shiny black and one shiny white. The cube should be filled with hot water. [O1]
- Students could record the temperature of one of the sides of the cubes as the water cools and then they could share their results to order the sides from 'emits most radiation' to 'emits least radiation'. [O1]
- Show students a second demonstration where a filament bulb is placed in between a black body and a light body (a can covered in matt black paper and one covered in aluminium foil would suffice). The students should be asked to predict which will absorb most radiation in a given time. Thermometers or temperature probes could be placed in the cans to test their predictions. [O1]
- Define a perfect black body as an object that absorbs all the radiation that strikes it. As all the radiation is absorbed, it does not reflect or transmit any radiation; be sure to emphasise that the better an absorber an object is, the better an emitter it is as well. [O2]

Consolidate and apply

The students should now be issued with the emission and absorption of infrared radiation worksheet appropriate to their ability:

- Low demand: Worksheet 8.8.1 [O1, O2, O3]
- Standard demand: Worksheet 8.8.2 [O1, O2, O3]
- High demand: Worksheet 8.8.3 [O1, O2, O3]

Extend

Ask students able to progress further to do one of the following:

- Suggest why the inside of a telescope is coated in dark, matt material.
- Investigate Stefan's Law, linking surface area, power output and surface temperature of stars.

Plenary suggestions

Ideas hothouse: Ask students to work in pairs to list points about what they know about a particular idea. Then ask the pairs to join together into fours and then into groups of six to eight to discuss this further and to come up with an agreed list of points. Ask one student from each group to report back to the class.

Learning triangle: At the end of the lesson ask students to draw a large triangle with a smaller inverted triangle that just fits inside it (so they have four triangles). Ask students to think back over the lesson and identify and write in the outer triangles:

- something they saw
- something they did
- something they discussed.

Ask students to write in the central triangle something they learned.

Ask me a question: Ask students to write a question about something from the topic and then devise a mark scheme for the answer. Encourage them to come up with questions worth more than one or two marks and to try them out on one another.

 © HarperCollins*Publishers* Limited 2016

Lesson 8.9: The Solar System

Lesson overview

OCR Specification reference

P8.3e (part)

Learning objectives

- Describe the orbits of planets and moons in the Solar System.
- Distinguish between planets, dwarf planets and moons.

Learning outcomes

- List the planets in the Solar System in order from the Sun. [O1]
- Describe some features of planets and moons in the Solar System. [O2]
- Explain why the definition of a planet was changed, causing Pluto to become a dwarf planet. [O3]

Skills development

- Understand how scientific methods and theories develop over time.
- Use models to solve problems.
- Interpret observations and other data.

Maths focus

- Recognise and use expressions in standard form.
- Use ratios, fractions and percentages.

Resources needed Solar System model, 10 A3 pieces of card, marker pens, tape measure (40 m); Worksheet 8.9, Practical sheets 8.9.1, 8.9.2 and 8.9.3, Technician's notes 8.9

Digital resources

Space song video, e.g. David Bowie's 'Space Oddity' https://www.youtube.com/watch?v=iYYRH4apXDo
Video of relative size of bodies in the Solar System, e.g. https://www.youtube.com/watch?v=7T1LO6nOUdw
Solar System body sizes and distances to scale video, e.g. https://www.youtube.com/watch?v=97Ob0xR0Ut8
Solar System animation, e.g. http://theplanets.org/orbit/ or http://www.solarsystemscope.com/

Key vocabulary Solar System, planet, orbit, natural satellite, dwarf planet

Teaching and learning

Engage

- As students arrive, play a space song (e.g. David Bowie's 'Space Oddity' https://www.youtube.com/watch?v=iYYRH4apXDo) to set the tone for the lesson.
- **Pair talk:** Challenge students' thinking and understanding of the Solar System by discussion. Students will come into the lesson with differing knowledge on this topic, so determine first what they know. Let students discuss their knowledge in pairs, then share their thoughts. Discussions might include:
  - Low demand: What makes up the Solar System? What is the order of the planets from the Sun? [O1]
  - Standard demand: As low demand, plus: describe the different bodies in the Solar System. [O1, O2]
  - High demand: As standard demand, plus: consider why Pluto is no longer considered a planet. [O1, O2, O3]
- Show a video showing the relative sizes of bodies in the Solar System. This can lead to a discussion of relative sizes and relative distances.

 © HarperCollins*Publishers* Limited 2016

Challenge and develop

- Ask students to carry out Practical 8.9 by issuing them with a Practical sheet at an appropriate level:
  - Low demand: Practical sheet 8.9.1 (gives the distances that the students need to measure) [O1]
  - Standard demand: Practical sheet 8.9.2 (distances that the students need to measure are determined from the distance in AU) [O1]
  - High demand: Practical sheet 8.9.3 (distances that the students need to measure are determined from the distance in km) [O1]
- **Ideas hothouse:** Ask students to work in pairs to list points about how the practical could have been made better. Then ask the pairs to join together into groups of four and then groups of six to eight to discuss this further and to come up with an agreed list of points. Ask one person from each group to report back to the class. [O1, O2]

Explain

- Challenge students' thinking and understanding of distances in the Solar System from the practical activities by using the evaluation sections of Practical sheet 8.9. [O1]

Consolidate and apply

Students should then complete a copy of Worksheet 8.9 at an appropriate level to consolidate this work:

- Low demand: Guide students through the tasks carefully. [O2, O3]
- Standard and high demand: Complete tasks independently, in pairs or in small groups. [O2, O3]

Extend

Ask each student/pair to design a new body in the Solar System, which they should then describe to the rest of the class. The class should then determine what type of body it is and give their reasoning. [O2, O3]

Plenary suggestions

Ask me a question: Ask students to write a question about something from the topic and then devise a mark scheme for the answer. Encourage them to come up with questions worth more than one or two marks and to try them out on each other.

Answers to questions

Worksheet 8.9

1.

Planet	Solid or Gas	Relative size	Type of planet	Moons?	Rings?	Other interesting features
Mercury	Solid	Small	Rocky dwarf	No	No	Day longer than a year
Venus	Solid	Small	Rocky dwarf	No	No	Rotates the 'wrong way'
Earth	Solid	Small	Rocky dwarf	Yes	No	Contains life
Mars	Solid	Small	Rocky dwarf	Yes	No	Tallest mountain
Jupiter	Gas	Large	Gas giant	Yes	Yes	Red spot hurricane
Saturn	Gas	Large	Gas giant	Yes	Yes	Largest rings
Uranus	Gas	Large	Gas giant	Yes	Yes	Rolls along orbit
Neptune	Gas	Large	Gas giant	Yes	Yes	Coldest

2.

a) Multiple large bodies discovered outside Pluto's orbit
b) 'Cleared its neighbourhood of most other orbiting bodies', which Pluto does not
c) Five: Pluto, Eris, Haumea, Makemake and Ceres
d) Hundreds or thousands

 © HarperCollins*Publishers* Limited 2016

Practical sheets 8.9.1, 8.9.2 and 8.9.3 Evaluation

1. From: relative distances, mostly empty space, planets orbit Sun, order of planets, etc.
2. As the distance from the Sun increases, the distance between bodies increases.
3. Mercury, Venus, Earth, Mars, Jupiter, Saturn, Uranus, Neptune (*not* Pluto)
4. Most of the Solar System is taken up by empty space.
5. Neptune; furthest from Sun
6. Any sensible answer and sensible reason why.

 © HarperCollins*Publishers* Limited 2016

Lesson 8.10: Orbits of planets, moons and artificial satellites

Lesson overview

OCR Specification reference

P8.3e (part) f, g

Learning objectives

- Compare the orbital motion of moons, artificial satellites and planets in the Solar System.
- Describe what keeps bodies in orbit around planets and stars.
- Explain how, for circular orbits, an object can have a changing velocity but unchanged speed.
- Explain why bodies must move at a particular speed to stay in orbit at a particular distance.

Learning outcomes

- [O1] Use information on the motion of the planets to compare each one.
- [O2] Describe the motion of planets, moons and artificial satellites.
- [O3] Explain how bodies remain in orbit.

Skills development

- Understand how scientific methods and theories develop over time.
- Use a variety of models to solve problems.
- Appreciate the power and limitations of science, and consider ethical issues.

Maths focus

- Visualise and represent 2D and 3D forms, including 2D representations of 3D objects.

Resources needed Moon and meteorite samples, equipment listed in Technician's notes; Worksheet 8.10, Practical sheets 8.10.1 and 8.10.2, Technician's notes 8.10.1 and 8.10.2

Digital resources

Moon song video, e.g. The Police's 'Walking on the Moon' https://www.youtube.com/watch?v=zPwMdZOIPo8
Size of planets animation, e.g. http://sciencenetlinks.com/interactives/messenger/psc/PlanetSize.html
Jupiter's moons video, e.g. https://www.youtube.com/watch?v=Q_AnmCTdNgk
Borrow Moon samples: http://www.stfc.ac.uk/public-engagement/borrow-the-moon/
Formation of Moon video, e.g. http://www.history.com/shows/the-universe/videos/creation-of-the-moon
Artificial satellites animation, e.g. http://stuffin.space/

Key vocabulary orbit, gravitational force, natural satellite, artificial satellite

Teaching and learning

Engage

- As students arrive, play a space song (e.g. The Police's 'Walking on the Moon' https://www.youtube.com/watch?v=zPwMdZOIPo8) to set the tone for the lesson.
- Compare the sizes of planets using students' previous learning and an animation (e.g. http://sciencenetlinks.com/interactives/messenger/psc/PlanetSize.html). Lead on to a video showing three of Jupiter's moons (e.g. https://www.youtube.com/watch?v=Q_AnmCTdNgk). Use the idea of moons as natural satellites of planets. [O1]
- Ideally, samples of Moon rock will have been borrowed from STFC's Loan Scheme (http://www.stfc.ac.uk/public-engagement/borrow-the-moon/). Use Practical sheet 8.10.1 to help students to examine these. [O2]

 © HarperCollins*Publishers* Limited 2016

- **Pair talk:** Challenge students' thinking and understanding of the Moon by discussion. Let students discuss their knowledge in pairs, then join up into fours to share their thoughts. Discussions might include:
    - Low demand: What is the Moon? How do we see the Moon? [O2]
    - Standard demand: As low demand, plus: describe the orbit of the Moon, including potentially the phases. [O2]
    - High demand: As standard demand, plus: consider how the Moon came into existence. [O2]
- Show a video showing the creation of the Moon (e.g. http://www.history.com/shows/the-universe/videos/creation-of-the-moon). Lead this on to a discussion of how the Earth would be different if the collision in the video had never happened. [O2]

Challenge and develop

- **Pair talk:** Move on from natural satellites to artificial satellites. Show an animation of artificial satellites (e.g. http://stuffin.space/), and lead this on to a discussion on the purpose of these satellites. Let students discuss their knowledge in pairs, then join up into fours to share their thoughts. Discussions might include:
    - Low demand: What are artificial satellites? What are their uses? [O3]
    - Standard demand: As low demand, plus: what features must the different orbits have for their particular use? [O3]
    - High demand: As standard demand, plus: describe the two main types of orbit and match the use to the type of orbit. [O3]
- Students should then complete Practical sheet 8.10.2 to investigate the path of an object in orbit. [O2, O3]

Explain

- Challenge the students' thinking and understanding of the Moon and satellites from the practical activities by using the evaluation sections of Practical sheets 8.10.1 and 8.10.2. [O2, O3]

Consolidate and apply

Students should then complete a copy of Worksheet 8.10 at an appropriate level to consolidate this work:

- Low demand: Guide students through the tasks carefully. [O1, O2, O3]
- Standard and high demand: Complete the tasks independently, in pairs or in small groups. [O1, O2, O3]

Extend

Ask students who are able to do the following:

- Investigate non-circular orbits. [O3]

Plenary suggestions

Ask me a question: Ask students to write a question about something from the topic and then devise a mark scheme for the answer. Encourage them to come up with questions worth more than one or two marks and to try them out on each other.

Where's the answer? As students enter the room, give them a card with a word written on it. At the end of the lesson, ask a question, and the students should say if they think they have the answer card and explain why.

 © HarperCollins*Publishers* Limited 2016

Answers to questions

Worksheet 8.10

1.

Nearest to furthest from Sun	Largest to smallest	Most moons	Most artificial satellites (currently)
Mercury	8th	None	3rd (1)
Venus	6th	None	3rd (1)
Earth	5th	6th (1)	1st (2465)
Mars	7th	5th (2)	2nd (5)
Jupiter	1st	1st (63)	None, 1 soon
Saturn	2nd	2nd (62)	3rd (1)
Uranus	3rd	3rd (27)	None
Neptune	4th	4th (13)	None

2.

a) Any sensible answer from:

Earth's spin and the Moon's orbit have similar orientations.

Moon samples indicate that the Moon once had a molten surface.

The Moon has a relatively small iron core.

The Moon has a lower density than Earth.

Evidence exists of similar collisions in other star systems (that result in debris discs).

Giant collisions are consistent with the leading theories of the formation of the Solar System.

The stable-isotope ratios of lunar and terrestrial rock are identical, implying a common origin.

b) Any sensible answer from:

The ratios of the Moon's volatile elements are not explained by the 'giant impact' hypothesis. If the 'giant impact' hypothesis is correct, they must be due to some other cause.

The presence of volatiles such as water trapped in lunar basalts is more difficult to explain if the Moon was caused by an impact that would entail a catastrophic heating event.

The iron oxide (FeO) content (13%) of the Moon, which is between Mars (18%) and the terrestrial mantle (8%), rules out most of the source of the proto-lunar material from the Earth's mantle.

If the bulk of the proto-lunar material had come from an impactor, the Moon should be enriched in *siderophilic* elements (like to bond to iron), when, in fact, it is deficient in those.

The Moon's oxygen isotopic ratios are essentially identical to those of Earth. Oxygen isotopic ratios, which may be measured very precisely, yield a unique and distinct signature for each Solar System body. If a separate proto-planet Theia had existed, it probably would have had a different oxygen isotopic signature from Earth, as would the ejected mixed material.

The Moon's titanium isotope ratio ($^{50}Ti/^{47}Ti$) appears so close to the Earth's (within 4 ppm), that little if any of the colliding body's mass could likely have been part of the Moon.

c) Any sensible answer.

Practical sheet 8.10.1 Evaluation

1. Any sensible answer, e.g. made from rock, not too different from parts of Earth, irregular yet fairly smooth meteorite surface.
2. Shooting stars are small pieces of asteroid or comet that glow as they burn up in the Earth's atmosphere.
3. Any sensible answer including astronaut going to Moon and bringing back. Note: only six missions, one country and 12 astronauts have been on the surface of the Moon.

 © HarperCollins*Publishers* Limited 2016

Lesson 8.11: Key concept: Gravity: the force that binds the Universe

Lesson overview

OCR Specification reference

KC

Learning objectives

- Understand that gravity provides the force that keeps planets and satellites in orbits.
- Understand that gravity is necessary at the start of a star's life cycle and to maintain equilibrium in a stable star.
- Describe how the weight of an object depends on the gravitational field strength.
- Recognise that there is still much about the universe that is not understood, e.g. dark mass and dark energy.

Learning outcomes

- [O1] State the force that keeps planets and satellites in orbits.
- [O2] Explain the role of gravity in the formation of stars and their life cycles.
- [O3] Describe the relationship between weight and gravitational field strength.

Skills development

- Use a variety of models to solve problems.
- Carry out and represent mathematical and statistical analysis.
- Interpret observations and other data.

Maths focus

- Change the subject of an equation.
- Substitute numerical values into algebraic equations using appropriate units for physical quantities.
- Solve simple algebraic equations.

Resources needed equipment listed in Technician's notes; Worksheet 8.11, Practical sheets 8.11.1 and 8.11.2, Technician's notes 8.11.1 and 8.11.2

Digital resources

Gravity song video, e.g. James Brown's 'G.R.A.V.I.T.Y.' https://www.youtube.com/watch?v=SshO5vpWt8E
Felix Baumgartner's supersonic freefall video, e.g. https://www.youtube.com/watch?v=FHtvDA0W34I
Gravity and orbits simulation, e.g. https://phet.colorado.edu/en/simulation/gravity-and-orbits
Gravity on different planets simulation, e.g.
http://highered.mheducation.com/olcweb/cgi/pluginpop.cgi?it=swf::800::600::/sites/dl/free/0072482621/78778/Gravity_Nav.swf::Gravity%20Variations%20Interactive

Key vocabulary gravitational field strength, dark matter, dark energy

Teaching and learning

Engage

- As students arrive, play a gravity song (e.g. James Brown's 'G.R.A.V.I.T.Y.' https://www.youtube.com/watch?v=SshO5vpWt8E) to set the tone for the lesson.
- Show a video of Felix Baumgartner's supersonic freefall (e.g. https://www.youtube.com/watch?v=FHtvDA0W34I). Discuss his jump. What caused him to accelerate? What would it have felt like? Why would someone do this?
- **Pair talk:** Challenge students' thinking and understanding of gravity by discussion. Pair students up to discuss their thoughts, then invite them to share their thoughts with the class. Discussions might include:

 © HarperCollins*Publishers* Limited 2016

  - Low demand: What does gravity act between? What happens to gravity when i) the mass of an object increases and ii) the separation between masses increases? [O1, O2]
  - Standard demand: As low demand, plus: discuss what happens to gravity when the mass of one object doubles. [O1, O2]
  - High demand: As standard demand, plus: consider what happens to gravity when i) the mass of each object doubles and ii) their separation doubles. [O1, O2]
- Carry out the demonstration as described in Practical sheet 8.11.1 according to students' ability:
  - Low demand: Ensure that the students know what each part is representing at all times. [O1, O2]
  - Standard and high demand: Invite students to explain what is represented by each part and what they could investigate. [O1, O2]

Challenge and develop

- Run the animation of gravity and orbits from https://phet.colorado.edu/en/simulation/gravity-and-orbits. Ideally, this will be run individually by students with a computer each (or at least in pairs), but it may be done as a class demonstration. Issue the students with copies of Practical sheet 8.11.2 and ask them to carry out the steps from the Method section appropriate to their ability:
  - Low demand: Carry out steps 1–9. [O1, O2, O3]
  - Standard demand: As low demand, and carry out at least one task from step 10. [O1, O2, O3]
  - High demand: As standard demand, and carry out at least one additional task from step 10. [O1, O2, O3]
- **Pair talk:** Pair up students to discuss how the animation could have been made better. Then ask students to join up into fours to share their ideas, then choose and share their best ideas with the class. [O1, O2, O3]
- Run an animation showing gravity on different planets (e.g. http://highered.mheducation.com/olcweb/cgi/pluginpop.cgi?it=swf::800::600::/sites/dl/free/0072482621/78778/Gravity_Nav.swf::Gravity%20Variations%20Interactive). If computers were used to run the previous animation, students can run this one individually as well. Otherwise, show as a class demonstration. Discuss what changes need to be made on different bodies (e.g. firing cannon in the animation given) and why. [O1, O2]

Explain

- Challenge students' thinking and understanding of gravity from the practical activities by using the evaluation sections of Practical sheets 8.11.1 and 8.11.2.
- Alternatively, provide students with the gravitational field strength on different planets and ask them to calculate how their weight would change as they moved from planet to planet. They could display their results graphically. (Give students a mass of 65 kg to use if they do not wish to use their own mass in the calculations.)
- Students may need to be reminded of the equation $W = mg$ from Chapter 2 Forces. [O1, O2, O3]

Consolidate and apply

- Students should now complete a copy of Worksheet 8.11 appropriate to their ability:
  - Low demand: Guide students through the tasks carefully. [O1, O2, O3]
  - Standard and high demand: Let students complete the tasks independently, in pairs or in small groups. [O1, O2, O3]
- The students should then complete questions 4 and 5 on page 283 in the Student Book. [O3]

Extend

- Ask students individually or in pairs to describe what the Universe would be like if gravity was i) a much stronger force or ii) a much weaker force. [O2, O3]

 © HarperCollins*Publishers* Limited 2016

- Alternatively, students could be asked to research the meaning of the terms 'dark energy' and 'dark matter'.

Plenary suggestions

Hot seat: Ask each student to think of a question, using material from the topic. Select someone to put in the hot seat; ask students to pose their questions and say at the end whether the answer is correct or incorrect.

Learning triangle: At the end of the lesson, ask students to draw a large triangle with a smaller inverted triangle that just fits inside it (so they have four triangles). Ask students to think back over the lesson and identify and write in the outer triangles:

- something they saw
- something they did
- something they discussed.

Ask students to write in the central triangle something they learned.

Answers to questions

Worksheet 8.11

1. a) Attractive force that acts between all masses
   b) Mass (of each body) and separation
   c) Force due to gravity triples
   d) Force due to gravity decreases
   e) Planets would no longer orbit the Sun and would fly off in (roughly) straight lines
   f) Led to formation of stars and galaxies (and therefore life!)

2.

Distance	Distance squared	Area over which gravity is spread out	Calculation to find strength of gravity	Strength of gravity	Relative strength of gravity
1	1	1	14 400 ÷ 1	14 400	1
2	4	4	14 400 ÷ 4	3600	$\frac{1}{4}$
3	9	9	14 400 ÷ 9	1600	$\frac{1}{9}$
4	16	16	14 400 ÷ 16	900	$\frac{1}{16}$
5	25	25	14 400 ÷ 25	576	$\frac{1}{25}$

 © HarperCollins*Publishers* Limited 2016

3.

Body in Solar System	Mass of object on body (kg)	Gravitational field strength of body (N/kg)	Force due to gravity on body (N)
Sun	20	274	5480
Mercury	20	4	80
Venus	20	9	180
Earth	20	10	200
Mars	20	4	80
Jupiter	20	25	500
Saturn	20	10.4	208
Uranus	20	8.9	178
Neptune	20	11	220
The Moon	20	1.7	34

Practical sheet 8.11.1 Evaluation

1. Reaction force from sheet (and/or friction)
2. Gravity
3. Increased
4. The closer the planet is to the Sun, the **faster** it moves in orbit.
5. Increased
6. The closer the planet is to the Sun, the **stronger** the gravity from the Sun upon it.
7. Increased force on marbles, spiral inwards in less time
8. Marbles lose energy due to friction
9. No friction – planets not losing energy

Practical sheet 8.11.2 Evaluation

1. Increases
2. Decreases
3. Adjust position of satellite slightly
4. Start planet very close to star
5. Start planet very far from star
6. Satellite no longer orbits and moves in a straight line

 © HarperCollins*Publishers* Limited 2016

Lesson 8.12: Temperature of the Earth

Lesson overview

OCR Specification reference

P8.3h

Learning objectives

- Describe how the atmosphere absorbs radiation in a way that varies with wavelength.
- List the factors affecting the temperature of the Earth.
- Explain how the temperature of an object is related to the radiation absorbed and radiation emitted.

Learning outcomes

- [O1] State the causes of an increase in the temperature of the Earth's surface.
- [O2] Carry out targeted research.
- [O3] Work as a group to develop knowledge.

Skills development

- Carry out valid research.
- Draw conclusions from research.
- Present research findings to peers.

Maths focus

Resources needed filament lamp, small drinks bottle filled with carbon dioxide, small drinks bottle filled with air, two temperature probes linked to a monitor, range of research material – ideally some books from the library and access to internet research sources; Worksheets 8.12.1, 8.12.2, 8.12.3 and 8.12.4

Digital resources two images of rainforest in Borneo – one taken in 1950 and another of the same area taken more recently

Key vocabulary absorption, emission, reflection, infrared radiation

Teaching and learning

Engage

- Show students two images: one of a tropical rainforest in Borneo in 1950 and one of the same area in 2010. [O1]
- **Pair talk:** Ask students to discuss what they think could be causing the loss of vegetation and what effects this could have. [O1]
- **Pairs to fours:** Ask the pairs to join into groups of four to develop their ideas further, before feeding back to the class. [O1]

Challenge and develop

- Demonstrate the difference in radiation absorption by carbon dioxide and air. This can be done by having two small drinks bottles – one filled with pure carbon dioxide and the other filled with air. Either use temperature probes linked to a monitor or insert a thermometer through the lid of each bottle and stand the two bottles an equal distance away from a filament lamp. [O1]
- Ask the students to record the temperature of each bottle after a few minutes. [O1]

Explain

- **Jigsaw:** Split the class into three. Ideally, students should be grouped by ability.

 © HarperCollins*Publishers* Limited 2016

- Group 1: Low demand, Worksheet 8.12.1 [O1, O2]
- Group 2: Standard demand, Worksheet 8.12.2 [O1, O2]
- Group 3: High demand, Worksheet 8.12.3 [O1, O2]

Consolidate and apply

- Students should be regrouped in such a way that they are working in groups of three, with someone from each of the original expert groups in each home group. [O1, O2, O3]
- Students should share their research findings and complete Worksheet 8.12.4 as a means of structuring their feedback. [O1, O2, O3]
- During their research, students may well have come across the fact that short wavelength radiation, such as UV, tends to travel straight through the atmosphere whereas longer wavelength radiation, such as infrared radiation, tends to get reflected. If not, this should be made clear to them at this stage. [O1]

Extend

Ask students able to progress further to do one of the following:

- Explain the difference between global warming and the greenhouse effect in no more than 30 words.
- Alternatively, answer question 8 on page 285 of the Student Book.

Plenary suggestions

Heads and tails: Ask each student to write a question about something from the topic on a coloured strip of paper and the answer on another coloured strip. In groups of six to eight, hand out the strips so that each student gets a question and an answer. One student reads out their question. The student with the right answer then reads it out, followed by their question, and so on.

Hot seat: Ask each student to think up a question, using material from the topic. Select someone to put in the hot seat, ask students to pose their question and say at the end whether the answer is correct or incorrect.

What do I know? Ask students to each write down one thing about the topic they are sure of, one thing they are unsure of and one thing they need to know more about, being specific. Ask them to work in groups of five or six to agree on group lists. Ask each group to say what they decided and agree as a class what they are confident about, what they are less sure of and what things they want to know more about.

 © HarperCollins*Publishers* Limited 2016

Lesson 8.13: Measuring wave speeds

Lesson overview

OCR Specification reference

P8.3i (part)

Learning objectives

- Explain how the speed of sound in air can be measured.
- Explain how the speed of water ripples can be measured.
- Describe the use of echo sounding.

Learning outcomes

- [O1] Recognise that the speed of sound through solids and liquids is faster than in air.
- [O2] Describe how to measure the speed of sound through air and the speed of water waves.
- [O3] Know how to calculate the depth of water using the speed of sound through water in echo sounding and the time between signal and echo.

Skills development

- Correct use of key terms.

Maths focus

Resources needed Worksheets 8.13.1, 8.13.2 and 8.13.3, Practical sheet 8.13, Technician's notes 8.13

Digital resources Video clip showing how to calculate the speed of sound in air, e.g. https://www.youtube.com/watch?v=KdaMiM-dJFc; BBC Ocean Giants: Voices of the Sea: Echolocation lifeline: http://www.bbc.co.uk/programmes/p00jz21r

Key vocabulary echo, echo sounding

Teaching and learning

Engage

- Watch a video clip showing how to calculate the speed of sound in air, e.g. https://www.youtube.com/watch?v=KdaMiM-dJFc . Pose the question to the students: "How would the speed of sound differ had they conducted the experiment through a liquid or a solid?" [O1, O2]

Challenge and develop

- Ask the students to investigate the speed of water waves. Practical sheet 8.13 gives details of how to carry out this investigation. [O2]

Explain

- Introduce the concept of echo sounding through a **snowball** to address the following question. Individuals consider their answer briefly, then pairs discuss the question, then double up to fours and continue the process into groups of eight. This allows for comparison of ideas or to sort out the best answer. Finally, the whole class is drawn together and a spokesperson for each group of eight feeds back ideas.
- Pose the question to the students: "Having seen that waves reflect off a surface, how might this be useful for navigators on ships to calculate the depth of the sea?" [O3]
- Sonar (SOund Navigation And Ranging) could also be introduced here.
- Show the BBC Ocean Giants: Voices of the Sea: Echolocation lifeline as an example of echolocation in action, where narwhals use sound to map air holes in ice.

 © HarperCollins*Publishers* Limited 2016

Consolidate and apply

- Issue the students with the measuring wave speed worksheet appropriate for their ability:
  - Low demand: Worksheet 8.13.1 [O2, O3]
  - Standard demand: Worksheet 8.13.2 [O2, O3]
  - High demand: Worksheet 8.13.3 [O2, O3]

Extend

Ask students able to progress further to do the following:

- **Pair talk:** Give the students the following situation: despite hunting in complete darkness, bats are able to catch their prey effectively. Explain how. [O3]

Plenary suggestions

- **Ask me a question:** Ask students to write a question about measuring wave speeds and then devise a mark scheme for the answer. Encourage them to come up with questions worth more than one or two marks and try them out on one another. [O1, O2, O3]
- Show a range of images of the use of echolocation in nature and ask the students to explain what is happening in the pictures. [O3]

Answers to questions

Worksheets 8.13.1 and 8.13.2

1. a) (i) 0.016 s (ii) 0.33 s b) (i) 900 m (ii) 3000 m

2. a) light travels faster than sound b) 1800 m c) 1200 m closer (at 600 m)

Time to hit sea bed = 1.2/2 = 0.6 s

1600 m/s × 0.6 s

= 960 m

Worksheet 8.13.3

1. a) (i) 0.016 s (ii) 0.33 s b) (i) 3000 m (ii) 60 m

2. a) light travels faster than sound b) 5 s c) 900 m closer (at 600 m)

Time to hit sea bed = 1.2/2 = 0.6 s

1600 m/s × 0.6 s

= 960 m

 © HarperCollins*Publishers* Limited 2016

Lesson 8.14: Seismic waves

Lesson overview

OCR Specification reference

P8.3i (part)

Learning objectives

- Describe how earthquakes are detected.
- Describe the properties of P waves and S waves.
- Explain how the properties of seismic waves allow us to investigate the inside of the Earth.

Learning outcomes

- [O1] Know how earthquakes can be detected.
- [O2] Use the terms focus and epicentre correctly.
- [O3] Use the properties of S waves and P waves to deduce useful information about the structure of the Earth..

Skills development

- Consider a range of evidence.
- Draw conclusions from evidence.
- Communicate ideas clearly.

Resources needed Worksheets 8.14.1, 8.14.2 and 8.14.3

Digital resources Clip of CCTV footage of April 2015 Nepal Earthquake
List of recent earthquakes: http://www.ncedc.org/anss/maps/alaska.M7.1900-2002.txt
Map template http://web.ics.purdue.edu/~braile/edumod/epiplot/epiplot.htm
World map with earthquake sites:
https://www.google.co.uk/url?sa=i&rct=j&q=&esrc=s&source=images&cd=&cad=rja&uact=8&ved=0CAcQjRxqFQoTCJKn2cW8g8kCFcOgDgodRm0JjA&url=http%3A%2F%2Fwww.regentsearth.com%2FIllustrated%2520ESRT%2FPage%25205%2520(Tectonic%2520Plates)%2FESRT%2520Page%25205%2520Index.htm&psig=AFQjCNG1A46x_2g0KhiBh0lTVkoqnWR4mg&ust=1447163032845018

Key vocabulary earthquakes, seismic waves, P waves, S waves

Teaching and learning

Engage

- Show the students a clip of CCTV footage from the Nepal earthquake in April 2015. During the clip, ask the students to note down how they think the people in the video were feeling at the time.
- Following the clip, locate Nepal on a world map (preferably one with the tectonic plate boundaries marked). [O1]

Challenge and develop

- Either provide the students with a list of recent earthquakes (such as the list found at the link given above) and ask them to plot these on a map template (see link above), or show an image of a world map with earthquake sites marked (see link above).
- **Pair talk:** What patterns can be seen in the picture? What conclusions can be drawn from the patterns that have been identified?
- **Pairs to fours:** Each pair should join up with another pair to share ideas and then feedback to the class. [O1]

 © HarperCollins*Publishers* Limited 2016

Explain

- **Class discussion:** How is the Earth's movement monitored? How are earthquakes detected? A seismometer in Nepal would detect more waves than one placed on the opposite side of the Earth (in Chile). Are all waves the same? How do we know?
- Remind the students of the difference between longitudinal and transverse waves (perhaps by modelling using a flexible spring) and ask the students to give an example of each type of wave. Remind the students of the materials that transverse and longitudinal waves will pass through.
- Refer to a picture of P and S waves transmitted through the Earth (Figures 8.48 and 8.49 in the Student Book). What can be deduced about the nature of each type of wave and therefore about the structure of the Earth? [O3]

Consolidate and apply

- Using the report template appropriate to their level, the students write a report for a journal on the earthquake in Nepal. The report should include: the cause of the earthquake, how the earthquake can be detected, different types of wave that the earthquake produced, what can be deduced about the nature of the structure of the Earth from looking at the patterns on seismometers. [O2, O3]
  - Low demand: Worksheet 8.14.1
  - Standard demand: Worksheet 8.14.2
  - High demand: Worksheet 8.14.3
- Alternatively, the class could be asked to produce a news report for the evening news bulletin. The class should be split into groups, where students act as either scientists, news reporters (to screen), journalists (who carry out interviews), local residents or politicians. The students could write a script and then film their report. [O2, O3]

Extend

- Students could be asked to research the aftershocks of a particular earthquake and then present their findings to the rest of the class.
- Alternatively, students could be asked to investigate the time difference between the detection of a P and an S wave and therefore decide the position of an earthquake, given the speeds of both types of wave.

Plenary suggestions

The big ideas: Ask students to write down three ideas they learned during the topic. Then ask them to share their facts in groups and to compile a master list of facts, with the most important at the top. Ask for ideas to be shared and find out which other group(s) agreed.

Ideas hothouse: Ask students to work in pairs to list points about what they know about a particular idea. Then ask the pairs to join together into fours and then into groups of six to eight to discuss this further and to come up with an agreed list of points. Ask one person from each group to report back to the class.

© HarperCollins*Publishers* Limited 2016

Lesson 8.15: Maths skills: Using scale and standard form

Lesson overview

Learning objectives

- Understand the scale of objects in the Universe.
- Use standard form.

Learning outcomes

- [O1] Compare the sizes of objects in the Universe to each other and to everyday objects.
- [O2] Convert very large and very small numbers into standard form, and vice versa.
- [O3] Carry out calculations using numbers in standard form.

Skills development

- Carry out and represent mathematical and statistical analysis.
- Use SI units (e.g. kg, g, mg; km, m, mm; kJ, J) and IUPAC chemical nomenclature unless inappropriate.
- Use prefixes and powers of 10 for orders of magnitude (e.g. tera, giga, mega, kilo, centi, milli, micro and nano).

Maths focus

- Recognise and use expressions in decimal form.
- Recognise and use expressions in standard form.
- Use ratios, fractions and percentages.

Resources needed Worksheet 8.15

Digital resources

Space song video, e.g. R.E.M.'s 'Man on the Moon' https://www.youtube.com/watch?v=1hKSYgOGtos
Animation 'Tediously Accurate Scale Model of the Solar System':
http://joshworth.com/dev/pixelspace/pixelspace_solarsystem.html
Scale of the Universe animation, e.g. http://htwins.net/scale2/

Key vocabulary scale, standard form

Teaching and learning

Engage

- As students arrive, play a space song (e.g. R.E.M.'s 'Man on the Moon' https://www.youtube.com/watch?v=1hKSYgOGtos) to set the tone for the lesson.
- Show the animation 'Tediously Accurate Scale Model of the Solar System': http://joshworth.com/dev/pixelspace/pixelspace_solarsystem.html (either individually or in pairs at computers, or as a class) to introduce the idea of scale and how small the Sun and planets are compared with the distances between them. Lead this on to a discussion of the amount of apparently empty space in the Solar System and how that makes the students feel. [O1]
- **Pair talk:** Challenge students' thinking and understanding of sizes within the Solar System by discussion. Pair up students to discuss their ideas, then ask them to share these with the class. Discussions might include:
    - Low demand: What makes up the Solar System? What are the largest and smallest bodies in the Solar System?
    - Standard and high demand: As low demand, plus: how many of our Moons do you think could fit inside the Earth? [about 50] Also how many Earths do you think could fit inside the Sun? [1.3 million]

 © HarperCollins*Publishers* Limited 2016

Challenge and develop

- Write down the following in full (i.e. not in standard form):
  average distance from Sun to Earth (1.50×10^{11} m)
  diameter of Earth (1.27×10^{7} m)
  the distance from John o'Groats to Land's End by road (9.70×10^{5} m)
  distance round an athletics track (400 m)
  diameter of an average atom (3×10^{-10} m).
  Discuss with students how better to write these numbers, as appropriate to their ability:
  - Low demand: Guide students through the above carefully. [O1, O2]
  - Standard and high demand: Students attempt all of the above by themselves initially. If they struggle, do the first one or two with them and then ask them to do the rest. [O1, O2]
- Show a scale of the Universe animation (e.g. http://htwins.net/scale2/) (individually or in pairs at computers, or as a class). Show both the very small and the very big. Highlight the use of standard form and the range of standard form used. Ask the students to choose some numbers to write out both in full and in standard form. Which do they prefer? [O1, O2]

Explain

- Challenge students' use of standard form by practising calculations where one or both numbers used are in standard form. Examples may include: i) calculate the distance when speed of light 3×10^{8} m/s travels for 1.2×10^{5} s [answer = 3.6×10^{13} m] ii) calculate the time taken for light to travel across the Milky Way (a distance of 9.6×10^{20} m, at a speed of 3×10^{8} m/s) [answer = 3.2×10^{12} s = 101 000 years]. [O1, O2, O3]

Consolidate and apply

Students should now complete a copy of Worksheet 8.15 appropriate to their ability:

- Low demand: Guide the students through the tasks carefully. [O1, O2, O3]
- Standard and high demand: Let the students carry out the tasks independently, in pairs or in small groups. [O1, O2, O3]

Extend

Ask students individually or in pairs to write a list of distances in the Universe in order of size from smallest to largest. Then ask them to estimate how many times bigger each one is compared with the others. [O1, O2, O3]

Plenary suggestions

Nominate students to ask standard form questions for the rest of the class to answer.

Answers to questions

Worksheet 8.15

1. a) Four times heavier
   b) Five
   c) 2×10^{23} = 200 000 000 000 000 000 000 000
   d) 2.8×10^{7} = 28 000 000
   e) 1×10^{23} = 100 000 000 000 000 000 000 000
2. a) i) 1×10^{3} m ii) 9×10^{4} kg iii) 5.4×10^{6} m^3 iv) 3×10^{8} m/s
   b) i) 4000 m ii) 6 000 000 000 kg iii) 83 000 000 m^3 iv) 0.000 000 002 m
3. a) i) 6.4×10^{6} ii) 3×10^{4} iii) 3×10^{8} iv) 500 or 5×10^{2}
   b) 8×10^{13} s = 2.5 million years
   c) 4.2×10^{25} m
4. a) i) 5 mm ii) 6 kN iii) 2 μm iv) 3 Tm
   b) i) 4×10^{9} V ii) 7×10^{-9} m iii) 9×10^{3} m iv) 8×10^{6} N

 © HarperCollins*Publishers* Limited 2016

When and how to use these pages: Check your progress, Worked example and End of chapter test

Check your progress

Check your progress is a summary of what students should know and be able to do when they have completed the chapter. Check your progress is organised in three columns to show how ideas and skills progress in sophistication. Students aiming for top grades need to have mastered all the skills and ideas articulated in the final column (shaded pink in the Student book).

Check your progress can be used for individual or class revision using any combination of the suggestions below:

- Ask students to construct a mind map linking the points in Check your progress
- Work through Check your progress as a class and note the points that need further discussion
- Ask the students to tick the boxes on the Check your progress worksheet (Teacher Pack CD). Any points they have not been confident to tick they should revisit in the Student Book.
- Ask students to do further research on the different points listed in Check your progress
- Students work in pairs and ask each other what points they think they can do and why they think they can do those, and not others

Worked example

The worked example talks students through a series of exam-style questions. Sample student answers are provided, which are annotated to show how they could be improved.

- Give students the Worked example worksheet (Teacher Pack CD). The annotation boxes on this are blank. Ask students to discuss and write their own improvements before reviewing the annotated Worked example in the Student Book. This can be done as an individual, group or class activity.

End of chapter test

The End of chapter test gives students the opportunity to practice answering the different types of questions that they will encounter in their final exams. You can use the Marking grid provided in this Teacher Pack or on the CD Rom to analyse results. This shows the Assessment Objective for each question, so you can review trends and see individual student and class performance in answering questions for the different Assessment Objectives and to highlight areas for improvement.

- Questions could be used as a test once you have completed the chapter
- Questions could be worked through as part of a revision lesson
- Ask Students to mark each other's work and then talk through the mark scheme provided
- As a class, make a list of questions that most students did not get right. Work through these as a class.

 © HarperCollins*Publishers* Limited 2016

Marking Grid for End of Chapter 8 Test

Student Name	Getting Started [Foundation Tier] Q. 1 (AO1) 2 marks	Q. 2 (AO1) 1 mark	Q. 3 (AO1) 2 marks	Q. 4 (AO1) 1 mark	Q. 5 (AO1) 1 mark	Q. 6 (AO1) 1 mark	Q. 7 (AO1) 2 marks	Q. 8 (AO1) 1 mark	Going Further [Foundation and Higher Tiers] Q. 9 (AO1) 2 marks	Q. 10 (AO1) 1 mark	Q. 11 (AO1) 1 mark	Q. 12 (AO3) 4 marks	More Challenging [Higher Tier] Q. 13 (AO1) 1 mark	Q. 14 (AO2) 2 marks	Q. 15 (AO1) 3 marks	Q. 16 (AO1) 1 mark	Q. 17 (AO1) 2 marks	Q. 18 (AO2) 6 marks	Most demanding [Higher Tier] Q. 19 (AO2) 3 marks	Q. 20 (AO2) 6 marks	Q. 21 (AO2) 2 marks	Q. 22 (AO3) 6 marks	Q. 23 (AO3) 6 marks	Total marks	Percentage

 © HarperCollins*Publishers* Limited 2016

Check your progress

You should be able to:

State that vehicle speed and reaction time affect the stopping distance of a vehicle	Describe factors that affect a driver's reaction time and a vehicle's braking distance	Calculate braking forces using ideas of stopping distance and energy transfer
State that various resources are used as fuels and to generate electricity	Describe the advantages and disadvantages of fossil fuel, nuclear and renewable energy resources	Evaluate and justify the use of various energy resources for different applications
Recall that the National Grid is a system of cables and transformers linking power stations to consumers	Describe how step-up and step-down transformers change the potential difference in the National Grid	Explain why electrical power is transmitted at high voltages in the National Grid
State that high potential differences are used to reduce power transmission losses	Explain how changing potential difference affects current for the same power	Explain how power transmission losses are related to the square of the current
Identify live, neutral and earth wires by their colour-coded insulation	Explain why a live wire may be dangerous even when a switch in the main circuit is open	Explain the dangers of providing any connection between the live wire and earth or our bodies
State that the hotter the body the more radiation it emits in a given time	Explain that a perfect black body absorbs all the radiation incident on it, and does not reflect or transmit any radiation	Explain how the temperature of a body is related to the balance between incoming radiation absorbed and radiation emitted
Describe how planets and dwarf planets are arranged in the Solar System	Describe how moons are arranged in the Solar System in relation to planets	Describe the relationship between a solar system and a galaxy
Recall what stars are formed from	Explain the role of gravity in star formation	Explain the balance of forces in a stable star
Recall that stars go through a life cycle	Describe the life cycle of stars similar in size to our Sun and much larger stars	Explain the role of fusion in the life cycle of a star
Recall that elements heavier than iron are produced in a supernova	Explain how heavy elements become distributed throughout the Universe	Explain the role of fusion in the formation of elements
Compare the orbital motions of planets, moons and artificial satellites	Explain the role of gravity in enabling objects to describe circular orbits	Explain the relationship between speed and radius in stable orbits
Recall that an observed increase in wavelength of light results in red-shift	Recall that more distant galaxies are travelling away faster	Explain what this evidence suggests about the Universe
Recall that the Big Bang suggests the Universe started from a small dense region	Explain how red-shift provides evidence for the Big Bang	Explain how scientists use evidence such as red-shift to support the Big Bang theory

 © HarperCollins*Publishers* Limited 2016

Worked example

1. **Describe the orbit of an artificial satellite.**

   It goes round the Earth.

   Yes satellites go round the Earth but the answer could be improved by describing what kind of orbit, i.e. circular.

2. **In what way do the observations made by Hubble support the Big Bang?**

   The observation shows there is red-shift.

   Just saying red-shift is not really enough. The answer should go into more detail to explain what is meant by red-shift, what Hubble observed, what it implied and why he called this effect red-shift.

3. **Explain the differences and similarities between moons and artificial satellites.**

   They both orbit planets, but one of them has been put into orbit.

   This is correct but does not give enough information. You should aim to state what the similarities are, and what the differences are – and which object the differences apply to, e.g. moons are natural satellites, but artificial satellites have been put into orbit by humans.

4. **What does it mean when a star is in its stable period?**

   It's in the main sequence.

   Stable period and main sequence are different names for the same thing.

   You need to show that you understand and describe what is happening inside the star when it is in its stable period.

5. **Every star goes through a cycle of birth, life and death. Explain how the mass of a star determines its life cycle.**

   A small star has a long life cycle and becomes a white dwarf. A big star has a much shorter life cycle and blows up.

   The answers needs to be more precise with the words used. Yes it has to do with size, but in what way?

   You should describe what happens to an average sized star and then describe the stages that a much more massive star goes through. Structure your answer carefully and use the correct scientific terms.

 © HarperCollins*Publishers* Limited 2016

Chapter 1: Matter

Lesson 1.1 Developing ideas for the structure of the atom

1 No it only includes the electrons. The rest of the atom was a positively charged sphere.

2 Yes – although it is unclear how they are balanced.

3 The results were very surprising and other scientists needed to peer review the work to check that they had not made any mistakes. The work is useful for other scientists to develop in order to produce further scientific theories.

4 The experimental results could not be explained by the current model of the atom.

5 Rutherford's model explained why most of the alpha particles went through the gold foil and only a few bounced back. It also went on to explain what was happening in radioactive decay.

6 More people do experiments to test whether the results agree with the predictions of the scientific theory. If the theory correctly predicts the results of many experiments over a long period of time it becomes gradually accepted. However just one experiment's results can force a theory to be changed as shown by Geiger and Marsden's experiment.

Lesson 1.2 Atomic structure

1 88 electrons

2 92 protons, 238 – 92 = 146 neutrons

3 The number of electrons is the same; the number of protons is the same; the number of neutrons is different.

4 Uranium: 92 protons, 238 – 92 = 146 neutrons

Thorium: 90 protons, 234 – 90 = 144 neutrons

5a 7 protons and 14 - 7 = 7 neutrons.

5b 92 protons and 235 - 92 = 143 neutrons.

6 Nuclear radiation can knock electrons off atoms. These atoms become positive ions since they have lost negative charge. Once the electrons have been knocked off, these can join onto other atoms. These atoms become negative ions because they have gained negative charge.

Lesson 1.3 Density

1a solid

1b gas

1c liquid

2 The particles in a solid are usually closer together than they are in a liquid or a gas. Therefore, the same mass of material will occupy a smaller volume which makes the density higher.

3 The particles in a gas are far apart. Therefore, the volume of a certain mass of gas is much bigger than the same mass of liquid and solid. This makes the density small.

4a $\rho = m \div V = 5400 \div 2 = 2700$ kg/m^3

4b $m = \rho V = 7700 \times 2 = 15\ 400$ kg.

4c Aluminium is less dense than steel. Therefore, aeroplanes made from aluminium are likely to be much lighter.

5 Volume = 5 × 4 × 3 = 60 m^3.

$m = \rho V = 1.3 \times 60 = 78$ kg.

6 Cork is less dense than water so it floats. Iron is denser than water, so it sinks.

7 The mass of the air stays the same but the volume of the air gets less. Since density = mass / volume, this means the density of the air will increase.

8 1 g / cm^3 means that each cm^3 of the substance will have a mass of 1 g. There are 100 × 100 × 100 = 1000 000 cm^3 in 1 m^3, so 1 m^3 of the substance will have a mass of 1000 000 g. 1000 000 g = 1000 kg, so 1 m^3 of the substance has a mass of 1000 kg – giving a density of 1000 kg / m^3.

Lesson 1.4 Key Concept: Particle model and changes of state

1 In a solid, the atoms and molecules vibrate around a fixed point. In a liquid the atoms and molecules can move past each other.

2 The particles vibrate with a larger amplitude. Therefore, their average separation increases.

3 The internal energy increases. This is because the potential energy increases from the particles getting further apart and the kinetic energy increases from the particles vibrating with a greater speed.

4

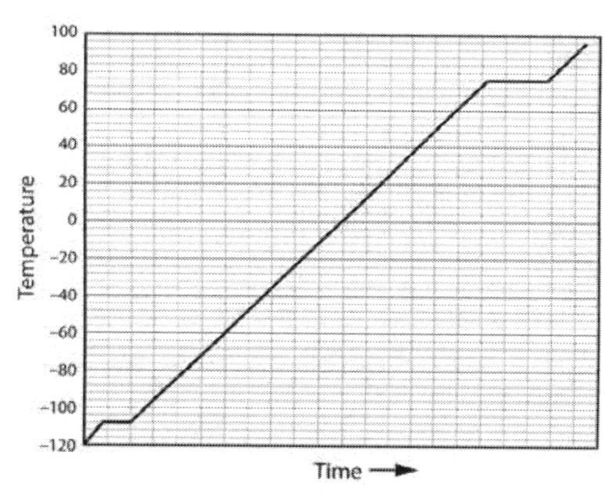

 © HarperCollins*Publishers* Limited 2016

5 The material is cooling down.

Lesson 1.5 Practical: To investigate the densities of regular and irregular solid objects and liquids

1 The balance would also be recording the mass of the measuring cylinder.

2 Subtract the mass of the empty measuring cylinder to get the mass of the liquid.

3 Density = mass / volume

Coconut oil: 18.5 / 20 = 0.925 g / cm^3

Acetone: 19.6 / 25 = 0.784 g / cm^3

Sea water: 51.3 / 50 = 1.026 g / cm^3

4 Volume of cork = 2.0 × 2.0 × 3.0 = 12 cm^3

5 Density of cork = mass / volume = 3 / 12 = 0.25 g / cm^3

6 Density of oak = mass / volume = 17 / (2.0 × 3.0 × 4.0) = 0.71 g / cm^3

Density of tin = mass / volume = 365 / (2.5 × 2.5 × 8.0) = 7.3 g / cm^3

7 The data is only measured to 2 significant figures. Therefore, the answer can only be given to two significant figures. It is incorrect to give any more significant figures as this suggests that the calculation is more accurate than it actually is.

8 You could half fill a measuring cylinder with water. Record the volume of the water. Then place the necklace into the water and make sure it is fully submerged. Record the new volume of the water. The volume of the necklace is the difference between the two volumes you measured. Then you could find the mass of the necklace by placing it on a balance. Repeat the measurements and find an average to reduce the effects of random errors.

9 You would need to measure the mass of the necklace and the volume of the necklace.

10 Density = mass / volume

11 There are many errors in the experiment such as not reading the measuring cylinder very accurately. Perhaps your eyes weren't lined up with the bottom of the meniscus or you weren't holding the measuring cylinder completely vertically. Also the volume of the necklace is quite small and the measuring cylinder would not be sensitive enough to measure small changes in the volume accurately.

Lesson 1.6 Changes of state

1 Freezing

2 You could place a block of ice in a container and then place the container on a balance. Record the mass and then wait for all of the ice to melt. Record the mass again and see whether the mass has changed.

3 e.g. Dry ice changing from a solid to a gas (sublimating). The material involved is carbon dioxide. (Dry ice is solid CO_2)

4 A freezing temperature is not necessarily a cold temperature. Some materials (e.g. tungsten) freeze at thousands of degrees Celsius. We are really only referring to the temperature at which water freezes.

5 This makes the surface area larger and so more evaporation can take place.

6 The fastest moving particles are the ones which evaporate. When they leave the liquid, the average speed of the remaining particles is less (since the fastest ones have left). The temperature is related to the average speed and so the temperature decreases.

7 When you are burned by steam, the steam transfers energy to your skin when it is condensing. This is extra to the energy transferred to your skin when the hot water cools down.

8 Sweat is no colder than your skin. The cooling effect occurs because the sweat evaporates. Not all of the water molecules in the sweat move at the same speed and it is the ones that move the fastest that evaporate. Therefore, the average speed of the molecules decreases as the sweat evaporates and this results in a lower temperature.

Lesson 1.7 Internal energy

1 They store kinetic energy because they are moving.

2 $E_k = \frac{1}{2} mv^2$ and the particles have the same kinetic energy at the same temperature. This means that the heavy particles are moving slower than the light particles at the same temperature.

3 The particles store potential energy because they are separated from each other.

4 The Pacific Ocean has more internal energy than the tea. Each particle in the tea does store more kinetic energy (the tea is hotter) and more potential energy (the particles are further apart). However, there are many more particles in the Pacific Ocean so the **total** of the kinetic and potential energies stored by the particles in the Pacific Ocean is a much bigger value.

5 The internal energy increases.

6 The water cools down, freezes and cools down again. All of this results in a decrease in internal energy.

 © HarperCollins*Publishers* Limited 2016

7a The internal energy of steam at 100 ^{0}C is much higher than that of water at the same temperature. The internal energy would also include all of the latent heat of vaporisation.

7b Steam is able to transfer much more energy than water at the same temperature as its internal energy is so much higher.

Lesson 1.8 Specific heat capacity

1 They move faster (gain kinetic energy) and they get further apart (gain potential energy).

2 There is a larger mass of water in the saucepan than there is in the cup. Therefore, more energy is needed.

3 Yes it will. The gain in internal energy of the milk is smaller when it heats up. Therefore, the decrease in internal energy will be smaller when it cools down, so the amount of energy transferred into the surroundings will be less.

4 $\Delta E = mc\Delta\theta = 0.1 \times 4200 \times (40 - 10) = 12\ 600$ J

5 e.g. in cooling systems. Water passing through a car engine can stop the engine from heating up by absorbing some of the thermal energy. The water can absorb lots of energy into its thermal energy store without heating up very much.

6a Energy needed = $mc\Delta\theta$ for the copper + $mc\Delta\theta$ for the water

= (0.5 × 380 × 10) + (1 × 4200 × 10) = 43 900 J

$t = E / P$ = 43 900 / 2000 = 21.95 s = 22 s (to 2 s.f.)

6b I have assumed that all of the energy from the heater has been transferred to the thermal energy stored in the water and in the copper kettle.

7 Copper is a very good conductor of heat – it has a very high thermal conductivity. It also has a low specific heat capacity so not much energy is needed to heat the saucepan up.

8a The temperature of the copper decreases and the temperature of the water increases. Energy is transferred from the mass to the water by heating. Therefore the mass loses internal energy and the water gains internal energy.

8b The final temperature is likely to be closer to 20 ^{0}C. The specific heat capacity of water is much larger than that of copper. Therefore for the same mass (both 50g) and the same amount of energy transfer by heating, the temperature change of the copper will be much larger than the temperature change of the water. So the final temperature will be closer to 20 ^{0}C.

Lesson 1.9 Specific latent heat

1 The material is changing state.

2 As water turns into steam the particles get further apart. The particles therefore gain potential energy and so they need energy to do this.

3 The particles are only gaining potential energy. The energy in their kinetic energy store remains constant so the temperature remains the same.

4 When particles move from a solid into a liquid they don't move apart from each other very much. However, when they move from the liquid state to a gas they move apart from each other a great deal and gain much more potential energy.

5 $E = mL$ = 0.1 × 340 000 = 34 000 J

6a Melt the ice at 0 ^{0}C: $E = mL$ = 0.2 × 340 000 = 68 000 J

Heat the water to 100 ^{0}C: $E = mc\Delta\theta$ = 0.2 × 4200 × 100 = 84 000 J

Boil the water at 100 ^{0}C: $E = mL$ = 0.2 × 2260000 = 452 000 J

Total energy transferred = 68000 + 84000 + 452000 = 604 000 J

6b

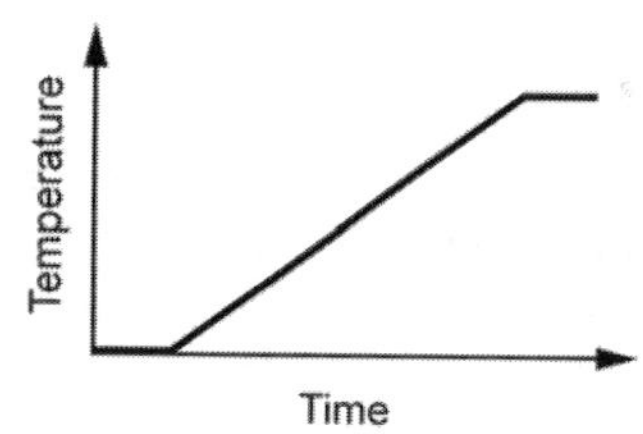

7 Energy is needed to heat the ice up to 0 ^{0}C, melt the ice, and then heat the water (the melted ice) up to 5^{0}C. Let the mass of the ice be m.

Therefore 7304 = (m × 2100 × 2) + (m × 340 000) + (m × 4200 × 5)

7304 = 4200m + 340 000m + 21 000m

7304 = 365 200m

So m = 7304 / 365200 = 0.02 kg

 © HarperCollins*Publishers* Limited 2016

Lesson 1.10 Maths Skills: Drawing and interpreting graphs

1 The maximum value of temperature is 95.0 ^{0}C and the minimum value is 25.0 ^{0}C.

2 & 3

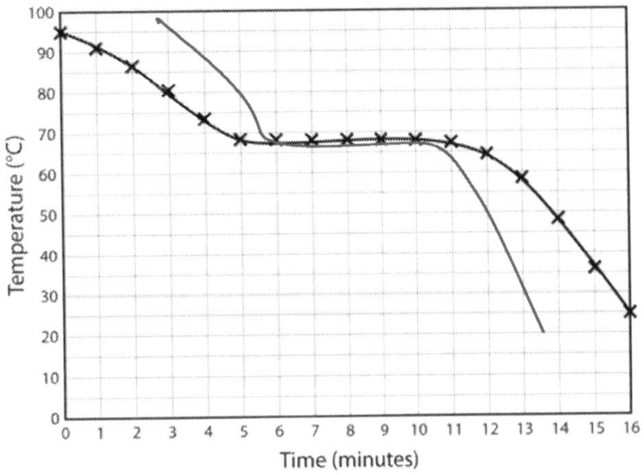

4 The line curves when the stearic acid cools from 95.0 ^{0}C to 68.1 ^{0}C. Then the line is horizontal at about 68 ^{0}C indicating that the temperature remains constant. Then the line begins to fall again after about 11 minutes.

5 The specific heat capacity of the wax must be changing.

6 The wax is melting.

7 About 44 ^{0}C.

8 The stearic acid was cooling in liquid form, then it changed state to a solid (as indicated by the horizontal portion of the graph). Once it had completely frozen it continued to cool towards room temperature.

9 The stearic acid is freezing.

10 About 68 ^{0}C

11 When the graph has a negative gradient, both the kinetic energy and the potential energy of the particles is decreasing. When the graph is horizontal only the potential energy of the particles is decreasing. However, the internal energy is decreasing at all parts of the graph.

12 For the stearic acid, both the potential energy and the kinetic energy is decreasing when the stearic acid is cooling. When the acid is freezing, then the kinetic energy remains the same while the potential energy continues to decrease.

For the wax, both the potential and the kinetic energy is increasing when the wax is warming up. When the wax is melting, then the kinetic energy remains the same while the potential energy continues to increase.

13 Same melting point would mean that the horizontal portion of the graph would still be at the same temperature. Lower specific heat capacity would mean less energy would be needed to raise the temperature so the temperature would take less time to change. Therefore the curved regions of the graph would be steeper. A larger latent heat would mean that more energy is needed to melt the substance so this would take longer. So the horizontal part of the graph is longer.

Lesson 1.11 Practical: Investigating specific heat capacity

1 Stir the water and measure the new temperature of the hot water + brass with the thermometer.

2 Subtract the new temperature of the hot water + brass from the initial temperature of the hot water (~80^{0}C)

3 Subtract the temperature of the ice-cold water + brass (0 ^{0}C) from the new temperature of the hot water + brass.

4 The temperature would decrease as thermal energy would be transferred to the surroundings.

5 The brass might warm up as you move it. You would assume that the brass changed temperature more than it actually did when you added it to the hot water. This would mean that your measurement of the specific heat capacity would be too small.

6 The temperature of the brass and hot water would steadily decrease as heat energy is being transferred to the surroundings.

7 You need to transfer the brass as quickly as possible and you also need to measure the new temperature of the hot water + brass as quickly as possible (making sure that all of the thermal energy transfer from the hot water to the brass had taken place).

8 In your calculation, you assume that all of the thermal energy transferred to the brass has come from the water rather than the surroundings.

9 $\Delta E = mc\Delta\theta = 0.25 \times 4200 \times (26 - 17) = 9450$ J

10 9450 J

11 The temperature would decrease from 100 ^{0}C to 26 ^{0}C, so $\Delta\theta = 100 - 26 = 74$ ^{0}C.

12 $c = \Delta E / (m\Delta\theta) = 9450 / (0.6 \times 74) = 213$ J/kg ^{0}C

13 The brass is likely to cool down as it moves from the boiling water to the cold water. Therefore the temperature change of the brass as it is heating the cold water up is likely to be less than 74 ^{0}C. This would result in the calculation in question 12 becoming a larger value.

 © HarperCollinsPublishers Limited 2016

Lesson 1.12 Particle motion in gases

1 The molecules move faster.

2 The temperature is related to the average kinetic energy of the molecules. The faster the molecules move, the higher the temperature.

3 The particles collide with the walls of their container. During the collision they exert a force on the walls. Since pressure = force / area, the force exerted by the particles produces a pressure on the container.

4 When you pump more air in a bicycle tyre there are more air particles. Therefore, there are more collisions between the particles and the walls of the tyre, which increases the pressure.

5 No gas can pass in or out of the container.

6 If the gas gets hotter, then the average kinetic energy of the particles increases. This means that the particles will move faster. This makes them collide with the container with a larger force and more often. Therefore, the pressure increases.

7 The temperature of a gas is a related to kinetic energy. So the particles in hot gases have a high kinetic energy. The particles also have a very small mass which means they must move very fast.

8 The pressure can drop by making the gas occupy a larger volume, cooling the gas down and reducing the amount of gas in a given volume.

9 The balloon will change volume until the pressure inside the balloon equals atmospheric pressure. When you heat the air inside a balloon the particles gain kinetic energy and will collide with the walls with more force and more frequently. Both of these increase the pressure inside the balloon. The pressure inside is now bigger than outside so there is a resultant force making the balloon expand. As the balloon expands, the particles inside the balloon collide with the walls of the balloon less often and the pressure drops. Once the pressure becomes the same as the outside pressure again, the balloon stops expanding.

Lesson 1.13 Increasing the pressure of a gas

1 The pressure decreases.

2 The pressure should double.

3 There are more collisions with the walls in a given time because the particles have a smaller distance to travel between each collision. This results in a larger average force on the walls and a higher pressure.

4 $p_1 \times V_1 = p_2 \times V_2$

$200 \times 4 = p_2 \times 3$

$P_2 = 800 / 3 = 267$ kPa

5 $p_1 \times V_1 = p_2 \times V_2$

$1.8 \times 10^5 \times 80 = 1.2 \times 10^5 \times V_2$

$V_2 = 144 \times 10^5 / 1.2 \times 10^5 = 120\ \text{cm}^3$

6 When you move the move the pump, the particles of gas colliding with the pump will end up moving faster. This increases the kinetic energy of the particles and so the internal energy increases.

7 The pump is doing work on the gas which heats up the gas. Since gas is not a very good conductor of heat, the region of the gas that gets hot remains near the pump.

8a The internal energy of the gas must decrease since the gas is doing work as it expands. Therefore, the temperature must decrease.

8b An expanding gas reduces the temperature. So you can make the gas expand inside a fridge to reduce the temperature of the fridge.

8c The compressor increases the temperature of the gas. This is used on the outside of the fridge (at the back). As the cold, expanded gas moves inside the fridge, it gains internal energy as the warmer food heats it up a little. Then the gas passes to the outside of the fridge where it is compressed and heats up. The gas is now hotter than the surroundings and it loses internal energy (which has originally come from the food) to the surroundings by heating.

Lesson 1.14 Pressure in a fluid

1 The pressure at the lower holes is greater so the water comes out faster.

2 The pressure squashing the can will increase. Eventually it will become so large that the can is crushed.

3 pressure due to the petrol = height of column × density × *g*

$= 5 \times 800 \times 9.8$

$= 39\,200$ Pa

4 pressure due to the water = height of column × density × *g*

$= 3 \times 1000 \times 9.8$

$= 29\,400$ Pa

5 Lead

6 ice / air/ petrol

7 No, ice is denser than petrol so it would sink.

 © HarperCollins*Publishers* Limited 2016

8 For most substances solids are denser than liquids, which are denser than gases. Ice (a solid) is not denser than water (a liquid), however, but this is unusual.

9a $p = h\rho g = 10 \times 1000 \times 10 = 100\ 000$ Pa

9b Top of cube is 9m below the surface of the water. $p = h\rho g = 9 \times 1000 \times 10 = 90\ 000$ Pa.

9c The water is trying to squash the cube. Therefore, the pressure acts upwards on the base of the cube and downwards at the top of the cube. The pressure upwards is bigger than the pressure downwards so there is an upthrust.

9d The force acting on the base of the cube due to the pressure of the water = $p \times A = 100\ 000 \times 1 = 100\ 000$ N.

The force acting at the top of the cube = $90\ 000 \times 1 = 90\ 000$N.

Therefore, the upthrust = 100 000 – 90 000 = 10 000 N.

This is smaller than the weight of the cube (= 12 000 N) so the cube will sink. (you could also work out the density of the cube = mass / volume = 1200 kg/m^3. This is larger than the density of water so the cube will sink).

Lesson 1.15 Atmospheric pressure

1a The pressure outside the bag is larger than the pressure inside the bag. Therefore, the bag will crush.

1b The sides have been pushed in by the atmospheric pressure outside the bag.

2a The helium atoms inside the balloon.

2b Atmospheric pressure.

3 There is a smaller height of air above you.

4 The air has a weight and the force of gravity pulls it downwards.

5 The helium balloon rises upwards. Therefore, atmospheric pressure gets less as the balloon rises but the pressure inside the balloon due to the helium is not affected. This means there is a bigger pressure inside the balloon than outside so the balloon expands.

6 Its molecules collide with the walls of its container. This creates a force when they rebound which creates a pressure.

7 The molecules in the atmosphere are colliding with the outside walls of the can and therefore create a force inwards. There is no force to balance this out (since there is a vacuum inside the can) so the can collapses inwards.

8 The air particles create a pressure by colliding with the suction cup. There are no air particles between the wall and the suction cup so the only pressure acting is on the outside of the cup. This creates a force which pushes the suction cup onto the wall. So the man is held onto the wall purely by lots of air particles colliding with the suction cup.

End of Chapter Questions

1 The left hand diagram is a gas, the middle diagram is a solid and the right hand diagram is a liquid. [1 mark]

2 Spacing: In a solid and a liquid the particles are close together and in a gas they are far apart. [1 mark]

Motion: In a solid the particles vibrate about a fixed position, in a liquid the particles slide past each other and in a gas the particles move freely until they collide with each other. [1 mark]

3 It condenses into a liquid [1 mark]

4 Latent heat [1 mark]

5 density = mass / volume [1 mark]

6 mass / volume = 100 / 25 = 4g/cm^{-3} [1 mark]

7 Latent heat of fusion of water is the energy needed to melt 1 kg of ice into water (or the energy given out when 1 kg of water freezes into ice). [1 mark for realising that the change of state is between ice and water and 1 mark for realising that it is for 1 kg]

8 $\Delta E = 2 \times 4200 \times 10 = 84\ 000$ J [1 mark]

9 Internal energy [1 mark]

10 There are two latent heats because the energy involved with melting/freezing is different to the energy involved with boiling/condensing. [1 mark]

11 It is the sum of all of the kinetic energies of the particles [1 mark] plus the sum of all of the potential energies [1 mark].

12a The particles are much further apart in steam so the same amount of mass occupies a much bigger volume. [1 mark]

12b Volume = mass / density = 2 / 0.59 = 3.4 m^3 [1 mark]

12c $p_1 \times V_1 = p_2 \times V_2$, so $1.5 \times 10^5 \times 3.4 = 1.0 \times 10^5 \times V_2$, so $V_2 = 5.1$ m^3 [1 mark]

12d $E = mL = 2 \times 2\ 260\ 000 = 4\ 520\ 000$ J [1 mark]

13 The internal energy decreases [1 mark]. When the water cools down to 0^0 C both the potential energy and the kinetic energy decreases. When the water is freezing at 0 ^{0}C only the potential energy decreases [1 mark].

 © HarperCollins*Publishers* Limited 2016

14a The mass of a substance is always conserved. [1 mark]

14b It can either increase in temperature or it can change state (melt, boil or sublimate) [1 mark].

15 The particles in the gas move faster [1 mark]. This makes them collide with the walls with a larger force and more often [1 mark]. Which results in a gain of pressure.

16 Level 3: A detailed description is made, which will lead to an accurate measurement of density. All necessary measurements are described together with the measuring instruments and apparatus needed. Consideration of at least two methods of measuring accurately are made. (5-6 marks)

Level 2: A description is made which will lead to a valid measurement of the duck. Measuring the mass and the volume is considered and an attempt at describing a displacement method is made. Some attempt is made to describe procedures to improve the accuracy. (3-4 marks)

Level 1: A simple response is made where it is recognised that the mass and the volume duck need to be measured. (1-2 marks)

Indicative content

- Measure mass of duck by using an electric balance
- Measure volume of duck by using a displacement method
- Displacement method could be by pushing the duck under water in a large measuring cylinder / filling a container full of water, pushing the duck under water and a method of measuring the volume of the water that has spilled / using a displacement/eureka/Archimedes can plus a measuring cylinder.
- Find density by dividing the mass by the volume

Methods to improve accuracy

- Repeat the experiment
- Discard any anomalous results
- Find the mean value of the density (which reduces random error)
- Make sure duck is completely submerged under the water
- When submerging the duck make sure nothing else (e.g. a finger) ends up under the water
- When reading the volume of water in a measuring cylinder make sure that eyes are level with the water.
- Make sure the measuring cylinder is completely vertical

When measuring the mass make sure there is no extra water on the duck.

17 $E = mc\Delta\theta$

So $c = E / (m \times \Delta\theta)$ [1 mark]

$= 8880 / (2 \times 10) = 444$ J / kg^{0}C [1 mark]

18 Since the temperature is constant the speed of the particles remains the same [1 mark]. However, if the volume increases then the particles travel a further distance between collisions with the walls which makes them collide less often [1 mark]. This reduces the pressure.

19 Level 3: A detailed description is made which includes ideas about changing state, latent heat, internal energy and pressure. All descriptions consider the action of the particles in terms of their motion, arrangement and energies and the description forms a logical sequence of steps which lead to an explanation of why the bottle explodes. (5-6 marks)

Level 2: A description is made which considers the action of particles when they are changing state and how they exert a pressure. There is some consideration of energy but it is not clear why the pressure increases and the sequence of steps do not necessarily follow logically on from each other. (3-4 marks)

Level 1: An attempt is made to describe the change of state from liquid to gas in terms of particles and that there is a build-up of pressure. (1-2 marks)

Indicative content

Changing state

- The temperature outside the bottle is much higher than the boiling point of nitrogen so the nitrogen will start boiling very vigorously.
- While the nitrogen is boiling the particles increase their potential energy but their kinetic energy stays the same.
- The nitrogen particles are close together in a liquid but far apart in a gas

Causing pressure

- The internal energy of the gas is much higher than the liquid due to the latent heat that is transferred as the nitrogen is boiling.
- The particles of gas are moving much quicker than the particles in the liquid because the gas is at a much higher temperature
- The gas particles collide with the walls of the bottle and exert a force
- The force creates a pressure

Why the pressure builds up

- As the liquid nitrogen continues to boil then more and more particles are released as gas

 © HarperCollins*Publishers* Limited 2016

- This means that there is an increasing number of collisions with the walls of the bottle
- So the pressure increases

Why the bottle explodes

- The pressure of the nitrogen gas pushes the bottle outwards
- Eventually the pressure inside the bottle becomes so great that the bottle cannot withstand the force and it explodes.

20 The pressure due to the water increases with depth [1 mark]. Since water is coming out of the jets, the water level inside the can is decreasing. This means the pressure at the holes is getting less [1 mark]. So the jets don't travel as far.

 © HarperCollins*Publishers* Limited 2016

Chapter 2: Forces

Lesson 2.1 Scalars and vectors

1 You would cover a distance equal to the circumference of the circle = 2πr = 2 × π × 10 = 63 m (to 2 significant figures)

2 You are back where you started.

3 4 m/s

4a acceleration, force and momentum are vector quantities

4b These quantities have a magnitude and a direction; the other quantities only have a magnitude.

5

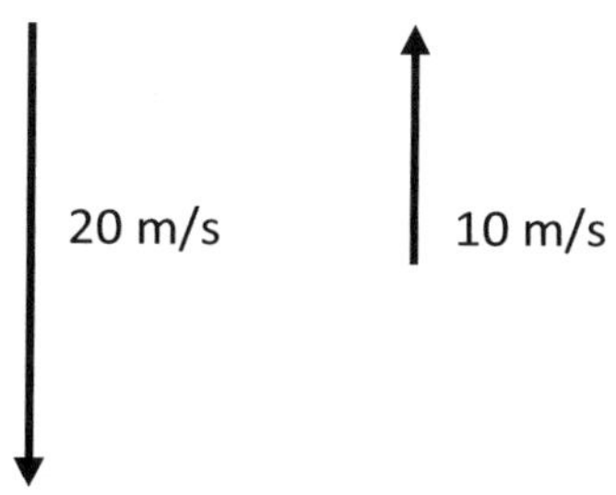

6 Final distance = 400m; Final displacement = 0m.

Lesson 2.2 Speed

1 For example, the traffic means you have to slow down if there is a car in front of you; there are different speed limits on different roads; you need to come to a stop when you reach a junction.

2 *average speed* = *distance* / *time* = 50 m / (2.5 × 60) s = 0.33 m/s

3 *average speed* = *distance* / *time* = 10 000 m / (1.75 × 60 × 60) s = 1.6 m/s

4 *distance travelled* = *speed* × *time* = 0.5 mm/s × (1 × 60 × 60) s = 1800 mm = 1.8 m

5 The gradient would increase / the line would become steeper.

6 The car is slowing down (because the gradient decreases as the time increases)

7 The gradient of the graph is getting steeper between 0 to 10 s. This tells us that the speed is getting faster. By measuring the gradient of the tangent you can see that the gradient gets steeper by the same amount each second. For example, when the time is 1 s the gradient is 4 m/s and when the time is 2 s the gradient is 8 m/s. Therefore, the speed is increasing by the same amount each second (constant acceleration).

8 Object travels a total distance of 200 m in 10 s. So the average speed for the entire journey = 200 / 10 = 20 m/s.

Lesson 2.3 Acceleration

1 Car A has the greater acceleration.

2 The ball has the force of gravity pulling it to the ground.

3 The air resistance has a greater effect on the feather's motion than it does on the ball.

4 The ball decelerates to a rest as it moves up to its highest point, then it accelerates downwards.

5 Acceleration = change in velocity / time taken = (40 m/s – 20 m/s) / 10 s = 2 m/s^2

6a The velocity increases by 2 m/s each second.

6b The velocity decreases by 2m/s each second.

7 Acceleration = change in velocity / time taken = (0 m/s – 24 m/s) / 3 s = - 8 m/s^2

8 Speed at the start = 72 km/h = 72 × (1000 / 3600) m/s = 20 m/s

Speed at the end = 108 km /h = 108 × (1000 / 3600) m/s = 30 m/s

Acceleration = change in velocity / time = (30 m/s – 20 m/s) / 5 s = 2 m/s^2

9 Every year the glacier moves 4 mm/s faster than it was moving a year ago.

Lesson 2.4 Calculations of motion

1a The velocity at the beginning of the motion.

1b The velocity at the end of the motion.

2 Acceleration is change of velocity / time so its unit needs to be the unit of velocity divided by the unit of time.

3 Rearrange $v^2 = u^2 + 2as$ to give $2as = v^2 – u^2$

Dividing by $2a$ gives $s = (v^2 – u^2) / 2a$

So length of runway = $(v^2 – u^2) / 2a$ = (60^2 – 0^2) / (2 × 2.5) = 720 m

4 Acceleration is 3 m/s^2 which means the velocity increases by 3 m/s each second. Therefore, after 4 s the velocity has increased from rest to 3 × 4 = 12 m/s.

From question 3, $s = (v^2 – u^2) / 2a$

So the distance the car travels = (12^2 – 0^2) / (2 × 3) = 24 m

5 $s = (v^2 – u^2) / 2a$ (a = - 2 m/s^2 since this is a deceleration)

Distance train travels = (0^2 – 40^2) /(2 × -2) = 400 m

 © HarperCollins*Publishers* Limited 2016

6 $v^2 = u^2 + 2as$

$0^2 = 10^2 + (2 \times -10 \times s)$

$0 = 100 - 20s$

$20s = 100$, so $s = 5$ m

7 $v^2 = u^2 + 2as$

$v^2 = 0^2 + (2 \times 10 \times 2.5)$

$v^2 = 50$

$v = 7.1$ m/s (to 2 significant figures)

Lesson 2.5 Velocity-time graphs

1

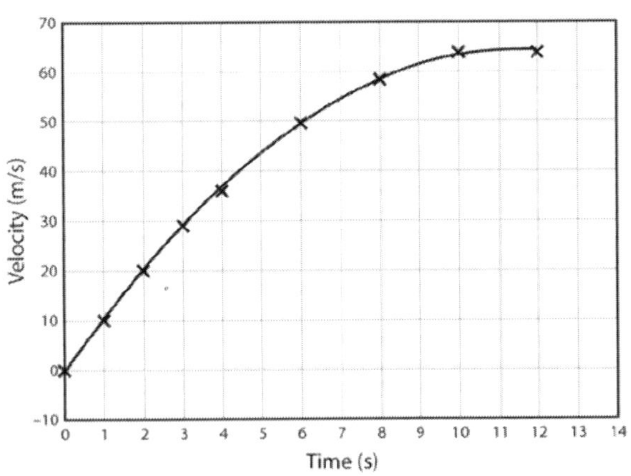

2 Graph (c) in Figure 2.8 shows a car braking and reversing. Graph (a) shows the car going at constant velocity; graph (b) shows the car constantly accelerating.

3

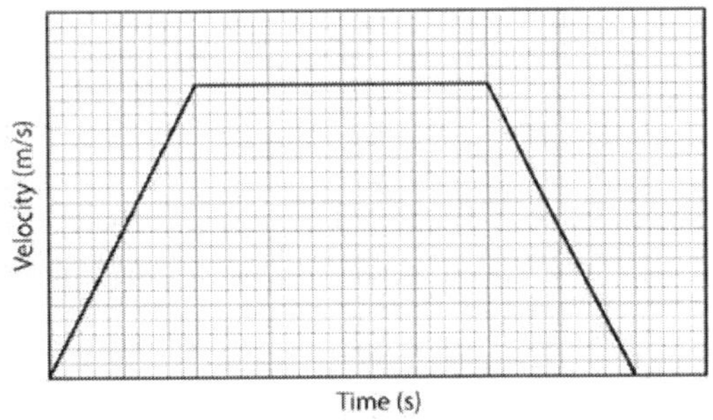

4 Graph (a) – acceleration = 25 / 5 = 5 m/s^2

Graph (b) – acceleration = -25 / 5 = - 5 m/s^2

Graph (c) – acceleration = 0 m/s^2

5 Acceleration = change in velocity / time = -20 / 5

= -4 m/s^2

6 Change of velocity = acceleration × time = -2 × 4 = - 8 m/s.

Truck moves at 25 m/s at the beginning and the velocity changes by -8 m/s so the velocity after braking = 25 – 8 = 17 m/s.

7a The train accelerates constantly for the first minute, then it travels at a constant velocity of 15 m/s for the next 3 and a half minutes and it then decelerates constantly for the final half a minute.

7b Distance train travels is the area under the graph.

Area of the left hand triangle = ½ × 60 s × 15 m/s = 450 m

Area of the rectangle = 15 m/s × (3.5 × 60) s = 3150 m

Area of the right hand triangle = ½ × 30 s × 15 m/s = 225 m

So the total distance = 450 + 3150 + 225 = 3805 m

8

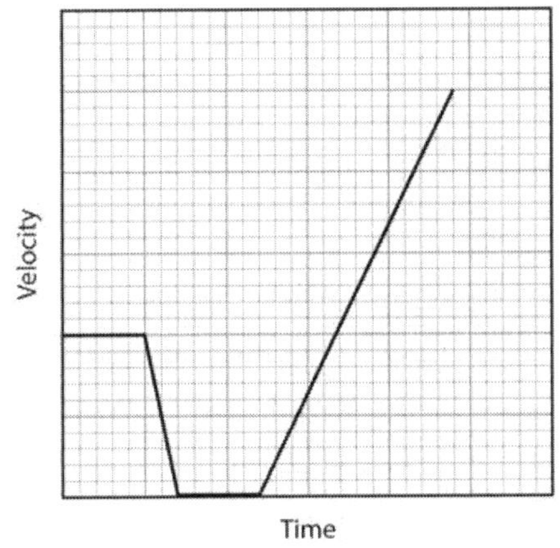

9 Graph (b) in Figure 2.9 could represent the motion of the stone as it moved upwards. This is because it is has an upwards velocity at the beginning of the throw which constantly reduces up to the stone's maximum height. For the stone moving downwards, the straight line could be continued so that it carries on going downwards below the axis. This would show that the stone is gaining speed but in the opposite direction. The whole graph should have a constant gradient, as the acceleration of the stone is the same throughout the motion (acceleration due to gravity).

Lesson 2.6 Maths Skills: Making estimates of calculations

1 Volume of this book is about 20 cm × 30 cm × 2 cm = 1200 cm^3 ≈ 1000 cm^3

 © HarperCollins*Publishers* Limited 2016

2 A typical adult would take about 2 minutes to run 500 m. So their running speed would be about 500 m / 120 s ≈ 4 m/s

3 Change in speed is about 20 m/s. The time is about 4 seconds. So the deceleration is about 20 / 4 which is about 5 m/s^2.

4 Braking distance = $(v^2 - u^2) / 2a \approx 20^2 / 14 \approx 400 / 10 \approx 40$m

Typical reaction time is about 0.5 s so thinking distance is about 20 × 0.5 = 10 m.

Therefore, total stopping distance ≈ 50 m.

5 Approximate mass of car = 1000 kg.

$F = ma \approx 1000 \times 5 = 5000$ N

6 Typical volume of room in orders of magnitude is $10^1 \times 10^1 \times 1 = 10^2$ m^3

Mass = density × volume ≈ 1 × 10^2 = 10^2 kg or 100 kg

7 Length of UK is about 10^3 km = 10^6 m. Speed of light is about 10^8 m/s.

So time = distance / speed = $10^6 / 10^8 = 10^{-2}$ s = 0.01 s.

Lesson 2.7 Forces explain how objects interact

1 e.g. friction between the tyres of a car and the road

2 e.g. the gravitational force pulling you down as you are falling from a tree

3 It would move to the right.

Lesson 2.8 Forces and motion

1 1 N

2 No, the object could also be moving at a constant speed (in a straight line).

3a The drag force increases.

3b No – the weight remains the same but the drag has increased, so there will be a resultant force upwards.

4a Their size must be the same.

4b They must be in opposite directions.

5 The forces must be the same size.

6 The weight acting downwards must have an equal size to the normal contact force acting upwards so that they balance out.

7 The normal contact force from the ground balances the weight of the boulder.

8 The driving force forward on the bicycle is balancing out the drag from the air.

9 The driving force will decrease but the drag force remains the same (for the same speed). This means that there is an unbalanced force opposing the motion and the bicycle slows down.

10 The Earth would move in a straight line in the direction it was moving in just before the Sun vanished. It would still be spinning on its axis though. (It would also continue to orbit around the Milky Way galaxy.)

11 Their direction is changing, so their velocity is changing (since velocity is a vector). Whenever there is a change in velocity then the object is accelerating.

Lesson 2.9 Resultant forces

1 Gemma and Alan produce a total force of 40 + 40 = 80 N. So the resultant force is 80 – 50 = 30 N.

2 Total force opposing the motion = 1000 + 500 = 1500 N. Total forward force = 1500 N so the forces are balanced and the resultant force = 0 N.

3

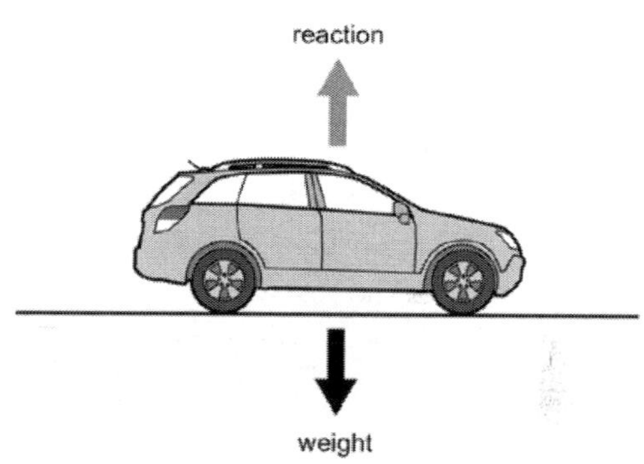

4

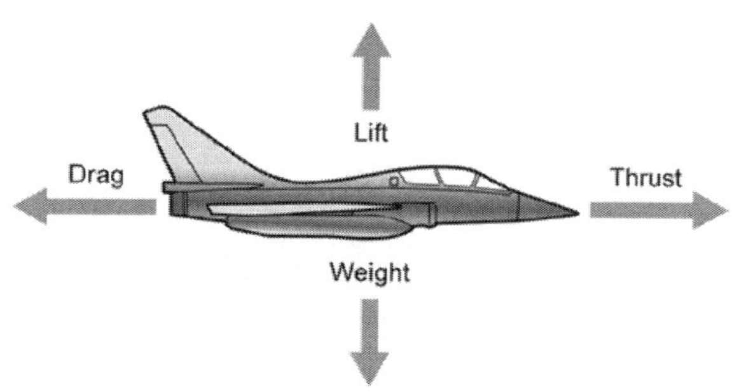

5 13N at an angle of 67^0 above the horizontal.

6 Draw a scale diagram where e.g. 1cm represents a force of 1N. Start with an arrow pointing to the right that is 12cm long. At the head of this arrow join the tail of a 16cm arrow that is pointing downwards.

The third force needs to end up where the 12cm arrow started so that there is no distance between

 © HarperCollins*Publishers* Limited 2016

the starting and ending points on the scale diagram (so there is no resultant force).

Therefore, draw the third arrow from the head of the 16cm arrow to the tail of the 12 cm one. Measure the length of this arrow (= 25 cm). So the size of the force needed = 25 N.

7 Draw a scale diagram with a line 14.1 cm long at an angle of 45^0 (using a protractor to measure the angle). Form a triangle with one side vertical and the other side horizontal. The lengths of the sides represent the magnitudes of the components. Both components are 10N.

8 The horizontal components of the two forces are both 10N to the right, so the total horizontal component = 20N. The vertical components are both 10N but one points upwards and the other points downwards so they cancel out. Therefore, the resultant force is 20N to the right.

Lesson 2.10 Forces and acceleration

1 In Figure 2.27 the resultant force is to the left; in figure 2.28a the resultant force is to the right and in figure 2.28b there is no resultant force so there is no direction.

2 The resultant force rises from zero and it acts to the left.

3 $a = F / m = 4200 / 1200 = 3.5$ m/s^2

4 $m = F / a = 1 / 10 = 0.1$ kg.

5 F = ma = 4000 × 2 = 8000 N. However this is the **resultant** force. The resultant force = the force needed – the weight of the rocket.

So 8000 = force needed – 40 000

Force needed = 8000 + 40 000 = 48 000N

6 Inertia is a measure of how difficult it is to change the velocity of an object (a reluctance to change motion). Inertial mass is calculated by dividing the force by the acceleration.

7 Inertial mass is a measure of how much force the object needs in order to get it to change its motion (accelerate) whereas gravitational mass is a measure of how strongly the object feels the force of gravity. More massive objects need a bigger force to accelerate them (due to their inertial mass) and they have a bigger weight (due to their gravitational mass).

8 Both inertial and gravitational mass = 2kg. For inertial mass, this is a measure of its reluctance to change its motion; for gravitational mass, this is a measure of how strongly it feels the force of gravity.

Lesson 2.11 Momentum

1 $p = mv = 1000 \times 20 = 20\,000$ kg m/s.

2 $v = p/m = 240 / 60 = 4$ m/s.

3 Force = change of momentum / time. The change of momentum is the same whether the air bag is there or not, but the air bag increases the time it takes to change momentum. Therefore, the force acting on the passenger is less and so there is less chance of harm.

4 If the car stops suddenly, the passengers in the back will continue moving forward (since there is no force to change their momentum). Therefore, they will crash into the back of the driver and the front seat passenger who will be crushed between them and the front of the car.

5 $F = m\Delta v / \Delta t$, so $\Delta t = m\Delta v / F = (800 \times 12) / 3000 = 3.2$ s.

6 Initial momentum = final momentum

$(1200 \times 30) + (m \times 0) = (1200 \times 4) + (m \times 4)$

$36\,000 = 4800 + 4m$

$4m = 31200$

$m = 7800$ kg.

7 Initial momentum = final momentum

$(2 \times 8) + (6 \times 0) = (2 + 6)\,v$

$16 = 8\,v$

$v = 2$ m/s

Lesson 2.12 Practical: Investigating the acceleration of an object

1 The trolley will accelerate (at a constant rate) to the right.

2 The force remains constant as the trolley is moving.

3 You could change the mass of the falling mass.

4 You can use the remaining masses in the trolley to add to the falling mass. This changes the force acting on the trolley but keeps the mass the same.

5 You could remove some of the remaining masses but not add them to the falling mass.

6 He should find that the acceleration increases as the force increases.

7 Yes they do.

 © HarperCollins*Publishers* Limited 2016

8

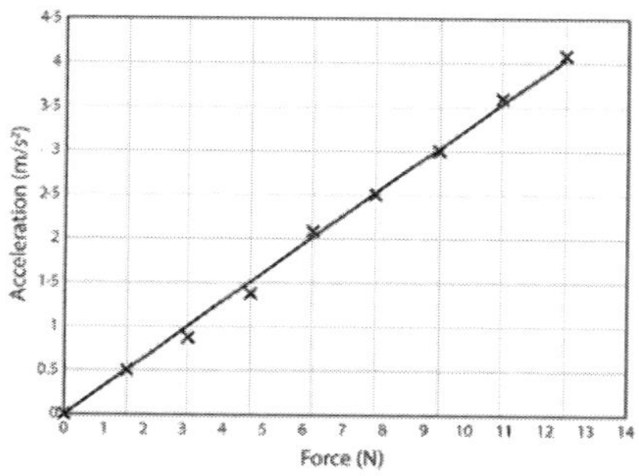

The graph is a straight line through the origin

9 This shows that acceleration is directly proportional to the force.

10 As the mass is increasing the acceleration is decreasing

11 The graph would have a negative gradient and it would be curved.

12 A graph of acceleration against force gives a straight line through the origin. If you look at the table with mass and acceleration you can see that if you double the mass (e.g. from 20 g to 40 g) the acceleration halves (e.g. from 5.0 m/s^2 to 2.5 m/s^2). This shows that acceleration is inversely proportional to the mass.

13 a = F / m. If F is tripled then the acceleration is tripled. However, if m is doubled then the acceleration is halved. The combined effect is that the acceleration is 3/2 = 1.5 times bigger.

Lesson 2.13 Newton's third law

1 They have the same magnitude but act in the opposite direction.

2 The road pushes on the tyres (due to friction).

3a They move apart.

3b There is a force pair. One person pushes on the other person with a force and the other person pushes on the first person with an equal force in the opposite direction.

4 Ben's body exerts a 50N force acts on Kim in the opposite direction. They move apart.

5a The resultant force will be the same on each vehicle (but acting in opposite directions) as the forces are a Newton's third law pair.

5b Since F = ma and the force is the same on each vehicle then the Lorry will decelerate at a smaller rate than the car as its mass is bigger.

6 Throw the backpack away from the ship. The backpack will exert an equal and opposite force on her which will push her back to the ship.

7a It is a gravitational force.

7b Since the Earth is pulling on the cat with a gravitational force downwards, the other force in the force pair is the cat pulling on the Earth with a gravitational force upwards.

8a The same as in question 7. The other force in the force pair is the gravitational force the cat exerts upwards on the Earth.

8b The force that the cat exerts downwards on the table and the force that the table exerts upwards on the cat (they're a pair of electrostatic forces).

9 The cat exerts a gravitational force upwards on the Earth (see question 7). This means the Earth will accelerate upwards towards the cat. The acceleration is very small though since the mass of the Earth is extremely large.

10 As you begin the jump, you push down onto the Earth. This means that the Earth will push upwards onto you. As you are in mid-air the Earth exerts a gravitational force on you and your body is exerting a gravitational force on the Earth. When you hit the ground you exert a force on the Earth and the Earth exerts an equal and opposite force on you, which decelerates you to rest.

Lesson 2.14 Work done and energy transfer

1 The amount of force and the distance that the force moves.

2 The gravitational force (weight) on the person as the person moves towards the ground.

3 $W = F \times s = 400 \times 1.5 = 600$ J

4 She is not doing any work as the force is not moving.

5 $W = F \times s$, so $s = W / F = 300 / 200 = 1.5$ m

6 $W = F \times s$, so $F = W / s = 3000 / 12 = 250$ N

7a $E_k = ½ mv^2 = 0.5 \times 800 \times 12^2 = 57\,600$ J

7b $F = W / s = 57\,600 / 8 = 7200$ N

8a $W = F \times s = 500 \times 4 = 2000$ J

8b 2000 J (assuming all of the energy in her E_p store is transferred to her E_k store)

8c E_k reduces by 2000 J so work done by the trainers = 2000 J.

$F \times s = 2000$ so $F \times 0.01 = 2000$.

$F = 2000 / 0.01 = 200\,000$ N.

Lesson 2.15 Understanding power

1 The kettle

2 Although the television is less powerful than the food blender, it might be operating for a much longer time.

 © HarperCollins*Publishers* Limited 2016

3 2 minutes = 2 × 60 = 120 seconds.

$P = W / t$ = 108 000 / 120 = 900 W

4 $W = P \times t$ = 2000 × 30 = 60 000 J

5a $W = F \times s$ = 4000 × 6 = 24 000 J

5b $P = W / t$ = 24 000 / 20 = 1200 W

6a $W = F \times s$ = 800 × 5 = 4000 J

6b $P = W / t$ = 4000 / 10 = 400 W

7 15 minutes = 15 × 60 = 900 s

Work done = gain of $E_p = mgh$ = 60 × 10 × 300 = 180 000 J

$P = W / t$ = 180 000 / 900 = 200 W

8 10 cm = 0.1 m

Work done for each step up = gain of $E_p = mgh$ = 60 × 10 × 0.1 = 60 J

So total work done = 60 × 20 = 1200 J

$P = W / t$ = 1200 / 30 = 40 W

9a 108 km /h = 108 000 m / 3600 s = 30 m/s.

So the distance the car travels in 1 second = 30 m.

9b Work done for 1 second = $F \times s$ = 1000 × 30 =

30 000 J.

P = W / t = 30 000 / 1 = 30 000 W = 30 kW.

Lesson 2.16 Key Concept: Forces and acceleration

1 The acceleration increases.

2 The roller coaster would decelerate but the passengers would continue to move forward at their original speed (and so come out of their seats).

3 You would fly off at a tangent to the circle.

4 The direction is changing to the velocity is changing. An acceleration is equal to the change in velocity / time rather than the change of speed.

5 Velocity includes a direction but speed doesn't.

6 The riders feel the force when the roller coaster accelerates and decelerates. This force is designed to make the ride a thrilling experience.

7 $F = ma$ = 2500 × 10 = 25 000 N

8 They would experience an acceleration of – 0.5g = -5 m/s^2 (taking g to be 10 m/s^2 in this question).

9 The water accelerates due to gravity towards the ground at 10 m/s^2 as normal. However the bottle is also accelerated at this rate as part of the ride.

Therefore, the water remains in the bottle even though the bottle is upside down.

Lesson 2.17 Forces and energy in springs

1 0N – there is no resultant force

2 3N

3 It would end up thinner than it was before it was used.

4 6.0 – 4.8 = 1.2 cm

5 $k = F/e$ = 4/0.08 = 50 N/m

6 Compression, e = 10 – 4 = 6cm = 0.06m. $F = ke$ = 2000 × 0.06 = 120N

7 $E_e = \frac{1}{2} ke^2 = \frac{1}{2} \times 30 \times 0.02^2$ = 0.006J

8 Work done = the gain of elastic potential energy $E_e = \frac{1}{2} ke^2$.

So 0.2 = ½ × 10 × e^2

0.2 / 5 = e^2

0.04 = e^2

e = 0.2 m

9 Assume that the limit of proportionality has not been exceeded. 10g = 0.01 kg, 10cm = 0.1m. Gravitational potential energy of the spring at 1m = mgh = 0.01 × 10 × 1 = 0.1J. Therefore elastic potential energy stored before the spring is released is 0.1J. $E_e = \frac{1}{2} ke^2$ so $k = 2E_e / e^2$ = 2× 0.1 / 0.12 = 20 N/m.

Lesson 2.18 Practical: Investigate the relationship between force and the extension of a spring

1 She could either use a spring balance / newton meter or she could hang known masses off the springs and calculate their weight.

2 She could use a ruler to measure the starting length of the spring. Then she could measure the total length of the spring when a force is added. The extension would be the total length – the starting length for each force that she uses.

3 She added all of the repeat readings together and then divided this total by the number of readings.

4 Yes – if the anomalies are discarded and further repeat readings are taken.

5 When you find a mean value, the random errors become smaller since the readings that are slightly too big tend to balance out the readings that are slightly too small. So this is a good way of getting close to the true values of the extension.

6 It shows that the extension is directly proportional to the weight in this region.

7 The spring has inelastically deformed.

 © HarperCollins*Publishers* Limited 2016

8 The extension was directly proportional to the force up to a point called the elastic limit but after that it was not proportional.

9 The experiment should stop at a weight less than 1.5 N (e.g. 1.4 N) to be sure that plastic deformation is prevented. The graph begins to curve at a weight of 1.5 N.

10 Spring B: $m = 10 / 8 = 1.25$

Spring C: $m = 6 / 8 = 0.75$

11 Spring B: $k = 8 / 10 = 0.8$ N/cm

Spring C: $k = 8 / 6 = 1.33$ N/cm

Spring C is the stiffest as it needs the greater force to stretch it by a certain amount.

12

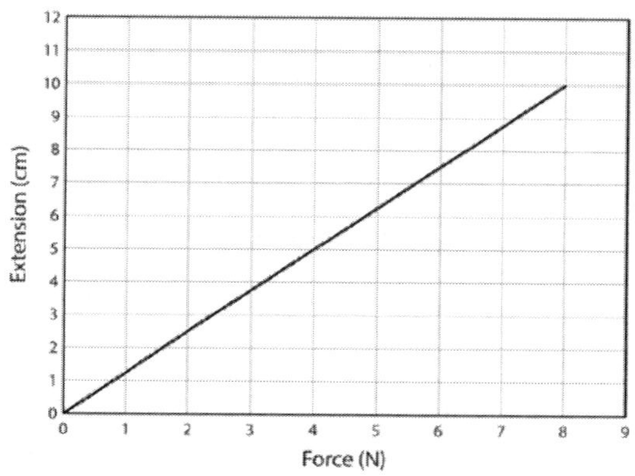

Gradient of this line = 8 / 10 = 0.8 N/cm, which agrees with the previous calculation.

13a Area under line A = a triangle = ½ × base × height = ½ × 8 × 0.16 = 0.64 Nm.

13b $k = F / e = 8 / 0.16 = 50$ N/m

$E_e = ½ke^2 = ½ \times 50 \times 0.16^2 = 0.64$ J

13c The answers are the same. This is because the area = ½ × F × e but $F = ke$. Therefore, area = ½ × ke × e = 1/2 ke^2. So the area = the elastic potential energy.

Lesson 2.19 Potential energy

1 You would feel that the force needed to stretch the rubber band increases, the more you stretch it.

2 As you turn the key, you stretch the spring more. This means that the energy stored increases.

3 Gravitational potential energy changes when the height changes. Since the aircraft is travelling horizontally, its height stays the same so there is no change in gravitational potential energy.

4 $E_p = mgh = 300 \times 10 \times 2 = 6000$ J

5 Mass of ball $m = 60 \div 1000 = 0.06$ kg.

$E_p = mgh = 0.06 \times 10 \times 2 = 1.2$ J

6 $E_e = ½ ke^2 = 0.5 \times 300 \times 0.1^2 = 1.5$ J

7 Extension = 25 – 20 = 5 cm = 0.05 m

$E_e = ½ ke^2 = 0.5 \times 500 \times 0.05^2 = 0.625$ J

8 $E_e = ½ ke^2$ so $k = 2E_e / e^2 = (2 \times 12) / 0.16^2 =$ 937.5 N/m

9 $E_e = ½ ke^2 = 0.25$J

So $0.5 \times 200 \times e^2 = 0.25$

$e^2 = 0.25/100 = 0.0025$

Therefore e = 0.05m = 5cm.

The total length of the spring = 20cm, which means that the unstretched length = 20 – 5 = 15cm.

Lesson 2.20 Heavy or massive

1 Yes they have lost weight as the gravitational force on them is smaller.

2 Yes, Denzil is correct. Weight is a force so a weighing scale is measuring a force.

3a 7 × 10 = 70 N

3b 0.5 × 10 = 5 N

3c 400 g = 0.4 kg, so weight = 0.4 × 10 = 4 N

4a $m = W / g = 30 / 10 = 3$ kg

4b The mass of the block remains at 3 kg.

$g = W / m = 11.1 / 3 = 3.7$ N/kg.

5 Yes Alex is right. Weighing scales measure the weight and convert this to a mass. However the conversion assumes that you are on the Earth.

6 Her weight is 0N as there is no gravitational force.

7 E.g. a balance where a beam is placed on a pivot. A known mass is placed on one side and the mass you need to measure is placed on the other. The masses are moved so that the beam balances and the reading is made. This would give identical readings on the Earth and on the Moon.

Lesson 2.21 Moments

1 5 × 1.5 = 7.5 Nm

2 If the distance is large, you only need a small force to produce a big moment so the door handle is easier to turn.

3 Clockwise moment: 1 × 60 = 60 Ncm

Anticlockwise moment: 2 × 30 = 60 Ncm

 © HarperCollins*Publishers* Limited 2016

The total anticlockwise moment is equal to the total clockwise moment so the beam is balanced.

4 $X \times 20 = 300 \times 40$

$20X = 12\ 000$

$X = 12\ 000 / 20 = 600$ N

5 Her father needs to move towards the pivot until the moment of his weight about the pivot is the same size as Jill's.

6 It is in the middle of the ruler because all of the clockwise moments and the anticlockwise moments from the gravitational forces on the particles in the ruler cancel out at that point.

7 You can consider the weight of the vase as a single force that acts on the centre of mass. This force creates an anticlockwise moment about the pivot and this turns the vase so it falls over.

8 The perpendicular distance between the weight and the pivot is very small when the vase starts to topple. Therefore there is only a small moment toppling it. However, as the vase turns, the perpendicular distance increases so the moment increases. This makes the vase topple faster which increases the moment yet further.

Lesson 2.22 Levers and gears

1 It acts in the same way as in Figure 5.45. The lid of the bottle is like the load that is being picked up and the pivot is the edge of the bottle.

2 The distance between the load force and the fulcrum is bigger than the distance between the effort force and the fulcrum. When the wheelbarrow is balanced, the moments are the same and so the effort force is less than the load force. Therefore, it is easier to pick up the handles of the wheelbarrow than it is to pick up the load directly.

3 They increase the distance that the cyclist pedals which reduces the amount of force that is needed.

4 The pedals rotate faster than the wheels. So if the wheels are rotating quickly then the pedals would rotate too fast for the cyclist to keep cycling.

5 It changes the direction of rotation without changing the force needed.

6 $P = F / A = 100 / 50 = 2$ N/cm^2

7 1 m^2 = 100 cm × 100 cm = 10 000 cm^2

$F = P \times A = 2 \times 10\ 000 = 20\ 000$ N

End of Chapter Questions

1 newtons / N [1 mark]

2 c (mass) [1 mark]

3 Contact force: e.g. air resistance / friction / normal contact force / tension [1 mark]

Non-contact force: e.g. weight / magnetic attraction and repulsion / electrostatic attraction and repulsion. [1 mark]

4 Pa means pascal (or N/m^2) – it is a measure of pressure. [1 mark]

5 A force has a direction, so it is a vector quantity. [1 mark]

6 $v = s/t = 100 /5 = 20$ m/s [1 mark]

7 $W = mg = 50 \times 10 = 500$ N [1 mark]

8 The object is not moving [1 mark] since the distance is staying the same / the gradient of the graph = 0 [1 mark] as shown by the horizontal line [1 mark].

9 It must be equal to the anticlockwise moment. [1 mark]

10 Constant speed (forwards). [1 mark]

11 Work done = force × distance [1 mark]

= 400 × 2 = 800 J [1 mark]

12a Moment = force × distance [1 mark] = 80 × 0.2 = 16 Nm [1 mark]

12b A longer spanner would require a smaller force [1 mark] to provide the same moment (needed to turn the nut). [1 mark]

13 The speed of the object **remains the same** and the velocity of the object **changes** [1 mark]

14 Force = 1500 N [1 mark]

(The car is travelling at a constant speed so the resultant force must be zero. Therefore the forward force must balance out the frictional force.)

15 On a graph of force against extension/compression [1 mark] the limit of proportionality is the point on the graph where it stops being a straight line [1 mark]

16a Ann's moment = 300 × 2 = 600 Nm anticlockwise

Beth's moment = 500 × 1.5 = 750 Nm clockwise

For the see-saw to balance, Cindy must provide a moment of 150 Nm clockwise [1 mark].

Cindy's weight is 300 N so the distance from the pivot needs to be 150 / 300 = 0.5 m [1 mark] on the same side as Beth [1 mark].

16b Total weight of the girls and the platform = 300 + 500 + 300 + 100 = 1200 N [1 mark]

$F = ke$ so $e = F/k = 1200 / 8000 = 0.15$ m [1 mark]

 © HarperCollins*Publishers* Limited 2016

Assumption is that the limit of proportionality has not been exceeded, as the formula wouldn't have been valid. [1 mark]

17 When two objects interact [1 mark] the force that one object exerts on the other is equal and opposite to the force the other object exerts on it [1 mark]]

18 First find the acceleration, *a* of the car. $a = F / m$ = 3000 / 1500 = (-) 2 m/s^2 [1 mark]

Then use $v^2 = u^2 + 2as$

$0^2 = 25^2 + (2 \times -2 \times s)$

$625 = 4s$

s = 156.25 m [1 mark] (or 160 m to 2 sig figs)

19 Level 3 (5 – 6 marks): A clear and logical description is made in a coherent sequence of steps. The answer includes a description of conservation of momentum and realises that the mass of the balls must be the same. Ideas about that this must be a closed system (since there are no external forces such as friction) are included. The answer uses sensible numbers for the values of the mass (10g – 1kg) and the velocity (0.1 m/s – 10 m/s).

Level 2 (3 – 4 marks): A clear description is made which uses ideas about the conservation of momentum. An attempt is made to illustrate the answer with sensible numbers.

Level 1 (1 – 2 marks): A basic description is made which identifies that the physics behind this is related to either Newton's second law (so the forces between the balls accelerate or decelerate them) or the conservation of momentum.

Indicative content

- Before the collision, the red ball had no momentum because it was stationary
- Before the collision, the total momentum equals the momentum of the white ball
- After the collision, the white ball has no momentum because it is stationary
- After the collision, the total momentum equals the momentum of the red ball
- This is a closed system since there is no friction
- Total momentum after the collision = total momentum before the collision
- So momentum of red ball after the collision = momentum of white ball before the collision
- Momentum = mass × velocity
- So if the red ball's velocity after the collision equals the white ball's velocity before the collision then the balls must have the same mass
- Since the velocity is the same, then the red ball must be moving in the same direction that the white ball moved

Numerical illustration

Estimated values: mass of white ball and red ball = 0.1 kg, original velocity of white ball = 1 m/s.

Let the velocity of the red ball = v.

Momentum before the collision = momentum after

Momentum of white + momentum of red before = momentum of white + momentum of red after

$(0.1 \times 1) + (0.1 \times 0) = (0.1 \times 0) + (0.1 \times v)$

$0.1 = 0.1v$

So v = 1 m/s = original velocity of the white ball.

 © HarperCollins*Publishers* Limited 2016

Chapter 3: Electricity

Lesson 3.1 Static Electricity

1 e.g. a polythene rod

1b e.g. a metal wire

2 If static charge begins to build up on a conductor when you are rubbing it with a cloth, charge will flow through the conductor through you to the ground.

3a attract

3b repel

3c attract

3d repel

4 Her hair builds up static charge. Since the type of charge is the same on each strand of hair (either all positive or all negative) then the strands of hair repel each other.

5 Electrons move from the duster to the polythene so the duster loses negative charge. Therefore, it builds up positive charge.

Lesson 3.2 Electric fields and currents

1 The direction of the arrows is the direction that the force acts on a positive charge. A positive charge placed in the field in Figure 2.5 would feel a force of repulsion, so it would always be acting away from the centre.

2 Positive charges would feel a force of attraction towards the negative charge in the middle. Therefore, the lines should all be pointing inwards to the centre.

3 The size of the force remains the same because the electron is always at the same distance from the charged sphere.

4a Anywhere close to the central charge.

4b Anywhere far away from the central charge.

5 The arrows show the direction that the electric field acts on a positive charge – which is useful. This could be misleading, however, if you are considering a negative charge in the field as the force would act the opposite way.

6 No – because a spark occurs when particles break apart due to an electric field and a vacuum doesn't have any particles in it.

7 The electric field between two charged objects is stronger if the objects are closer together. Sparks are more likely to occur in stronger electric fields.

8 The rate that lightning strikes should increase. When the cloud passes over a hill, the distance between it and the ground becomes smaller – this creates a stronger electric field.

9 $I = Q / t = 80 / 16 = 5$ A

10 $t = Q / I = 96 / 6 = 16$ s

Lesson 3.3 Electric circuits and potential difference

1 e.g.

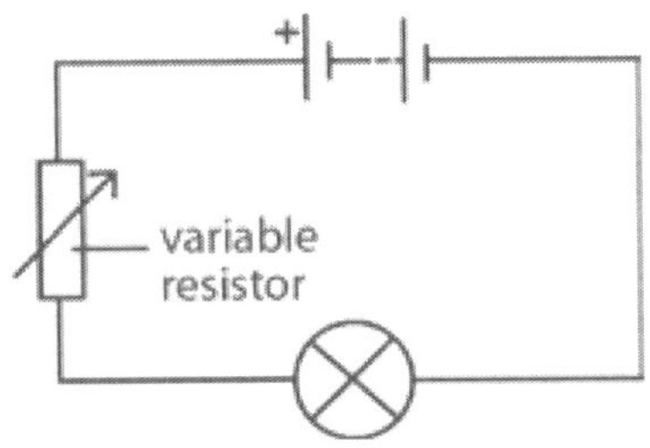

2 LDR stands for "Light Dependent Resistor" – it's resistance changes with the amount of light shining on it.

3 There needs to be a source of potential difference within a complete loop in a circuit.

4 6V battery transfers 6J of energy per coulomb of charge. Number of coulombs of charge, $Q = It = 1 \times 60 = 60$ C.

Therefore total energy transferred = 6 × 60 = 360 J

5 Electrons transfer energy to the kinetic energy stored in the metal ions. This makes them move faster and therefore become hot.

6 If you increase the resistance of the variable resistor, the current decreases. The lamp will get dimmer if the current decreases.

7 $R = V / I = 12 / 3 = 4\ \Omega$

8 $V = IR = 1.5 \times 6 = 9$ V

Lesson 3.4 Series and parallel circuits

1 Close the switches

2 They both get dimmer and they are the same brightness as each other (and the third lamp)

3 In the series circuit both light bulbs would not be shining; in the parallel circuit the unscrewed bulb would not be shining but the other bulb would continue to shine.

4a 0.4 A – the current through components in series is the same

4b 10 Ω + 20 Ω = 30 Ω

4c 8 V – the potential difference of the power supply (12 V) is shared between the components (4 V and 8 V)

4d The current would be half as big because the total resistance has doubled from 30Ω to 60Ω.

5a 0.6 A – the current from the supply (1.8A) is the sum of the currents through the motor (1.2A) and

 © HarperCollinsPublishers Limited 2016

the lamp. So the current through the lamp = 1.8 – 0.6 = 1.2 A.

5b 12 V – the potential difference across each component in a parallel circuit is the same and is equal to the p.d. of the power supply.

5c It must be smaller than 10 Ω. Connecting resistors in parallel decreases the total resistance, so connecting the lamp to the motor in parallel would make the resistance less than the resistance of the motor.

5d Adding the resistor decreases the total resistance, therefore the battery will be providing a bigger current.

Lesson 3.5 Resistance

1 They are directly proportional to each other.

2 The voltmeter is connected in parallel.

3 The I-V graph is not a straight line through the origin.

4 Tungsten obeys Ohm's law when the temperature remains constant. In a filament lamp the temperature increases as more current flows through it (to thousands of degrees), so the lamp does not obey Ohm's law.

5a At 1 V: $R = V / I = 1 / 0.2 = 5\ \Omega$

At 6 V: $R = V / I = 6 / 0.4 = 15\ \Omega$

5b Graph should show the resistance starting at a low value (not 0 Ω) and then rapidly rising to a higher, steady value as the filament lamp heats up to a constant temperature.

Lesson 3.6 Practical: Use circuit diagrams to set up and check appropriate circuits to investigate the factors affecting the resistance

1 You would expect the current to increase as the p.d. across it is increased. This is because $I = V / R$.

2 You could either vary the setting on the power supply between 0 V and 12 V or you could alter the resistance of the variable resistor (usually by moving a slider or by rotating a dial).

3 The temperature is also likely to increase.

4 If the current becomes too high, then the lamp becomes so hot that the filament melts. The lamp blows and no longer works.

5 The graph would be a straight line through the origin until the current reaches 13A. If the current goes above 13A then the fuse blows and breaks the circuit. The current will then become 0A for all values of potential difference.

6

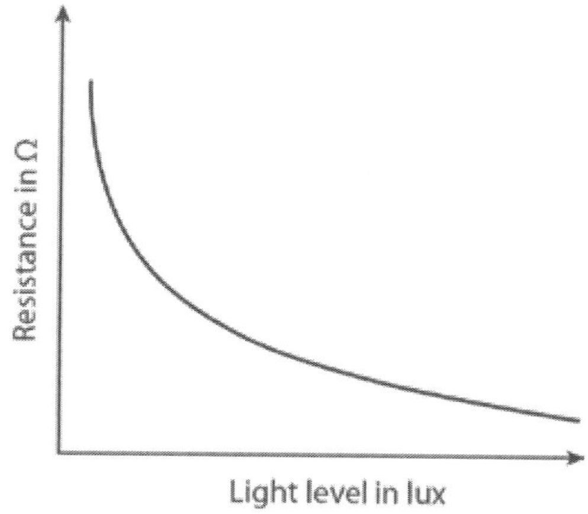

7 e.g. a temperature measuring circuit (such as a temperature sensor for a data logger) and in a temperature control circuit – where a heater in a circuit automatically acts to keep the temperature constant (such as a thermostat).

8 When the potential difference is positive then the current is very small for values up to 0.5 V but then becomes very large. This suggests that the diode has a very low resistance in this direction once the potential difference reaches 0.5 V. When the potential difference is negative the current is small until the potential difference reaches a high negative value. Then the current becomes very large. This suggests that the diode has a large resistance in this direction until the potential difference becomes very large, at which point the diode breaks down and allows a current to pass through.

Lesson 3.7 Investigating circuits

1 If the component or the cable is able to conduct electricity, then a current flows in the circuit when it is connected between the terminals. This makes the buzzer sound.

2a $R = V / I = 12 / 2 = 6\Omega$

2b You can adjust the variable resistor (either by moving a slider or by rotating a dial) so that its resistance increases.

3 The voltmeter needs to be connected in parallel to the resistor. Its high resistance means that no measurable current flows when it is connected in series.

4 $R = V / I = 12 / 0.6 = 20\ \Omega$

5 Total resistance, $R = 5 + 7 = 12\ \Omega$

$I = V / R = 6 / 12 = 0.5$ A

 © HarperCollins*Publishers* Limited 2016

6 Total resistance, $R = 3 + 6 = 9\ \Omega$

Current, $I = V / R = 12 / 9 = 1.33$ A

Therefore p.d. across 3 Ω resistor = $IR = 1.33 \times 3 = 4$V

7 The potential difference across the 3Ω resistor = $I \times R = 3$ V.

The current through the 6 Ω resistor is also 1 A so the p.d. across the 6 Ω resistor = 1 × 6 = 6 V.

Therefore, the p.d. of the battery = 3V + 6V = 9V.

8 The current through the 4 Ω resistor = $V / R = 8 / 4 = 2$ A.

Therefore, the current through the other resistor = 2 A.

The potential difference across the other resistor = 12 – 8 = 4 V.

So the resistance = $V / I = 4 / 2 = 2\ \Omega$.

9 The total p.d. stays the same (12 V) and the current increases. Since $R = V / I$ this means that the equivalent resistance of the circuit gets less.

10 The same current will pass through a 9 V battery connected to a 9 Ω resistor.

Therefore, the current = $V / R = 9 / 9 = 1$ A.

Lesson 3.8 Control circuits

1 The resistance of the LDR.

2 When the temperature increases, the resistance decreases.

3 The I-V graph is not a straight line through the origin.

4 The temperature increases when the current increases. You can work out the resistance by reading off the graph and dividing the p.d. by the current. At higher values of current (and therefore higher temperatures) the resistance is a lower value.

5 When you cover the LDR, the resistance increases so the current decreases. When you uncover the LDR the opposite happens so the current increases.

6 The current passing through the diode is very small when the p.d. is negative. Therefore, a diode has a very high resistance when the p.d. is negative.

7 The resistance of a diode is very small when the current goes through it in the allowed direction. This might allow the current to become dangerously high and damage the diode. An extra resistor prevents the current from becoming too high.

Lesson 3.9 Power and energy transfers

1 A kettle transfers energy from the mains electricity store into the thermal energy stored in the water by heating.

2a The electric current turns the motor in the drill. This means that it transfers energy from the mains electricity store into the kinetic energy stored in the drill. Energy is also transferred to the thermal energy stored in the surroundings.

2b The power of the drill = 400 W.

3 $E = Pt = 2500 \times (45 \times 60) = 6\ 750\ 000$ J (or 6.75 MJ)

4 $P = E / t = 10\ 000 / 5 = 2000$ W (or 2 kW)

5 $E = QV = 30 \times 230 = 6900$ J

6 $V = E / Q = 1800 / 75 = 24$ V

7a Energy transferred in 5 minutes, $E = 1150 \times 5 \times 60 = 345\ 000$ J

$Q = E/V = 345\ 000 / 230 = 1500$C.

7b $Q = E / V = 345\ 000 / 33\ 000 = 10.5$ C

Lesson 3.10 Calculating power

1a $P = VI = 230 \times 4 = 920$ W

1b $R = P / I^2 = 920 / 4^2 = 57.5\ \Omega$

(You could also calculate this by using $R = V / I = 230 / 4 = 57.5\ \Omega$)

2a $I = P / V = 36 / 12 = 3$ A

2b $R = P / I^2 = 36 / 3^2 = 4\ \Omega$

(Again, you could calculate this by using $R = V / I = 12 / 3 = 4\ \Omega$)

3 It could be transferred to the thermal energy stores in the kettle and the surroundings.

4a Volume = 15 × 10 × 2 = 300 m^3

Mass = density × volume = 1000 × 300 = 300 000 kg

4b Energy transferred = $mc\Delta\theta = 300\ 000 \times 4200 \times (22 - 17) = 6\ 300\ 000\ 000$ J (or 6300 MJ)

4c $t = E / P = 6\ 300\ 000\ 000 / 2000 = 3\ 150\ 000$ s = 875 hours (over 36 days!)

5a The microwave oven transfers 800 J to the thermal energy stored in the food every second.

5b A vacuum cleaner transfers 1600 J to the kinetic energy stored by the dust, and the thermal energy stored in the surroundings every second.

 © HarperCollins*Publishers* Limited 2016

Lesson 3.11 Practical: Investigating series and parallel circuits

1 In order to measure the resistance of the wire you would need to measure the p.d. across the wire by using a voltmeter connected in parallel and you would need to measure the current flowing though the wire by connecting an ammeter in series. You would then calculate the resistance by dividing the p.d. by the current. The wire would also need to be in a series circuit with a power supply and a variable resistor.

Calculate the resistance for different lengths of wire, which you can measure with a ruler.

2a The resistance should be directly proportional to the length.

2b A graph of resistance on the y-axis against length on the x-axis should be a straight line (with a positive gradient) through the origin.

3 To keep the temperature constant, you could measure the temperature of the wire using a thermometer and adjust the p.d until the temperature is the same each time.

4a

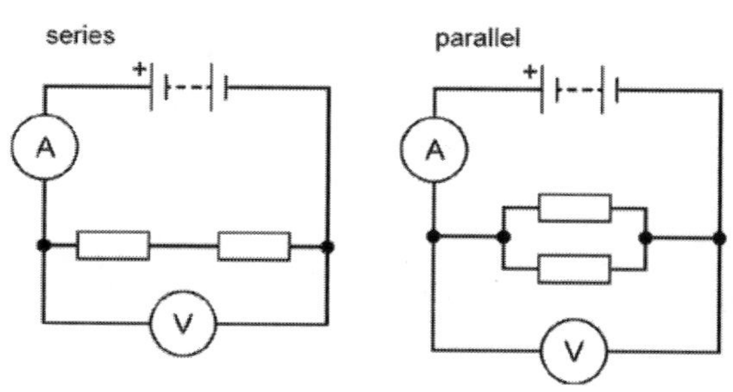

4b In series, the combined resistance should be the sum of the individual resistances. In parallel the combined resistance should be less than the smallest individual resistance.

5a First resistance = $V / I = 6 / 0.1 = 60\ \Omega$

Second resistance = $3 / 0.2 = 15\ \Omega$.

Therefore the expected value of the total resistance = $60 + 15 = 75\ \Omega$.

The student obtained a value of $3.9 / 0.05 = 78\ \Omega$ which is higher than expected.

5b Higher resistance is unlikely to be due to a gain in temperature since the current is actually lower than before. Extra resistance is likely to be from the contact between the resistors. Another possible reason is from inaccurate measurements such as rounding error in the current reading.)

Lesson 3.12 Key Concept: What's the difference between potential difference and current?

1 Your head is rubbing against the nylon jumper and electrons pass from one object to another. This means your head becomes oppositely charged to the jumper and there is an electric field between them. If the electric field is strong enough, the charges in the air particles get pulled apart – which creates the spark.

2 Connect two 12 V car batteries in series.

3 No – it also depends on the current that the device is able to provide.

4 The higher the p.d. (the volts) the larger the electric shock. However, the energy deposited to your body also depends on the current. So a large p.d. is safe if the current through you is very small. However, a current as large as about 0.1 A can kill you.

5 $P = VI = 230 \times 0.05 = 11.5$ W

6 $R = P / I^2 = 11.5 / 0.05^2 = 4600\ \Omega$

Lesson 3.13 Maths Skills: Using algebra in electric circuit calculations

1 One minute = 60 s.

$Q = It = 4.5 \times 60 = 270$ C

2 $I = Q / t = 35 / 14 = 2.5$ A

3 $R = V / I = 12.5 / 2.5 = 5\ \Omega$

4 $I = V / R$

5 $I = V / R = 12 / 200 = 0.06$ A

6 $E = Pt = VIt = 12 \times 2 \times (5 \times 60) = 7200$ J

7 You could have used $E = QV$ and $Q = It$

This gives $E = ItV$ which gives the same answer (as you can multiply values in any order).

Lesson 3.14 Maths Skills: Using formulae and understanding graphs

1a $V = IR = 10\text{ A} \times 100\ \Omega = 1000$ V

1b $V = IR = 5\text{ A} \times 3000\ \Omega = 15\,000$ V

2 $V = IR$

Divide both sides by I

$R = V / I$

3 $P = VI = 230 \times 5 = 1150$ W

4 $P = I^2R$

Divide both sides by r

 © HarperCollins*Publishers* Limited 2016

$I^2 = P / R$

Square-root both sides

$I = \sqrt{(P / R)}$

5 $P = I^2R = 5^2 \times 1000 = 25\ 000$ W

6 If the straight line goes through the origin, then the slope indicates the **constant of proportionality** or (**1/constant of proportionality**) depending on which way round the axes are plotted.

7 The relationship is not proportional because the graph is not a straight line through the origin.

End of Chapter Questions

1

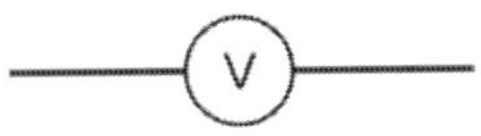

[1 mark]

2 Series [1 mark]

3 $I = Q / t$ [1 mark] = 100 / 20 = 5 A [1 mark]

4 You could rub it on e.g. your hair / your jumper [1 mark] making sure that there is no direct electrical connection to the Earth [1 mark].

5

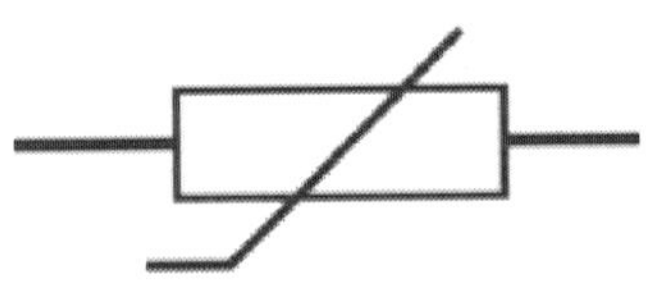

[1 mark]

6 A region of space where a <u>force acts</u> [1 mark] on an electric <u>charge</u> [1 mark].

7 The balloons will repel each other.

8 The brightness of <u>all</u> of the lamps decreases [1 mark]. The current decreases [1 mark].

9 $I = P / V$ [1 mark] = 2000 / 230 = 8.7 A [1 mark]

10 The resistance increases. [1 mark]

11

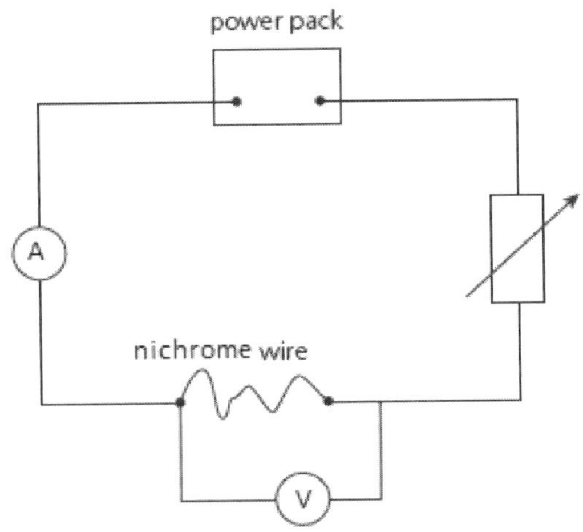

[1 mark for the ammeter connected in series and the voltmeter connected in parallel. 1 mark for power supply/battery/cell and variable resistor included in series with the wire]

12 $I = V / R$ [1 mark] = 12 / 8 = 1.5 A [1 mark]

13 C was the filament lamp [1 mark]

D was the high-valued resistor [1 mark]

14 Power = current^2 × resistance (or $P = I^2R$) [1 mark]

15 Adding resistors in series increases the total resistance of the circuit as this makes it harder for the charge to flow through [1 mark]. Adding resistors in parallel decreases the total resistance of the circuit as this makes it easier for the charge to flow through due to the extra path they can follow [1 mark].

16 $E = P \times t$ [1 mark] = 60 × (1 × 60) = 3600 J [1 mark]

17 Power rating of torch = IV = 5 × 12 = 60 W

Power rating of food mixer = IV = 6 × 230 = 1380 W [1 mark]

$E = Pt$

Energy transferred by kettle = 3000 × 20 = 60 000 J

Energy transferred by microwave = 920 × 60 = 55 200 J

Energy transferred by torch = 60 × 200 = 12 000 J

Energy transferred by food mixer = 1380 × 50 = 69 000 J

So the food mixer transferred the most amount of energy and it transferred 69 000 J [1 mark]

 © HarperCollins*Publishers* Limited 2016

Chapter 4: Magnetism and magnetic fields

Lesson 4.1 Magnetism and magnetic forces

1 They repel each other.

2 You need to see which way the compass points at different places so that you can plot the field lines.

3 When the pins and tacks are inside a magnetic field they become magnets themselves. When they are removed from the magnetic field, the tacks lose their magnetism straight away because they are made of iron but the pins remain magnetised because they are made from steel.

4 The iron tacks become magnetised so they are magnetic materials whereas the pins remain permanently magnetised. Therefore, the tacks will have no effect on other tacks; the tacks and pins will always attract each other; the pins will either attract or repel other pins depending on which way round they are.

5 A magnet will be repelled by another magnet. A material like iron can only be attracted by another magnet.

6 The magnetic compass experiences forces of attraction and repulsion from the Earth which would suggest that the core of the Earth is magnetic.

Lesson 4.2 Compasses and magnetic fields

1a

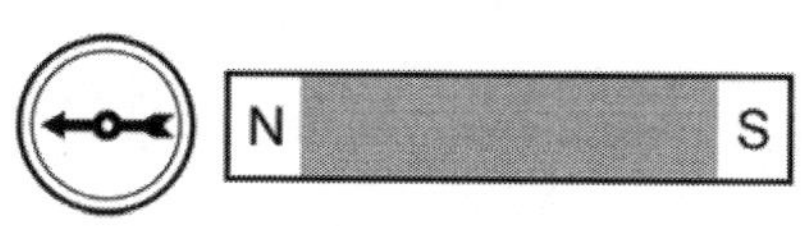

1b

2 A magnetic compasses point away from the geographic south pole so the polarity of the geographic south pole is actually north.

3 It seeks out (i.e. points towards) the Earth's North Pole.

4

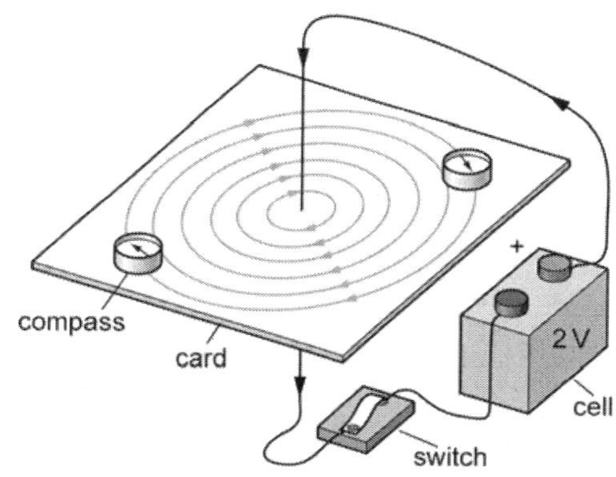

5

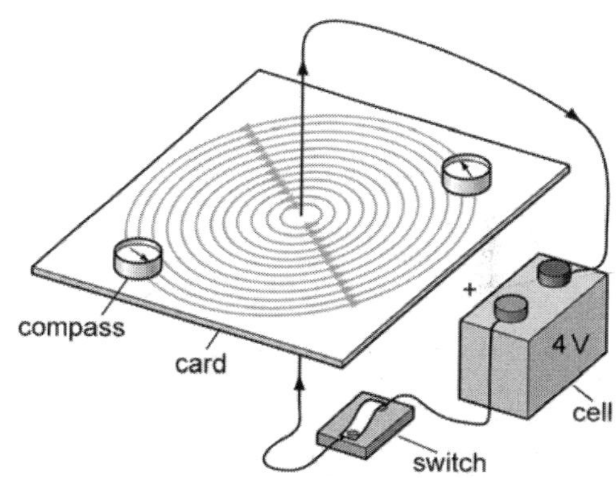

6 The magnetic field produced by the current is very weak and it is even weaker the further you are from the wire. Therefore, a compass towards the edge of the card will be affected by the Earth's magnetic field as much as the wire's.

Lesson 4.3 Magnetic effects

1 The magnetic field is in the opposite direction.

2 The south pole.

3 All of these will make the magnetic field stronger: Increase the number of turns of wire; increase the current; place iron inside the coil.

4a To the right

4b Assuming the current is in the original direction, the wire moves to the right. If the current remains reversed, then the wire would move to the left.

5 You could reduce the current or reduce the strength of the magnetic field by using a weaker magnet.

6 Applying Fleming's left hand rule: the first finger needs to point from N to S; the second finger needs to point down along the wire. This means that the thumb is pointing out of the magnet towards the switch. So the vertical wire will move

 © HarperCollins*Publishers* Limited 2016

to the right towards the switch (whilst remaining vertical).

Lesson 4.4 Calculating the force on a conductor

1 The magnetic field inside the solenoid is uniform (the same everywhere). This can be seen by the field lines that are parallel and equally spaced. The strength of the field is indicated by how close the lines are together and they remain the same distance apart throughout.

2a The force doubles.

2b The force is four times bigger.

3a $F = BIl = 0.05 \times 0.5 \times 2 = 0.05$ N

3b $F = BIl = 0.1 \times 2 \times 0.5 = 0.1$ N

4 $I = F / Bl = 0.03 / (0.02 \times 0.3) = 5$ A along the wire.

5 $B = F / Il = 0.05 / (3 \times 0.25) = 0.067$ T

6 No. When the length of wire is parallel to the magnetic field then no force acts; a maximum force acts when they are at right angles to each other.

7a Force needed to lift wire = 0.30 N

$F = BIl$

So $0.30 = 0.00003 \times I \times 2.0$

$I = 0.30 / (0.00003 \times 2.0) = 5000$ A.

7b No, it wouldn't be possible. The magnetic field lines near a pole would be vertical. The force needs to be at right angles to the field lines (from Fleming's left hand rule) so there is no orientation of the wire that can produce a force upwards.

Lesson 4.5 The generator effect

1 By changing the magnetic field – either by varying the strength of the magnets or by moving the wire relative to the magnets.

2 0 V – since the magnetic field doesn't change.

3 You can move the magnet more slowly, use a weaker magnet or use a solenoid with fewer turns.

4a The induced current is flowing through the solenoid. This current produces the second magnetic field.

4b On the left hand diagram the left hand end of the coil is a north pole and the right hand end is a south pole. On the right hand coil, the left hand end is a south pole and the right hand end is a north pole. You can work this out by knowing that the coil opposes the motion of the magnet (so it must repel the magnet in the left diagram and attract the magnet in the right diagram).

4c The magnetic field produced by the current in the coil attracts and repels the magnet, making it harder for the magnet to move. Therefore, you need to apply a force to balance out this magnetic effect.

4d For example on the diagram on the left – this time the coil would attract the magnet. The magnet would speed up and therefore there would be more energy in its kinetic energy store. This energy would have come from nowhere and so energy has not been conserved. Therefore it is impossible for the induced p.d. to be in this direction.

5 There would not be a complete circuit so no current flows. Therefore, the meter would always be pointing to 0 A.

Lesson 4.6 Electric motors

1 e.g. washing machine, fan, vacuum cleaner

2 The magnetic field produced by the electric current interacts with the magnetic field produced by the magnets. This creates forces on the coil which push one side up and the other side down. These forces turn the coil.

3 The split-commutator connects the coil to the power supply. However, the connections swap to the other side of the coil every time the coil makes half a turn. This ensures that the direction of the current is correct for the forces to continue to spin the coil in the same direction.

4 When a current is parallel to the magnetic field from the magnets then no force acts.

5 You can turn the dynamo's coil faster; you can use a coil with more turns; you can use a stronger magnet.

6 The dynamo and the motor look exactly the same (in fact you can use the same piece of equipment for both devices). The difference is that you physically turn the dynamo's coil, which generates an electrical current; but you connect up the motor's coil to a power supply, which makes a current flow in the coil and this turns the coil.

Lesson 4.7 Generators and transformers

1 The coil moves through a magnetic field which induces the potential difference from the generator effect. The potential difference is alternating because one side of the coil is alternately moving upwards then downwards through the magnetic field as the coil turns.

2 The induced p.d. will be larger and the output will vary quicker. This means that the graph will have higher peaks and lower troughs and the horizontal

 © HarperCollins*Publishers* Limited 2016

distance between the peaks and troughs will be smaller.

3 The magnetic field in the iron core is changing. So the secondary coil is in a changing magnetic field which means there is a p.d. induced in the secondary coil due to the generator effect.

4 $V_p / V_s = n_p / n_s$

$240 / 12 = 720 / n_s$

$20 = 720 / n_s$

$n_s = 720 / 20 = 36$

5 $V_p / V_s = n_p / n_s$

$240 / 20 = n_p / 100$

$12 = n_p / 100$

$n_p = 12 \times 100 = 1200$

6 $V_p / V_s = n_p / n_s$

$240 / V_s = 100 / 10$

$V_s = 240 / 10 = 24$ V.

7 There is no heating effect due to friction.

Lesson 4.8 Loudspeakers and microphones

1 Sound waves are produced by vibrating objects, which in turn make the air vibrate.

2 The alternating current keeps changing direction. This produces a force on the loudspeaker cone which keeps changing direction – so the loudspeaker cone vibrates.

3 The louder speaker could have more turns in its coil and/or it could be using a stronger magnet. Also the coil might be made of lighter wire so it can move more easily.

4 This is an accurate statement. The headphones are just very small loudspeakers.

5 A microphone has the same components as a loudspeaker – a diaphragm / cone, a magnet and a coil.

6 The pressure variations in the sound waves make the diaphragm vibrate which makes the coil vibrate. The moving coil in the magnetic field induces a current in a coil due to the generator effect. So the pressure variations produce variations in the current flowing through the coil.

7 This makes the microphone more sensitive to vibrations. The large number of turns means that a large current is produced when there is a small movement and the thin wire means that the coil is light enough to move easily when the sound waves hit the diaphragm.

Lesson 4.9 Key Concept: The link between electricity and magnetism

1 e.g. an electromagnet, a bell, a relay, a motor, a loudspeaker

2 e.g. a motor, a loudspeaker

3 e.g. a dc or ac generator, a microphone, an electric guitar pickup

4 A motor consists of a coil of wire which carries a current and the coil is inside a magnetic field. The current in the coil produces its own magnetic field which interacts with the magnetic field from the magnet. This interaction produces forces on the coil. The forces act upwards on one side of the coil and downwards on the other side which makes it turn. The coil is connected to a power supply using a split ring commutator. The commutator ensures that the current goes through the coil in the correct direction to make the coil keep turning the same way. It does this by reversing the direction of the current every half turn.

5 You can use Fleming's Left Hand Rule. Place your thumb, first finger and second finger of your left hand at right angles to each other. The first finger needs to point in the direction of the magnetic field produced by the magnets (north pole to south pole) and the second finger needs to point in the direction of the current. Your thumb will point in the direction of the magnetic force. In Figure 7.31 you can see that the force would push the wire down towards the bottom of the magnet.

6 Motors and generators both consist of a rotating coil inside a magnetic field. However, motors require a power supply and an electric current is used to produce motion. Generators, however, do not require a power supply but they do require something to rotate the coil. They use the rotation of the coil to generate an electric current.

Lesson 4.10 Maths Skills: Rearranging equations

1 $F = BIL = 3 \times 10^{-5} \times 3 \times 25 = 0.00225$ N $= 2.25 \times 10^{-3}$ N

2a Method 1

$F = BIL$

$0.125 = 1.0 \times 10^{-4} \times I \times 250$

$0.125 = 0.025\ I$

$I = 0.125 / 0.025 = 5$ A

Method 2

$F = BIL$

$I = F / BL = 0.125 / (1.0 \times 10^{-4} \times 250) = 5$ A

2b Method 2 is quicker but you need to be good at algebra. Most people find method 1 easier.

 © HarperCollins*Publishers* Limited 2016

3 e.g.

$L = F / BI$

$= 0.05 / (3 \times 10^{-5} \times 2.5)$

= 670 m (to 2 significant figures)

End of Chapter Questions

1 a) Near the poles [1 mark]

2 b) Stripping the plastic insulation off the wire. [1 mark]

(Using a thicker wire reduces the resistance so the current would be larger)

3 Permanent magnet remains magnetised when removed from another magnetic field but an induced magnet loses its magnetism. [1 mark]

A permanent magnet attracts and repels other magnets but an induced magnet is only attracted to other magnets. [1 mark]

4 c) They repel

5 e.g. hang both iron bars from a piece of cotton [1 mark]

The iron bar that rotates and points due North is the magnetised one [1 mark]

(Allow 1 mark for the description of the procedure and 1 mark for explaining how the magnetised bar can be found)

(Finding out which bar attracts the other one is incorrect as you still would not know which one was magnetised).

6 c) Seeing what potential difference the power supply is set to.

This doesn't directly indicate the strength of the field but changing it to a higher value would probably make the magnetic field stronger.

7 c) Attract the striker to hit the gong.

8 d) Towards the magnet's south pole [1 mark]

9 d) Pressure differences due to sound make the coil move. [1 mark]

10 A Increase the potential difference. [1 mark]

11 Any two out of:

- Have a large number of turns in the solenoid
- Use a large current / large potential difference across it
- Put iron in the middle

[2 marks]

12a $V_p / V_s = n_p / n_s$ [1 mark]

$240 / 12 = 500 / n_s$

$n_s = 500 / 20 = 25$ [1 mark]

12b Power supplied to engine = VI

$= 12 \times 1.5 = 18$ W [1 mark]

So power supplied to transformer = 18W

$I = P / V$

$= 18 / 240 = 0.075$ A [1 mark]

13 The student needs to identify variables other than the length, which would also affect the magnetic force and needs to makes sure they are kept constant [1 mark]. These variables include the current and the magnetic flux density/the distance from the magnet [1 mark for identifying one of these variables].

14 b) How hard the person pushes the wire [1 mark]

15 c) There should be fewer coils on the secondary coil than on the primary [1 mark]

16 The current through the coil and the magnetic field from the permanent magnet produce a force on the cone. [1 mark]

The alternating current means that the force on the cone goes backwards and forwards so the cone vibrates which produces sound. [1 mark]

[1 mark is for a description of the motor effect and 1 mark is for linking this to the motion of the cone]

17 Level 3: A clear and logical description is made in a coherent sequence of steps. The description includes why a potential difference is induced in the solenoid. It also links the shape of the graph to the motion of the magnet including a discussion about the why the potential difference starts positive and ends up negative. The fact that the magnet speeds up is identified and this is linked to both the size of the maximum positive and negative potential difference and the difference in the time when the pd is positive compared to when it is negative. (5-6 marks)

Level 2: A clear description is made which links the fact that there is an induced potential difference to the motion of the magnet. An attempt is made to explain why the potential difference changes sign and that the negative pd has a greater magnitude by linking this to the motion of the magnet. (3-4 marks)

Level 1: A basic description about why a potential difference is induced is made and there is some comment about the shape of the graph. (1-2 marks)

Indicative content

 © HarperCollins*Publishers* Limited 2016

- A potential difference is induced when the magnet moves through the solenoid
- This is due to the fact that the solenoid is cutting through the field lines or that the magnetic field is changing
- The induced potential difference is positive when the magnet enters the solenoid and is negative when the magnet leaves the solenoid
- This is because the magnetic field is changing the opposite way: it gets bigger as the magnet enters the solenoid and it gets smaller as it leaves the solenoid
- Another way of thinking about the direction is that any current induced in the solenoid would act to stop itself being formed so the polarity of the solenoid would repel the magnet as it was entering and attract the magnet as it was leaving – and so the pd must swap direction halfway through. This would exert a force on the magnet to reduce its motion.
- The magnet is leaving the solenoid at a quicker speed than it is entering, so it must have accelerated inside the solenoid.
- You can see this because the time it takes when the pd is positive is longer than when it is negative due to the wider peak.

  Also the magnitude of the pd when it is negative is higher than when it is positive and moving the magnet faster increases the size of the pd.

18 You can reverse the direction of motion in the coil in the generator [1 mark].

You can reverse the polarity of the magnetic field [1 mark].

19 When the coil is vertical, its edges (A and B) are moving parallel to the magnetic field so there is no potential difference induced. [1 mark]

When the coil is horizontal, its edges are moving at right angles to the magnetic field so there is a maximum potential difference induced. [1 mark]

The induced p.d. changes direction because each side passes up and then down through the magnetic field. [1 mark]

[Maximum of 2 marks]

 © HarperCollins*Publishers* Limited 2016

Chapter 5: Waves in matter

Lesson 5.1 Describing Waves

1 2 waves

2 As the amplitude increases, the amount of energy transferred by a wave increases.

3a $f = 1 / T = 1 / 0.1 = 10$ Hz

3b $f = 1 / T = 1 / 0.25 = 4$ Hz

4 $v = f\lambda = 2$ Hz × 0.1 m = 0.2 m/s

5 The wavelength will halve (assuming the speed stays the same).

6 Since $v = f\lambda$ then v is 6 times bigger.

Lesson 5.2 Transverse and longitudinal waves

1

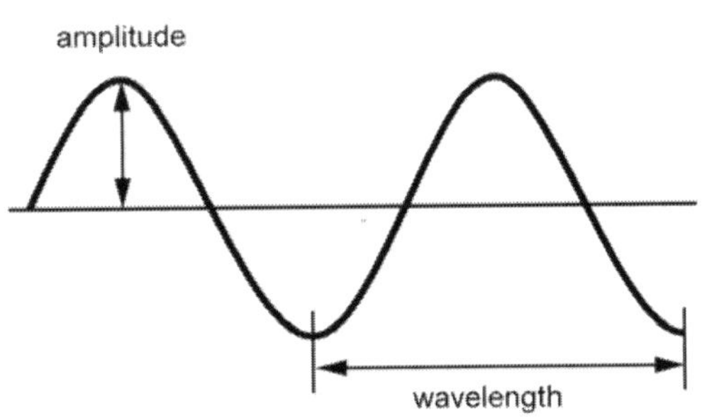

2 Objects floating on water bob vertically up and down as waves pass along the water.

3 Alex is wrong because in a transverse wave, the vibrations are at right angles to the direction the wave is moving, so the water doesn't move along with the wave. The objects might be moving over the water because the wave has transferred energy to their kinetic energy store – but they are not being carried along with the water.

4 D (transverse)

5 You could lay a slinky spring next to a ruler on a bench. Set up a video camera to record the wave and set it recording before the wave is produced on the slinky. Produce a longitudinal wave on the slinky and then stop recording. Play back the wave in slow motion. Use the ruler to record the location of a link on the slinky before the wave is produced. Then observe the motion of that link as the wave passes and record the maximum distance it travels away from its original position. This maximum distance is the amplitude of the wave.

6 The sound wave shown by the blue graph has the highest frequency.

7 15 cm

8 The sound waves travel through the iron railing and also through the air. Sound travels faster through the iron railing so Ann hears the sound wave that has passed along the iron railing first and then she hears the sound wave that has passed through the air.

9 Speed of sound in air at 50 ^{0}C = 331 + (0.6 × 50) = 361 m/s.

$\lambda = v / f = 361 / 1000 = 0.361$ m.

Lesson 5.3 Reflection and refraction of waves

1 When light reflects you would see an image / reflection; when sound reflects you would hear an echo; when a water wave reflects you would see the wave moving back in the opposite direction. We are assuming that the surface that they are reflecting off is smooth.

2a The sound would disappear and you wouldn't hear it.

2b You would hear an echo but it would not be as loud as the original sound.

3 40^0. The ray diagram should look like figure 6.18.

4 The angle of incidence = $90^0 - 35^0 = 55^0$, so the angle of reflection is 55^0.

5 When light speeds up (e.g. when it passes out of a glass block into the air).

Lesson 5.4 Sound waves

1 $f = v / \lambda$ so if the wavelength increases, the frequency decreases (the speed stays the same). This means that the sound will be lower in pitch.

2 16 cm

3 The air particles vibrate left and right about a fixed position.

4a No

4b Yes

5 The particles are closer together so they can transfer their energy quicker.

6 Sound travels faster in glass than it does in air so it bends the other way than light does. The sound will bend upwards towards the top of the block and then bends towards the normal as it leaves the block (almost vertically).

Lesson 5.5 Exploring ultrasound

1 It has a frequency that is too high for humans to hear.

2 Sound has a longer wavelength than ultrasound and this means that it can't be focused into a narrow beam. This would blur the image.

3 X-rays are particularly dangerous to bodies that are growing rapidly (such as a developing baby or a young child). So using ultrasound is much safer.

© HarperCollinsPublishers Limited 2016

4 e.g. checking the health of unborn babies; checking that materials used in aircraft don't have any cracks in them.

5 An ultrasound image can detect cracks in pipes so this will determine whether there is a leak.

6 If the speed changes by a large amount then most of the ultrasound is reflected. This might be a problem because if most of the ultrasound is reflected then the ultrasound that can pass to deeper structures in the body becomes very weak and these structures might not be able to be imaged.

7 To prevent too much reflection so most of the ultrasound passes into the body.

8 A short pulse of ultrasound waves is transmitted into the body. When the pulse hits a boundary (e.g. between one tissue and another tissue) some of it reflects back and some of it continues onwards. By measuring the reflected pulses, a computer can form an image of the structures inside the body.

9 The second pulse has travelled further than the first pulse. The extra distance it goes is twice the size of the head (because it passes through the head and reflects back again).

Let the size of the head = d

Distance = speed × time

$2d = 1500 \times 140 \times 10^{-6} = 0.21$ m

So $d = 0.21 / 2 = 0.105$ m or 10.5 cm

10 Lungs are full of air. Ultrasound travels at a very different speed in air than in body tissue so most of the ultrasound would reflect back from the lungs. Therefore, any pulses passing to the other side of the lungs would be far too weak to be used to form an image.

Lesson 5.6 Practical: Measuring the wavelength, frequency and speed of waves in a ripple tank and waves in a solid

1 It shakes the rod, which produces the ripples in the tank.

2a m or cm

2b Hz

3 You could use a transparent ruler taped onto the bottom of the tank of water. The shadow of the ruler can be used to measure the distance between the ripples on the viewing screen.

4 This makes the measurement more accurate.

5 She would divide the number of waves she counted by ten.

6 They could look at a single ripple and time how long it took to move a certain distance between to markers on the viewing screen.

7 They could use a strobe light to freeze the pattern. The frequency of the waves would equal the frequency of the strobe light. They could find the wavelength by measuring the distance between two ripples with a ruler.

8 If both methods give the same answer then it is likely that their conclusions are valid. If you only use one method you can't be sure that the measurement is correct as there is nothing to check it.

9 You can use a metre ruler by taping it to the bench.

10 Your reaction time may be too slow. The stop clock may not be precise enough.

11 You need to place your head directly above the ruler to measure the distances accurately which would be difficult to do perfectly. The pattern might not be stable, so that it is moving around slightly. The ruler might not be completely parallel to the string.

12 You could measure the speed for the different patterns that are formed. If all of the speeds are nearly the same you could be confident that your measurement accurately measured the true speed of the waves on the string.

Lesson 5.7 Maths Skills: Using and rearranging equations

1a $T = 1 / 100 = 0.01$ s

1b $T = 1 / 1000 = 0.001$ s

1c $T = 1 / 15\,000 = 0.000067$ s $= 6.7 \times 10^{-5}$ s

2a $f = 1 / 5 = 0.2$ Hz

2b $f = 1 / 10 = 0.1$ Hz

2c $f = 1 / 150 = 0.0067$ Hz $= 6.7 \times 10^{-3}$ Hz

3a $v = f\lambda = 100 \times 2 = 200$ m/s

3b $v = 100 \times 2 = 200$ cm/s = 2 m/s

3c $v = 100 \times 2 = 200$ mm/s = 0.2 m/s

4 $f = v/\lambda$

5a $f = 25 / 0.5 = 50$ Hz

5b $f = 250 / 0.05 = 5000$ Hz

5c $f = 2500 / 0.005 = 500\,000$ Hz = 500 kHz

6a $\lambda = v / f$

6b $\lambda = 330 / 500 = 0.66$ m

 © HarperCollins*Publishers* Limited 2016

7 Speed of red light = distance / time = 3.0 / 1.0 × 10^{-8} = 3.0 × 10^{8} m/s.

$f = v/\lambda$ = 3.0 × 10^{8} / 6.5 × 10^{-7} = 4.6 × 10^{14} Hz.

8 Frequency of sound wave in air = v / λ = 330 / 0.25 = 1320 Hz

In titanium, $\lambda = v / f$ = 6100 / 1320 = 4.6 m (to 2 significant figures).

Lesson 5.8 Key Concept: Transferring energy or information by waves

1 Sound travels much more slowly than light does.

2

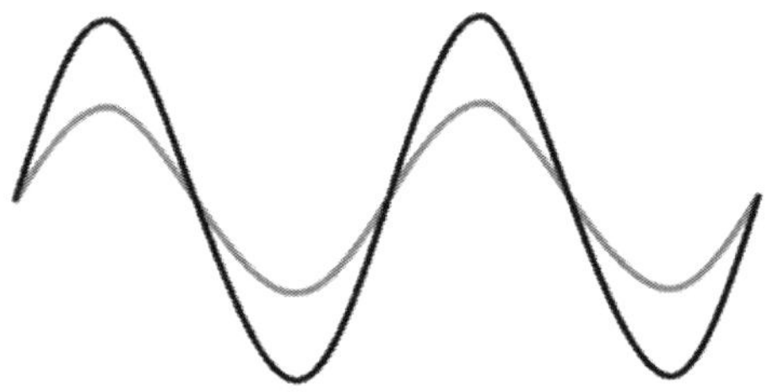

3 If the amplitude of a sound wave increased, you would hear a louder sound.

4 Data is sent faster and over greater distances through fibre optic cables than through copper wires.

5 e.g. Some food absorbs microwaves

Ultraviolet light can be reflected by snow

X-rays can be reflected in x-ray telescopes

Gamma rays are absorbed by lead

6 The Sun is replenishing its energy from fusion reactions inside its core. (It is actually losing energy but this results in its mass decreasing rather than its temperature.)

Lesson 5.9 The electromagnetic spectrum

1a Amplitude is the maximum distance the particles in the wave move from their rest position.

1b The wavelength is the distance from one point on a wave to the same corresponding point on the next wave.

2 e.g. a sound wave.

3 A transverse wave vibrates at right angles to the direction that it travels and a longitudinal wave vibrates along the same direction that it travels.

4 They don't need a material to carry them; they can travel in a vacuum.

5 They are all transverse waves and they all travel at the same speed in a vacuum.

6 Gamma rays

7 $f = v / \lambda$ = 3.0 × 10^{8} / 3.5 × 10^{-7} = 8.6 × 10^{14} Hz

8 $f = v / \lambda$ = 3.0 × 10^{8} / 0.2 = 1 500 000 000 Hz = 1.5 × 10^{9} Hz

9 Ultraviolet waves have a higher frequency than radio waves. Therefore, they can transfer more energy. This increases their ability to do harm.

Lesson 5.10 Gamma rays and X-rays

1 It is long enough to be useful but not too long so that the patient is exposed where there are no more benefits.

2 They should be gamma emitters; they should not be dangerously toxic.

3 Usually radiotherapy needs X-rays which transfer a higher energy than the X-rays used for imaging. This means that X-rays used for radiotherapy will tend to have shorter wavelengths.

4 Gamma rays can pass straight through bone and tissue so there wouldn't be enough contrast in the image to enable any diagnosis to take place.

5 Similarities: they are both electromagnetic wave.

Differences: Gamma rays tend to have a higher frequency than X-rays (although there is some overlap); gamma rays come from the nucleus of an atom whereas X-rays are produced by the elctrons.

6 Gamma rays come from a radioisotope which doesn't need electricity to work. X-rays are produced by accelerating electrons so they collide with a target with high energy. You need an electricity supply in order to accelerate the electrons.

7 They are dangerous and can give you radiation sickness / cancer.

8 The X-ray image is different if there is a fault in a weld so an expert can detect the fault.

9 The chest X-ray has a very small dose compared to what an average person would receive anyway so this procedure is relatively safe and the benefits for most people would easily outweigh the risks. The CT scan has a much bigger dose but this is still below what an average person would receive from natural sources. So if you only have one CT in a year then it is relatively safe although the doctors need to decide whether the risk is worth it and CT scan will help them to make the patient better.

Lesson 5.11 Ultraviolet and infrared radiation

1 e.g. security marking; tanning; producing vitamin D

 © HarperCollins*Publishers* Limited 2016

2 You can use an invisible ink that glows when ultraviolet light is shone on it.

3 So they can check the bank notes to see if they are genuine.

4 Ultraviolet lights often do have a purple colour but this is due to the visible light they produce in addition to the ultraviolet. This is a safety measure to show the user that they are switched on. Ultraviolet light is invisible to our eyes so it has no colour.

5 Small doses help your body to make vitamin D. Large doses can cause sunburn or skin cancer.

6 Snow reflects ultraviolet light so skiers will be exposed to more ultraviolet than people on the beach if the sunlight is the same in both situations.

7 Infrared radiation can't pass through the chicken. You need to wait for the inside of the chicken to be heated by other processes.

8 Conduction.

9 e.g. they can detect the presence of a warm body and set the alarm off; they can be used with a beam of infrared and set the alarm off if the beam is blocked by something going past.

10 Infrared radiation reflects off surfaces, so the remote can point towards a surface and the infrared beam can reflect off this towards the TV.

Lesson 5.12 Microwaves

1 Mobile phones / Satellite communication / cooking

2 This is quite a long wavelength in the microwave region. Therefore, these microwaves have low energy and are likely to be used for communication.

3 Microwaves only cook water or fat molecules and they penetrate about 1cm beneath the surface of the food. Conventional ovens use infrared to cook food. Infrared red only reaches the outer surface of the food but it does cook all materials. Both processes rely on conduction for the remainder of the food to cook.

4 It is quicker because the microwaves penetrate into the food. Food is cooked more evenly because the microwaves heat up the food from the inside rather than just at the surface.

5 Usually bread is cooked using infrared. This makes the outside of the toast crispy and the inside moist. Making toast in a microwave oven would have the opposite effect. The inside would be baked hard (or burnt) and the outside will still be soft bread.

6 Distance microwave travels = speed × time = $3.0 \times 10^8 \times 64 \times 10^{-6}$ = 19 200 m. The microwave travels to the target and back so the target is 19 200 / 2 = 9 600 m away.

7 The signal gets weaker the further it goes so the equipment won't be sensitive enough. Also there are more reflections from other objects which confuse the signal, the further away the target.

Lesson 5.13 Radio and microwave communication

1 Radio waves travel very fast (the speed of light) which means that communication can be very fast. Radio waves can travel large distances and they can pass through most walls.

2 $f = v / \lambda = 3.0 \times 10^8 / 100 = 3.0 \times 10^6$ Hz = 3.0 MHz.

3 A radio receiver has a wire or a coil. Radio waves induce oscillations in the wire or coil which results in a changing current. This current is the radio signal that the receiver picks up and can be turned into sound.

4 These radios use the energy transmitted by the radio waves for their power.

5 Microwaves don't pass through the walls of houses very effectively.

6 Buildings absorb microwave signals so you need lots of aerials for the signals to make sure that users can pick up a signal wherever they are.

7 They refract in the ionosphere and return to Earth.

8 Waves with this frequency don't return to Earth but are transmitted into space.

Lesson 5.14 Reflection, refraction and wave fronts

1 Aluminium foil is shiny so it is a good reflector of infrared radiation. Therefore, the heat is reflected back into the room rather than absorbed by the walls. This reduces heat transferring through the walls and so reduces the heating bills.

2 e.g. a thermos flask, white buildings in hot countries

3 $\lambda = v / f = 3.0 \times 10^8 / 10 \times 10^6 = 30$ m

4 $f = v / \lambda = 3.0 \times 10^8 / 100 = 3.0 \times 10^6$ Hz = 3.0 MHz

5

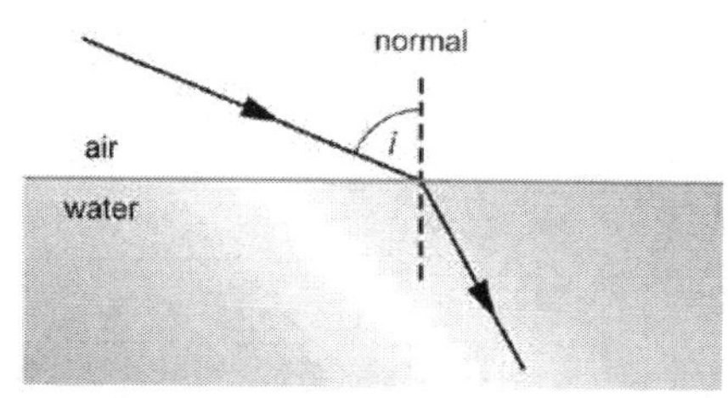

6 Under the right conditions, radio waves can reflect off the atmosphere. This enables them to reach

 © HarperCollinsPublishers Limited 2016

places that are out of line of sight due to the curvature of the Earth.

7a The wavefronts get closer together and they change direction.

7b They travel faster in deeper water. The diagram shows that they slow down when they enter shallow water because they become closer together (like cars do in a traffic jam).

Lesson 5.15 Practical: Investigate how the amount of infrared radiation absorbed or radiated by a surface depends on the nature of that surface

1 The experiment shows how quickly a surface can absorb infrared radiation.

2 The stopper connected to the blackened metal plate should drop first. This is because black surfaces absorb infrared more effectively than shiny surfaces.

3 The thickness of the plates should be the same, the mass of the stoppers and the amount of wax used to stick them should be the same, the heater needs to give the same heat transfer in both directions and the plates need to be placed the same distance either side of the heater.

4 This ensures that the temperature of each surface is the same.

5 To make it a fair test and ensure that the effect of the area on the infrared radiation is kept the same.

6 The infrared detector needs to be placed the same distance from each of the sides of the container.

7 The blackened bulb will absorb the infrared effectively and so the thermometer can measure the amount of infrared being produced.

8 The infrared sensor.

9a A firefighter's suit needs to be white or shiny.

9b Solar panels needs to be matt black.

10a It needs to be shiny (so that it doesn't emit much infrared radiation).

10b They need to be matt black (so they emit lots of infrared radiation).

11 Although they are emitting different intensities of infrared light, the sides of the cube are all at the same temperature. This is because of conduction of heat through the walls from the water. Therefore, the thermometer would give the same reading on all of the sides.

Lesson 5.16 Practical: Investigate the reflection of light by different types of surface and the refraction of light by different substances

1 e.g. unpolished shoes; rock; wood

2 A car driver might be dazzled if the light from their headlights is reflected from the bicycle in front of them straight back into their eyes.

3 They could use a ray box to produce the light ray, a mirror, a protractor to measure the angles, a piece of white paper, a pencil and a ruler. Place the mirror vertically on a piece of white paper and draw a line along its edge. Remove the mirror and draw in a normal line at 90^0 to the line that is already on the paper using the protractor and pencil. Replace the mirror and shine a ray of light from the ray box so that it hits the mirror at the normal line. Use the pencil and the ruler to draw lines along the incident ray and the reflected ray. Measure the angle of incidence and the angle of reflection between the rays and the normal line by using the protractor.

4 They should measure them from the normal line.

5 e.g. the mirror could move; the pencil lines could be too thick to measure the angles from. Fix the mirror in place with tape; use a sharp pencil; take repeat readings and find a mean value.

6 You could plot a graph with the angle of incidence on the x-axis and the angle of reflection on the y-axis. The shape of the line of best fit shows you the relationship between the angles.

7 No – the angle of reflection always equals the angle of incidence but in diffuse reflections the normal lines are in many different directions because the surface is rough.

8a The angle of refraction is smaller than the angle of incidence – this happens when the light slows down.

8b The angle of refraction is larger than the angle of incidence – the light has sped up.

9 It should leave the block in the same direction as it entered.

10 No – because at some angles the light would come out of the right hand side of the block rather than the bottom edge.

11 They should put the water into a transparent container with very thin walls. The container should be the same shape as the glass block. The walls should be thin to minimise the effect of refraction in the container so that nearly all the measured refraction is due to the water. They should then carry out the same procedure as they did with the glass block.

Lesson 5.17 Lenses

1a It is the place where rays that are parallel to the axis meet once they have passed through the lens.

 © HarperCollins*Publishers* Limited 2016

1b Yes, for example parallel rays that enter the lens at an angle to the principal axis will be focused onto another point.

2 The principal axis.

3 To reduce the focal length you would have to bend the rays more so they were focused closer to it. Therefore, the lens needs to have a fatter shape.

4 The light bends towards to the normal when it enters the lens and away from the normal after it leaves the lens. Due to the shape of the surface of the lens – this means that the light diverges.

5 If you draw rays from a point on an object near a lens, the rays that can enter the lens will spread apart from each other (diverge) quite steeply. Moving the object further away, the rays that can enter the lens will still be diverging but not as steeply – so they are more parallel to each other. The further away you move the object, the more parallel the rays get.

6 You can consider the centre of a thin lens to be the same shape as the middle of a rectangular block. This displaces the ray sideways but it still continues in the same direction (see section 6.6). If the lens is thin then the displacement is very small and so it looks like the ray does not refract at all.

Lesson 5.18 Images and magnification

1

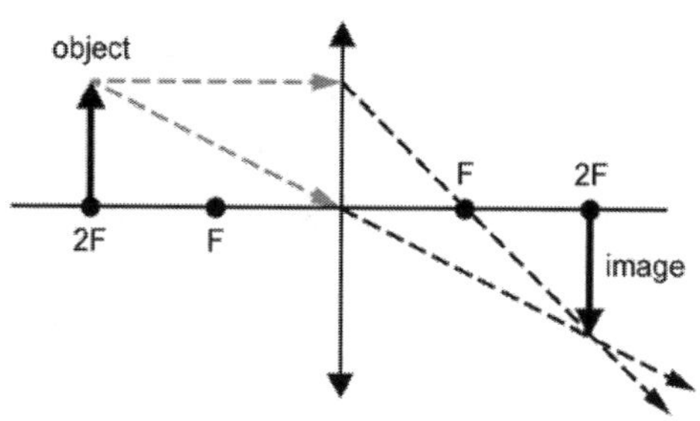

The image is inverted, it is real and it is the same size as the object.

2

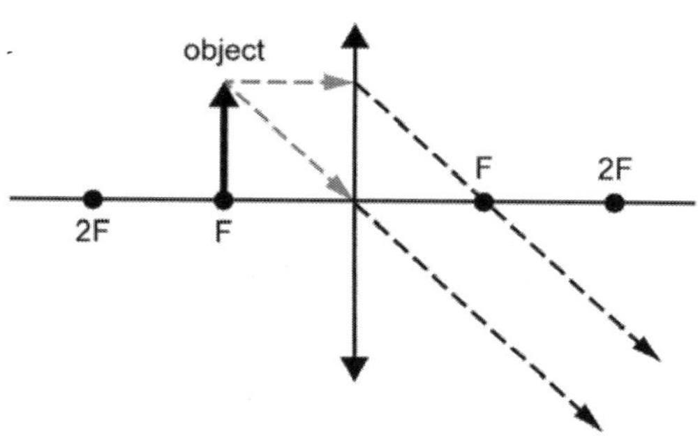

The image can't be formed because the rays entering the eye are parallel to each other. (It is also correct to say the image is virtual, an infinite size, the right way up and an infinite distance away)

3

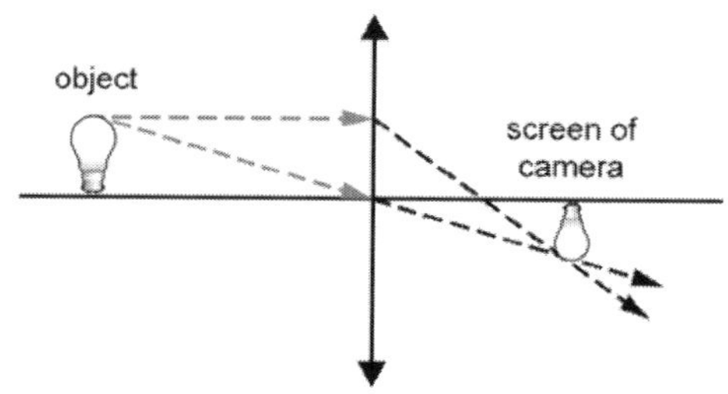

4

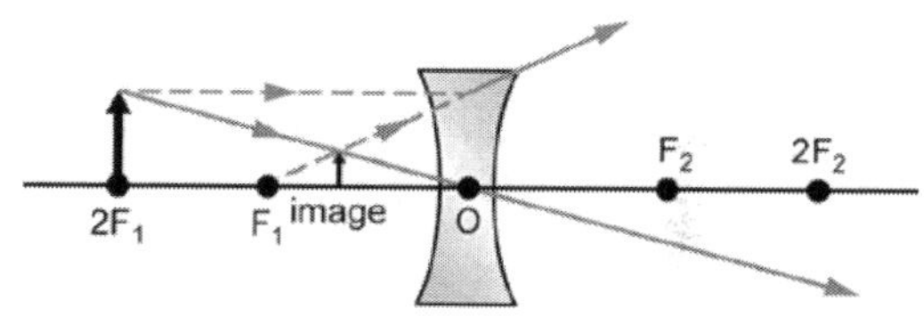

The image is upright, virtual and smaller than the object.

5 1.2 cm = 12 mm. Magnification = 12 / 1.6 = 7.5

6 1.76 m = 176 cm. Magnification = 1.9 / 176 = 0.011 (to 2 significant figures)

7 A ray diagram drawn to scale will produce an image that is 2cm high and is 6cm to the right of the lens. Therefore, the magnification = 2 cm / 1 cm = 2.

Lesson 5.19 Colour

1 No colours are reflected by a black object.

2 All the colours apart from yellow.

3 The colour of an object is the colour of light that it reflects.

4 All of the colours apart from blue.

5a red

5b black

5c black

5d black

6a black

 © HarperCollins*Publishers* Limited 2016

6b green

6c black

6d green

7 A grey colour is the same as white only darker. Therefore, a grey object will partially absorb all the colours and partially reflect all of them.

End of Chapter Questions

1 a) Sound wave [1 mark]

2 a convex lens [1 mark]

3 c) hertz [1 mark]

4 It is the maximum distance [1 mark] that a point moves from its undisturbed position [1 mark]

5 Gamma rays [1 mark]

6 $T = 1 / 4 = 0.25$ s [1 mark]

7 A radio wave [1 mark]

8 This is called diffuse reflection [1 mark]. The light has reflected in all directions [1 mark].

9 A = infrared and B = X-rays [1 mark for both correct]

10 Microwaves and infrared [1 mark]

11a Cooking and communication [1 mark]

11b Killing cancer cells and being used in a tracer [1 mark]

12 Magnification = image height / object height [1 mark] = 2.5 / 7.5 = 1/3 (or 0.33) [1 mark]

13 $v = f\lambda$ [1 mark] = 2 × 8 = 16 cm/s (or 0.16 m/s) [1 mark]

14 White is a poor absorber infrared radiation [1 mark] so the houses do not heat up as quickly by absorbing infrared radiation from the Sun [1mark].

15 Because its frequency is higher than 20 000 Hz which is the highest frequency humans can hear [1 mark].

16 Any colours apart from white and blue. [1mark]

17 White surface are both poor emitters and poor absorbers of radiation [1 mark]. Therefore, in hot countries the house remains cool as it absorbs infrared radiation slowly and in cold countries the house remains warm as it only emits infrared radiation slowly [1 mark].

18a

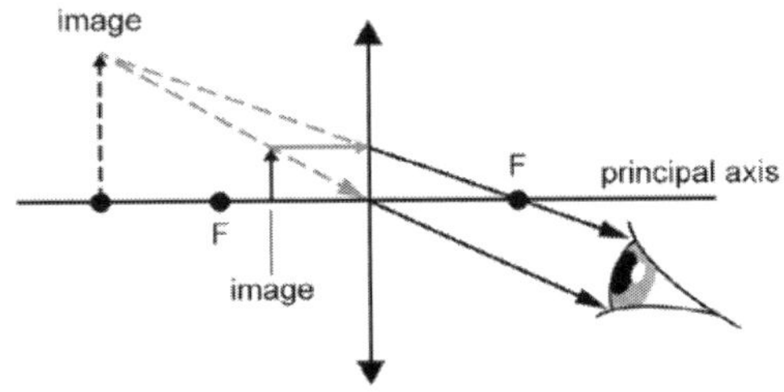

[1 mark for both of the rays drawn correctly, 1 mark for the correct location of the image.]

18b The image is virtual [1 mark], larger than the object [1 mark] (and the right way up).

18c The long sighted person looks at objects that are closer to the lens than one focal length [1 mark]. This makes the image further away and larger which both help the long sighted person to see the object clearly [1 mark].

19 As the ship is moving along the sea bed gradually becomes deeper, then stays at a constant depth and then gradually gets shallower [1 mark]. The anomalous result could be from a fish swimming underneath the ship but close to the surface.

20 Time between the pulses = 0.0600 – 0.0500 = 0.0100 ms

Distance travelled in this time = speed × time = 4000 m/s × 0.0100 × 10^{-3} s = 0.0400 m [1 mark]

This is the distance that the ultrasound travelled from the front of the stone to the back of the stone and then from the back of the stone to the front again.

So the thickness of the stone = 0.04 / 2 = 0.0200 m [1 mark].

 © HarperCollins*Publishers* Limited 2016

Chapter 6: Radioactivity

Lesson 6.1 Atomic structure

1 88 electrons

2 92 protons, 238 – 92 = 146 neutrons

3 The number of electrons is the same; the number of protons is the same; the number of neutrons is different.

4 Uranium: 92 protons, 238 – 92 = 146 neutrons

Thorium: 90 protons, 234 – 90 = 144 neutrons

5a 7 protons and 14 - 7 = 7 neutrons.

5b 92 protons and 235 - 92 = 143 neutrons.

6 Nuclear radiation can knock electrons off atoms. These atoms become positive ions since they have lost negative charge. Once the electrons have been knocked off, these can join onto other atoms. These atoms become negative ions because they have gained negative charge.

Lesson 6.2 Radioactive decay

1 150 × 20 = 3000 counts

2 They are all unstable and form radioisotopes.

3 You can't predict when a particular nucleus is going to decay.

4 The nitrogen nucleus has one more proton and one fewer neutron than the carbon nucleus.

5a The nucleus loses 2 protons and 2 neutrons.

5b One of the neutrons in the nucleus becomes a proton.

6 This is beta decay.

7 A nucleus has a lot of energy after it has undergone alpha or beta decay. This energy is enough for gamma radiation to be released.

Lesson 6.3 Nuclear equations

1 They both have to be balanced.

2 A chemical equation only has one set of numbers that need to be balanced, a nuclear equation has two sets of numbers.

3a ${}^{226}_{88}\text{Ra} \rightarrow {}^{222}_{86}\text{Rn} + {}^{4}_{2}\text{He}$

3b ${}^{219}_{86}\text{Rn} \rightarrow {}^{215}_{84}\text{Po} + {}^{4}_{2}\text{He}$

4a ${}^{90}_{38}\text{Sr} \rightarrow {}^{90}_{39}\text{Y} + {}^{0}_{-1}\text{e}$

4b ${}^{32}_{15}\text{P} \rightarrow {}^{32}_{16}\text{S} + {}^{0}_{-1}\text{e}$

5a ${}^{24}_{11}\text{Na} \rightarrow {}^{24}_{12}\text{Mg} + {}^{0}_{-1}\text{e}$

5b The particle emitted is a beta particle (an electron).

6a An alpha particle is emitted.

6b A = 228, Z = 88

7a platinum-190 → osmium-186 + alpha particle

${}^{190}_{78}\text{Pt} \rightarrow {}^{186}_{76}\text{Os} + {}^{4}_{2}\text{He}$

7b rhenium-187 → tantalum-183 + alpha particle

${}^{187}_{75}\text{Re} \rightarrow {}^{183}_{73}\text{Ta} + {}^{4}_{2}\text{He}$

7c copper-66 →zinc-66 + beta particle

${}^{66}_{29}\text{Cu} \rightarrow {}^{66}_{30}\text{Zn} + {}^{0}_{-1}\text{e}$

7d nickel-66 → copper-66 + beta particle

${}^{66}_{28}\text{Ni} \rightarrow {}^{66}_{29}\text{Cu} + {}^{0}_{-1}\text{e}$

7e rhodium-105 → palladium-105 + beta particle

${}^{105}_{45}\text{Rh} \rightarrow {}^{105}_{46}\text{Pd} + {}^{0}_{-1}\text{e}$

7f osmium-186 → tungsten-182 + alpha particle

${}^{186}_{76}\text{Os} \rightarrow {}^{182}_{74}\text{W} + {}^{4}_{2}\text{He}$

Lesson 6.4 Radioactive half-life

1 Because radioactive decay is random.

2 It is the time it takes for the activity of a sample to fall to half of its current amount.

3a 40 counts per minute

3b 20 counts per minute

3c 10 counts per minute

3d 5 counts per minute

4 8 minutes

5a & 5b

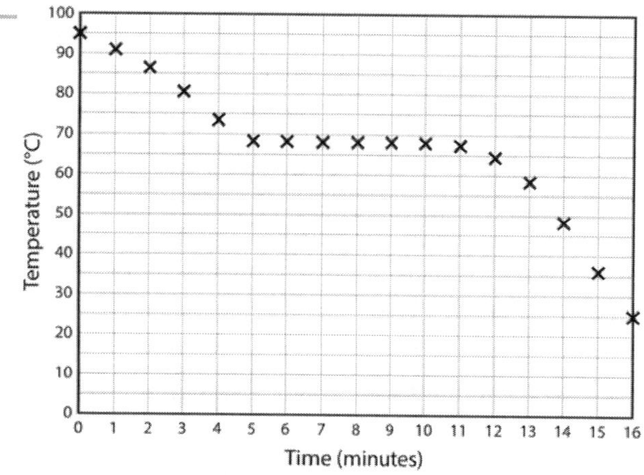

 © HarperCollins*Publishers* Limited 2016

5c Time taken to halve from 100 Bq to 50 Bq = 1 minute

Time taken to halve from 40 Bq to 20 Bq = 2.4 – 1.4 = 1 minute

Time taken to halve from 32 Bq to 16 Bq = 2.7 – 1.7 = 1 minute

So the average of these three measurements = (1 + 1 + 1) / 3 = 1 minute.

6 It takes 1 half-life to decrease from 100 Bq to 50 Bq and a further half-life to decrease from 50 Bq to 25 Bq. Therefore 4 hours is 2 half-lives which means that 1 half-life = 2 hours.

Lesson 6.5 Background radiation

1 Radon and thoron from soil, rocks and building materials; gamma rays from rocks and soil; radiation from living things and food; cosmic rays from outer space; medical; fallout from nuclear weapons testing and other forms such as air travel; work related; nuclear power industry.

2 All of the forms of background radiation apart from medical, work related, nuclear fallout, nuclear industry and air travel are natural. Therefore, the total percentage = 37 + 19 + 17 + 14 = 87 %.

3a The types of rock are different in different areas so some areas are exposed to more background radiation than others.

3b For example, nuclear fallout will vary depending on which way the wind is blowing. Cosmic background radiation increases when the sun is particularly active.

4 Beta particles can pass through a few metres of air and paper but they are stopped by a few mm of low density metals such as aluminium. Beta particles are more ionising than gamma rays but are less ionising than alpha particles.

5 Gamma radiation

6 Radon is an alpha emitter and alpha particles are very ionising and therefore dangerous if the inside of the body is exposed to them. Radon is also a gas so you can breathe it in.

7 It is likely that they would receive a higher amount of radiation. Pilots and cabin crew spend much of the time in the air at a higher altitude. The atmosphere will not have absorbed as much cosmic radiation at this altitude so the pilots and cabin crew will receive more cosmic radiation than people who spend more time at ground level.

Lesson 6.6 Maths Skills: Using ratios and proportional reasoning

1 200 Bq → 100 Bq: 70 – 10 = 60 seconds

100 Bq → 50 Bq: 145 – 70 = 75 seconds

50 Bq → 25 Bq = 230 – 145 = 85 seconds

Average = (60 + 75 + 85) / 3 = 73 seconds

2a

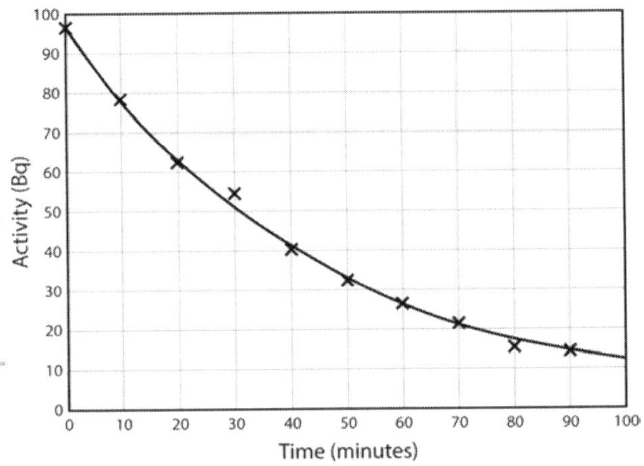

2b 92 Bq → 46 Bq: 34 – 2 = 32 minutes

60 Bq → 30 Bq: 53 – 22 = 31 minutes

40 Bq → 20 Bq: 70 – 40 = 30 minutes

Average = 31 minutes.

3a Number of half-lives = 4.

So the activity = 640 ÷ 2 ÷ 2 ÷ 2 ÷ 2 = 40 Bq

3b Net decline = 40 / 640 = 0.0625 = 1/16

4a Number of half-lives = 32 / 8 = 4

So count rate = 1800 ÷ 2 ÷ 2 ÷ 2 ÷ 2 = 112.5 Bq

4b Net decline = 112.5 / 1800 = 0.0625 = 1/16

5 1/16 decrease in activity corresponds to 4 half-lives.

So 1 half-life = 24 / 4 = 6 hours.

6 The number of atoms of polonium that have decayed equals the number of atoms of lead that are formed. After two half-lives ¼ of the original polonium atoms would be remaining and ¾ would have turned into lead atoms. Therefore at this time there will be three times as many lead atoms as polonium atoms. So it takes two half-lives = 2 × 138 = 276 days.

Lesson 6.7 Hazards and uses of radiation

1 When a radioactive material is somewhere where it isn't wanted.

2 Radioactive materials produce ionising radiation which is harmful to health. Ionising radiation can kill cells and can cause cancer.

3 Gamma (since it has the weakest ionising power)

 © HarperCollinsPublishers Limited 2016

4 Alpha particles are very ionising which means that they are more likely to kill cells or cause cancer. However, they are not very penetrating so if they are outside the body they can't get in and cause the damage.

5a You need to make sure that your measurements are a true measure of the activity of the tracer rather than that of the background radiation.

5b Background radiation is random and its activity can be higher at some times than at others. You need to use an accurate average.

6 The isotope with the half-life of 6 hours. 6 seconds would mean that the isotope has decayed to unmeasurable levels before the tracer can be monitored and 6 days would mean the patient would remain radioactive long after the procedure had taken place which would add unnecessary risk.

7 A gamma emitter. Gamma radiation has the weakest ionising power so it does the least harm to the body. It is also the most penetrating so it can pass from the inside of the body where the tracer is, to the outside where it can be detected.

8 Inject the tracer into the blood. Leave enough time for the tracer to be carried by the blood through the body. Examine the patient with a gamma camera. If there is a large signal at a particular place in the vessel then it is likely that there is a blockage in that place.

Lesson 6.8 Irradiation

1 Irradiation is when you expose an object to nuclear radiation.

2 We receive much more irradiation from the food than from the air (it's about 500 times more).

3 Accurate repair, cell death, misrepair

4 If a sperm cell or an egg cell is misrepaired then this change of genetic material could be passed onto offspring.

5 Irradiation is exposing someone to nuclear radiation. Contamination is when radioactive material is actually present on the person (which will continue to irradiate them).

6 The people doing the experiments might have made mistakes. If other people carrying out an experiment agree with the findings, then the findings are more likely to be true.

7 Once the pigeons move away from something that is irradiating them then they are no longer exposed to the radiation. However, if they are contaminated with radioactive material then they will continue to be irradiated for as long as the material in them remains radioactive. This is much more likely to cause them serious harm.

Lesson 6.9 Uses of radiation in medicine

1 They are both types of electromagnetic waves.

2 Many manufactured radioisotopes are used in medicine. Some of these are used to destroy cancerous cells from the radiation they produce; others are used as tracers to help diagnose problems with a patient in order to treat them.

3 X-rays are only produced when needed. You can control the energy of the x-rays that you produce.

4 Bracytherapy uses the radioactive source right next to (or inside) the tumour rather than the radiation coming from outside the body.

5 You need to make sure that you can extract all of the radioactive source so you don't contaminate the patient. Placing the radioactive source near the tumour might need invasive surgery which can lead to problems with infection.

6 Alpha particles wouldn't penetrate very far into the tumour (and probably wouldn't even be able to leave the protective casing). Many gamma rays would pass through the tumour so there would be a large dose applied to tissue outside of the tumour. Beta particles would penetrate into the tumour and deposit most of their energy within the tumour. Therefore, the tumour would get the biggest dose with a beta emitter.

Lesson 6.10 Using nuclear radiation

1 The patients have cancer and are therefore likely to suffer much more if they are left untreated. The benefits of the cancer treatment outweigh the risks of side effects from the treatment.

2 Compare the amount of radiation produced from the radioactive iodine with an image produced by a normal thyroid gland. If the patient is suffering from hypothyroidism then the image will show less radiation since the gland would not have as absorbed as much iodine.

3 Radiotherapy has a risk of harm to the patient. Although surgery is also dangerous, it is the safer option once the tumour has been reduced in size.

4 Not all of the cancer cells would have been removed from the tumour and these could go on to cause further cancer. The radiotherapy is used to kill all of the remaining cancer cells.

5a Activity is the number of decays per second.

5b Dose is a measure of harm that a particular procedure does to the patient.

6 It is more important to consider the dose as this is a measure of the direct harmful effect that the procedure has on the patient. A high activity of one source might actually be less harmful than a low activity of another source.

 © HarperCollins*Publishers* Limited 2016

7a The heart scan.

7b Over this time interval you would receive the same amount of radiation from background radiation as you would from the scan.

7c The patient might have a had a radiation dose previously and this scan might make them go over the maximum allowed radiation dose.

Lesson 6.11 Nuclear fission

1 Nuclear fission is the splitting up of a large nucleus into two smaller daughter nuclei and a few neutrons.

2 Fission is made to happen when a neutron is absorbed whereas radioactive decay happens spontaneously. The nucleus splits into two approximately equally sized parts in fission whereas only a small part of the nucleus is ejected (if any) in radioactive decay.

3 Fission can be induced by absorbing a neutron. Fission also releases a few neutrons which can then go on to cause further nuclei to fission. Thus a chain reaction can build up.

4 The similarities are that they both heat up water to produce steam, which turns a turbine and a generator. The steam turns back into water using cooling towers. A nuclear power station differs from a coal-fired one in that it uses uranium as a fuel rather than coal. Also you only have to replace the uranium every few years whereas you have to continually feed a coal fired power station with coal. The amount of fuel you need in a coal fired power station is much more than you need in a nuclear power station.

5 A coolant (e.g. a gas) is made to flow through the nuclear reactor. This heats up the gas. The hot gas then passes by some cold water and energy is transferred to the water by conduction. This turns the water into steam.

6 The moderator slows down the neutrons. Slower moving neutrons are more likely to cause fission.

7 Raising the control rods increases the temperature and lowering them decreases the temperature. When the rods are raised, more neutrons per fission go on to cause further fissions so the number of fissions happening at a particular time increases and energy is released more quickly. When the control rods are lowered then fewer neutrons go on to cause further fission so the number of fission reactions at any given time decreases. When, on average, one neutron in a fission reaction goes on to cause a further fission then the temperature remains constant.

Lesson 6.12 Nuclear fusion

1 They both release energy from changes in the structure of a nucleus. However, fission involves a large nucleus splitting into smaller ones whereas fusion is the joining of two small nuclei to form a larger one.

2 High temperatures and high pressures. The nuclei need to get very close together before fusion can take place. A large force is needed to do this because the nuclei repel each other (since they are both positively charged). Both high pressure and high temperature provide the conditions for the nuclei to get close enough to fuse.

3 There is plenty of available fuel and the waste products do not harm the environment.

4 The work is shared so that the work can be checked to see if everyone agrees with the results. Also other people can continue the work and use it in further developments in science.

5 You don't need to generate electricity from a hydrogen bomb. Therefore, you don't need to control the rate that energy is released from a hydrogen bomb. You can create the conditions needed for nuclear fusion using a fission bomb within a hydrogen bomb – it would not be possible to do this safely in a domestic fusion reactor.

6 There is plenty of fuel for fusion and the waste produces are clean so this would solve many of the future problems that we are currently facing in energy production. Also the fuel is available everywhere so this limits potential for wars over gaining resources in the future. However, fusion is very difficult to achieve and we might spend lots of money trying to develop fusion power with no success. This money could be spent on developing energy resources that we know will work. Also fusion power will turn buildings radioactive so there will be hazards associated with fusion power.

End of Chapter Questions

1

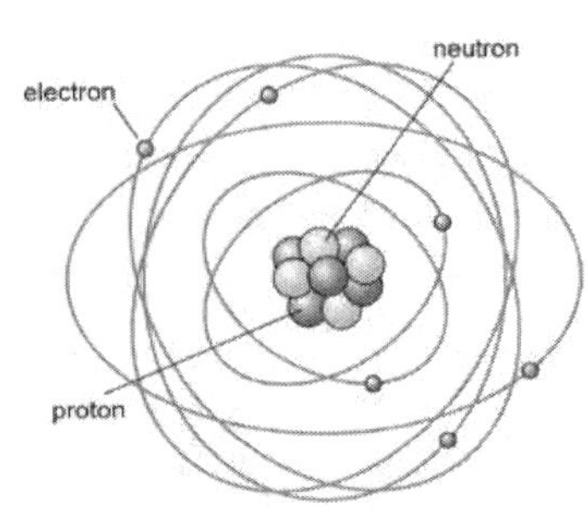

An atom contains a nucleus with protons and neutrons in it. Electrons are located further out. [1 mark]

2 Alpha, beta and gamma radiation. [1 mark for one of these correct, 2 marks if all three are correct.]

3 The time it takes for the activity to fall to half of its present value. [1 mark]

 © HarperCollins*Publishers* Limited 2016

4 Any **two** from rocks/building materials, living things/food/soil, cosmic rays/outer space/the Sun, nuclear weapons/nuclear power stations. [2 marks]

5 A large proportion of background radiation comes from rocks [1 mark]. The type of rock changes from place to place and some are more radioactive than others [1 mark].

6 X-rays and gamma rays are both ionising forms of radiation, which means they can kill cells or turn them into cancer cells [1 mark]. However, x-rays and gamma rays can also be used to treat cancer by killing cancer cells / they can help to diagnose illnesses that can then be treated more effectively [1 mark].

7 It is the number of protons in the nucleus. [1 mark]

8 A photographic plate / a fluorescent screen / a Geiger counter / a gamma camera [1 mark]

9 Fusion is the joining of two small nuclei to form a larger one [1 mark]. Fission is the splitting of a large nucleus to form two smaller ones (and some neutrons) [1 mark].

10a 23 is the mass number (or the nucleon number) which is the total number of protons and neutrons in the nucleus. 11 is the atomic number which is the total number of protons in the nucleus. [1 mark]

10b An isotope of an element contains the same number of protons but a different number of neutrons (from other isotopes of the same element). [1 mark]

11a Iodine-131 because it has a shorter half-life [1 mark]. After a few weeks it will no longer be contaminating the body (whereas the activity of the Carbon-14 would not have decreased very much at all) [1 mark].

11b The benefits of using a tracer is that it might diagnose a condition that can be treated and so make the patient better. The risks of using a tracer is that it might harm the patient by causing cancer. [1 mark]

The benefits must be likely to greatly outweigh the risks before a tracer is used. [1 mark]

12 Ionising radiation that is around us all the time. [1 mark]

13 Irradiation is exposing someone to nuclear radiation. Contamination is when radioactive material is actually present on the person (which will continue to irradiate them). [1 mark]

14 ${}^{219}_{86}\text{Rn} \rightarrow {}^{215}_{84}\text{Po} + {}^{4}_{2}\text{He}$ [1 mark for all the top numbers correct and 1 mark for all the bottom numbers correct.]

15 The radioactive sample took 3 half-lives to decay from 400 Bq to 50 Bq (400 Bq → 200 Bq → 100 Bq → 50 Bq). [1 mark]

Therefore 60 minutes = 3 half-lives. So the half-life = 60 / 3 = 20 minutes. [1 mark]

16a Radioactivity is a random process [1 mark]. Therefore, all of the points would show considerable scatter and you would have to draw a curve of best fit [1 mark].

16b Time to halve from 2000 counts/min to 1000 counts/min = 8 minutes

Time to halve from 800 counts/min to 400 counts/min = 18.5 – 10.5 = 8 minutes

Therefore average half-life = 8 minutes.

[1 mark for obtaining a value of the half-life to be between 7.5 and 8.5 minutes, 1 mark for finding more than one half-life and obtaining an average.]

17 Any two marks from:

Alpha particles consist of 2 protons and 2 neutrons; beta particles are electrons; gamma rays are electromagnetic rays. [1 mark]

Alpha particles are heavily ionising; beta particles are less ionising and gamma rays are very weakly ionising. [1 mark]

Alpha particles are stopped by paper / a few cm of air; Beta particles are stopped by a few mm of aluminium / a few m of air; Gamma rays can even pass through a few cm of lead. [1 mark]

18 Any two marks from:

- Geiger and Marsden used a radioactive source to fire alpha particles at a thin piece of gold foil
- Most of the particles passed straight through
- A very few of them bounced straight back

Any one mark from:

- This showed that most of the atom consisted of empty space
- This showed that there must be a small / dense, positively charged nucleus at the centre

Any one mark from:

- The evidence contradicted the predictions from Thomson's plum pudding model
- Rutherford proposed a new nuclear model of the atom which was consistent with the evidence from the experiment.

19

- The tracer needs to be a gamma emitter (so it can be detected easily) [1 mark]

 © HarperCollins*Publishers* Limited 2016

- It needs to have a sensible half-life (e.g. a few hours / days) so that it lasts long enough to be useful but it doesn't contaminate the water after the procedure is over. [1 mark]
- You need to place the tracer in the pollution that you suspect might be contaminating the water. [1 mark]
- You can collect samples from the water source and examine them using e.g. a gamma camera to see if there are any signs of the tracer. [1 mark]

 © HarperCollins*Publishers* Limited 2016

Chapter 7: Energy

Lesson 7.1 Investigating kinetic energy

1 The amount of energy in the kinetic energy store depends on both the mass and the speed. An adult has more mass than a child so they would have more kinetic energy even though the speed is the same.

2 The fuel is the food that the child has eaten.

3a $E_k = \frac{1}{2} mv^2 = 0.5 \times 50 \times 2^2 = 100$ J

3b Her kinetic energy would be four times as much.

4 Kinetic energy stored at 10 m/s = $0.5 \times 1200 \times 10^2$ = 60 000 J

Kinetic energy stored at 30 m/s = $0,5 \times 1200 \times 30^2$ = 540 000 J

So kinetic energy increases by 540 000 – 60 000 = 480 000 J

5 240 km / h = (240 × 1000 m) / 3600 s = 66.7 m/s

6a When the ball is at the highest point.

6b As the ball hits the surface for the first bounce

7 After you let go of the ball and the ball is moving upwards, energy stored in the ball's kinetic energy store is being transferred to energy stored in the gravitational potential energy store.

8 E_p stored by the ball at 20m = mgh = 2 × 10 × 20 = 400 J

So E_k ball stores at the ground = 400 J

$\frac{1}{2}$ mv² = 400

So $0.5 \times 2 \times v^2 = 400$

$v^2 = 400$

so v = 20 m/s

Lesson 7.2 Work done and energy transfer

1 The amount of force and the distance that the force moves.

2 The gravitational force (weight) on the person as the person moves towards the ground.

3 $W = F \times s = 400 \times 1.5 = 600$ J

4 She is not doing any work as the force is not moving.

5 $W = F \times s$, so $s = W / F = 300 / 200 = 1.5$ m

6 $W = F \times s$, so $F = W / s = 3000 / 12 = 250$ N

7a $E_k = \frac{1}{2} mv^2 = 0.5 \times 800 \times 12^2 = 57\,600$ J

7b $F = W / s = 57\,600 / 8 = 7200$ N

8a $W = F \times s = 500 \times 4 = 2000$ J

8b 2000 J (assuming all of the energy in her E_p store is transferred to her E_k store)

8c E_k reduces by 2000 J so work done by the trainers = 2000 J.

$F \times s = 2000$ so $F \times 0.01 = 2000$.

$F = 2000 / 0.01 = 200\,000$ N.

Lesson 7.3 Specific heat capacity

1 The material with the bigger increase of temperature does not need as much thermal energy transferred to it to increase its temperature by 1 °C.

2 $\Delta E = mc\Delta\theta = 1 \times 380 \times 20 = 7600$ J

3 $\Delta E = mc\Delta\theta = 2 \times 450 \times 30 = 27\,000$ J

4 Water has a very high specific heat capacity. This means that the water can transfer lots of thermal energy to the bed as it cools down – making the bed warmer.

5 It is the thermal energy needed to raise the temperature of 1 kg of a substance by 1 °C.

6 $\Delta E = mc\Delta\theta = 3 \times 450 \times 15 = 20\,250$ J

7 $\Delta\theta = 30 - 10 = 20$ °C

$\Delta E = mc\Delta\theta = 50 \times 800 \times 20 = 800\,000$ J

Concrete is chosen as it has a relatively high specific heat capacity so it can store more thermal energy than most other solids.

8 Change in thermal energy of the steel =

$1 \times 450 \times (80 - \theta) = 36\,000 - 450\,\theta$

Change in thermal energy of the water = $0.5 \times 4200 \times (\theta - 10) = 2100\,\theta - 21\,000$

So $36\,000 - 450\,\theta = 2100\,\theta - 21\,000$

So $2550\,\theta = 57\,000$

$\theta = 22.4^0$

Lesson 7.4 Electrical energy and power

1 Energy transferred from fuel in power station to kettle due to electric current.

Energy transferred from electric current to water by heating.

Energy transferred from water to kettle by heating.

Energy transferred from kettle to surroundings by heating.

Energy transferred from water to surroundings by sound.

 © HarperCollins*Publishers* Limited 2016

2 You either need to know the power of the kettle and the time that it is switched on for or you need to know the mass of the water, the specific heat capacity of the water and the temperature change (= 90°C).

3 It is transferred to the potential energy of the water particles as it is changing state into steam.

4 $I = P / V$ = 1800 / 230 = 7.82 A (to 3 significant figures)

5 $E = Pt$ = 1800 × 90 = 162 000 J

6 Some of the energy is not transferred to the water (e.g. some of it is transferred to the thermal energy store of the kettle).

7 $E = Pt$

1kW = 1000 W and 1 hour = 60 × 60 = 3600 s

So E = 1000 × 3600 = 3 600 000 J

Lesson 7.5 Dissipation of energy

1 Lubricate the wheels to reduce the effect of friction.

2 Newspaper reduces the amount of thermal energy that is transferred through it in any direction. Therefore, it reduces the thermal energy transferred to a cold ice cream as much as the thermal energy transferred away from the fish and chips.

3 Most thermal energy is lost through the windows. You could draw the curtains / use double glazing as this would reduce the rate that the thermal energy conducts.

4 The eco home needs to be well insulated so that it reduces the amount of thermal energy it loses to a bare minimum. It can do this by having thick walls with cavity wall insulation, thick loft insulation, triple glazed windows.

5 The thermal energy dissipates into the surroundings. This means that it spreads out and becomes diluted. The energy is too thinly spread for us to collect it again to re-use.

6 Energy is transferred to the surroundings and the car itself as thermal energy. Energy is also transferred into sound energy and gravitational potential energy if the car is going uphill.

7 Energy is being transferred to the thermal energy stored in the surroundings. Here the energy is dissipated which means it becomes very spread out and the temperature of the surroundings does not increase by very much (shown by the dark blue colour of the sky). Energy is being transferred more quickly through the windows than through the walls of the buildings because the windows are not as thick and they have a higher thermal conductivity. The ground floors of the buildings are transferring thermal energy more quickly than the higher floors. This is probably due to the ground floors storing more thermal energy as they have been heated more than the upper floors (since people have been there during the day).

Lesson 7.6 Energy efficiency

1 It heats up the surroundings. However, the temperature increase of the surroundings is very small as the energy has become very spread out.

2 An electric motor does not get as hot as a petrol or diesel engine as there is no need to convert energy from the fuel into thermal energy. Therefore, much less energy is wasted by transferring energy into the thermal energy stored in the surroundings.

3 Efficiency = useful energy output / total energy input = 80 / 100 = 80 %

4a Efficiency = useful energy output / total energy input = 135 / 500 = 0.27 = 27 %

4b Not all of the coal is burned.

5 Some of the energy is transferred to the thermal energy stored in the kettle and the surroundings rather than the thermal energy stored in the water.

6 Efficiency = useful energy output / total energy input

0.65 = useful energy output / 200

So useful energy output = 0.65 × 200 = 130 J.

7 Energy cannot be created or destroyed, only transferred from one store to another.

8 The metal expands so some of the energy is transferred to the elastic potential energy stored in the metal.

9 Some of the energy is transferred to the thermal energy store of the eardrums; some of it is transferred to the thermal energy stored in the air; some of the sound reflects back off your eardrums; the energy spreads out so not all of the energy transferred by the sound ends up at your eardrums.

10 First calculate the total energy input to the car:

Total energy input = useful energy output (× 100) / efficiency = 100 × 100 / 85 = 117.6 J

Now calculate energy input at the power station using the same formula:

Total energy input = 117.6 × 100 / 35 = 336 J.

 © HarperCollins*Publishers* Limited 2016

Lesson 7.7 Practical: Investigating ways of reducing the unwanted energy transfers in a system

1 Tazim is correct to say that only metals are good thermal conductors but other materials are also able to conduct a little. Different materials conduct differently so it is likely that all the materials will have different insulating effects. So Tazim's statement is incorrect.

2 The type of insulating material in the box; the amount / thickness of insulating material; the size / surface area of the box; the colour of the surfaces of the box; the material the box is made out of; the temperature of the surroundings; initial temperature of the liquid.

3 e.g. the better that an insulating material traps air then the better insulator it is and the hotter the liquid will be after a certain period of time. So expanded polystyrene would be the best insulator and air would be the worst.

4 Variables that need to be kept the same: size and shape of the box; thickness of insulating material (by making sure the box is completely packed full of the insulating material); the initial temperature of the liquid; the volume of the liquid; the time interval between measuring the initial temperature and the final temperature of the liquid.

Variables to change: the insulating material

Variables to measure: the initial temperature of the liquid, the temperature of the liquid after a certain amount of time.

Method:

Pour 300 ml of boiling water into the cooking pot and wait for the temperature to decrease to 90.0 ^{0}C. Start a stopwatch to measure the time.

Then place the lid on the pot and place the pot in the box as quickly as possible (for the control, take the measurements without putting the pot in the box).

Pack in the insulating material as quickly as possible and make sure that sure the box is fully packed.

Every hour (up to 5 hours) take the pot out of the box and record the temperature of the water with a thermometer. Put the pot back into the box and repack the insulating material as quickly as possible.

5

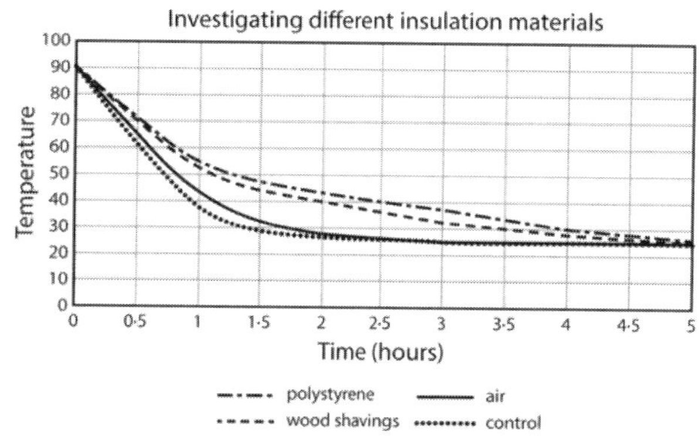

6 The gradient of the graph – the line with the shallowest slope indicates the best insulator.

7 The liquid becomes almost the same temperature as the surroundings so it only cools very slowly.

8 The rate of cooling is the gradient of the graph.

9 The rate of cooling depends on the difference in temperature between the liquid and the surroundings. During the first hour, the temperature difference was large so the rate of cooling was high; during the fifth hour the temperature difference was small so the rate of cooling was low.

10a No matter how good an insulating material is the pot would end up as cold as the room eventually. It is the time it takes to do this that matters. The experiment showed that the different insulators had different rates of cooling so it was able to test the hypothesis.

10b This is correct. A larger mass of liquid would mean that the liquid takes longer to cool down but the experiment is only meant to investigate how the type of insulating material affects the rate of cooling.

10c The box is trapping the air inside so this is a different situation from when the pot is open to the atmosphere where the air can move around freely. You can see from the graph that the air was able to insulate the pot a little bit better when it was trapped inside the box.

10d Some lines (such as the control) cooled down to their final temperature much sooner than five hours. So if you drew a straight line from 90 ^{0}C to 25 ^{0}C at the time when the graph reached this temperature then the lines would have different gradients – showing that the materials have different insulating properties.

11a The energy transferred, $\Delta E = mc\Delta\theta$. Since m and c remain the same then the amount of energy transferred depends on the temperature change, $\Delta\theta$.

 © HarperCollinsPublishers Limited 2016

For air: $\Delta\theta = 90.0 - 43.5 = 46.5\ ^0C$

For polystyrene: $\Delta\theta = 90.0 - 55.0 = 35.0\ ^0C$.

46.5 / 35.0 = 1.33

11b For air: $\Delta\theta = 43.5 - 28.0 = 15.5\ ^0C$

For polystyrene: $\Delta\theta = 55.0 - 43.5 = 11.5\ ^0C$

15.5 / 11.5 = 1.35

Lesson 7.8 Key Concept: Energy transfer

1 All of these are storing energy. The ball is storing kinetic energy, the string is storing elastic potential energy, the hot object is storing thermal energy and the mixture of oxygen and fuel is storing chemical energy.

2a The ball could transfer kinetic energy by colliding with another object. This has limited use but is used in e.g. marble runs, an aid to ten pen bowling, Rube Goldberg machines.

2b The spring could transfer energy from its elastic potential energy store to its kinetic energy store. This is a useful energy transfer in some mechanical clocks and watches or in other clockwork devices.

2c The hot object could transfer energy to the thermal energy stored in the surroundings – for example in a hot water bottle or the heating element in a toaster or kettle.

2d The mixture of oxygen and fuel can transfer energy from its chemical energy store to a thermal energy store. This is used in engines such as rockets.

3

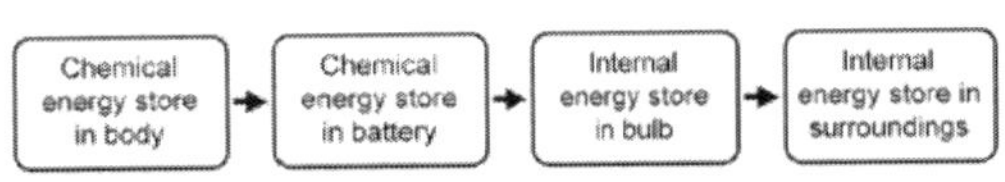

4a Chemical energy store in body decreases; kinetic energy store of bicycle increases (if bicycle is accelerating); gravitational potential energy store of the bicycle increases (if bicycle is going uphill); internal energy of the surroundings increases (since the cyclist needs to do work against air resistance and friction).

4b Chemical energy store in match decreases; thermal energy store of surroundings increases.

5 The amount of energy is represented by the number of bricks. No matter what the child does to the bricks and if one becomes lost – the total number of bricks remains the same.

6 The system must include the surroundings as all of the chemical energy in the coal ends up in the internal energy of the surroundings once the train has stopped.

Lesson 7.9 Maths Skills: Calculations using significant figures

1 $E_p = mgh$

$E_p = 70.0\text{ kg} \times 9.8\text{ N/kg} \times 2.0\text{ m}$

$= 1372\text{ J} = 1400\text{ J}$ (to 2 significant figures)

2 $E_p = mgh$

$E_p = 0.0500\text{ kg} \times 9.8\text{ N/kg} \times 10.0\text{ m}$

$= 4.9\text{ J} = 4.90\text{ J}$ (to 3 significant figures)

3 $E_e = \frac{1}{2}ke^2$

$= 0.5 \times 350\text{ N/m} \times 0.09^2\text{ m}^2$

$= 0.14175\text{ J} = 0.14\text{ J}$ (to 2 significant figures)

End of Chapter Questions

1 d) The second one will transfer energy more quickly. [1 mark]

2 efficiency = useful output energy transfer (× 100) / total input energy transfer [1 mark] (Allow answer without the (× 100) and/or using the word "power" instead of "energy")

3 An energy resource that will eventually run out. [1 mark]

4 c) dissipating [1 mark]

5 elastic potential energy is the energy stored when something is stretched or compressed [1 mark] whereas gravitational potential energy is the energy stored when something is moved vertically upwards (and work is done against the pull of gravity) [1 mark]

6 c) They have the same amount of stored thermal energy. [1 mark]

7 150 W, 0.1 kW, 0.08 kW, 60 W, 15 W. [1mark]

8 a) Its colour. [1 mark]

9 Air is a poor conductor of heat / good insulator [1 mark]. The bubble wrap has lots of trapped air in it [1 mark].

10 Energy is stored in the thermal energy store of the fish and chips [1 mark]. Since the fish and chips are hotter than the surroundings then some of this energy will transfer to the thermal energy store of the surroundings [1 mark].

11 The temperature of the roof is higher than most of the rest of the house [1 mark]. This is probably because the energy is transferring upwards (since heat rises) [1 mark].

12 It transfers 2000 J of energy every second. [1 mark]

 © HarperCollins*Publishers* Limited 2016

13 ΔE stands for the change in energy / the energy transferred to the object's thermal energy store; m stands for the mass of the object; $\Delta\theta$ stands for the change in temperature of the object. [1 mark]

14 The night storage heaters need to be able to store a large amount of thermal energy so they can keep on heating the house throughout the night [1 mark]. Objects with a high specific heat capacity store a large amount of thermal energy at a particular temperature [1 mark].

15 Level 3: A clear and logical description is made in a coherent sequence of steps. The procedure would lead to a successful conclusion and methods for ensuring a fair test are discussed. There is some valid comment about safety precautions that should be carried out. (5-6 marks)

Level 2: An experiment is outlined that would lead to a successful conclusion although detail is missing such as what measuring instruments should be used. The need for a fair test and safety precautions are covered although they are not coherently linked to the procedure. (3-4 marks)

Level 1: Some of the ideas are covered but there is not a full description and there is not a logical sequence of ideas. (1-2 marks)

Indicative content

- A method for heating one end of the rod (e.g. by using a Bunsen burner)
- A method for measuring the temperature of the other end of the rod (e.g. by using a thermometer or using a pin stuck on to the end of the rod with wax)
- A method for timing to reach a certain temperature (or for the pin to fall off due to the wax melting)
- The material with the shortest time would have the highest thermal conductivity
- A method to ensure a fair test (e.g. rods same thickness and length / temperature of Bunsen flame is kept the same)

Precautions to ensure that the experiment is safe (e.g. tie hair back, wear safety spectacles, avoid touching the rods if they are hot or use heat proof gloves).

16 Temperature is a measure of how hot something is [1 mark]. Thermal energy is the capacity of the object to do work by transferring this energy into other forms [1 mark].

So temperature is a measure of the *ability* of an object to transfer thermal energy and do work whereas thermal energy is the *amount* of work the object can do by transferring its energy.

17 Energy cannot be created or destroyed [1 mark]. So energy lost in one energy store is transferred to another one [1 mark].

18 Level 3: The data is linked to a numerical calculation of efficiency to show that a higher bounce height results in a higher efficiency. Several limitations of the experiment are evaluated including experimental techniques and the quality of the data and it is clearly justified whether the student's conclusion is valid. (5-6 marks)

Level 2: Several limitations of the experiment are evaluated and it is justified whether the student's conclusion is valid – however no attempt to link bounce height to a calculation of efficiency is made. (3-4 marks)

Level 1: Some limitations of the experiment are described and a comment about whether the conclusion is valid is made. (1-2 marks)

Indicative content

Making the conclusion

- A calculation of the efficiency at two different temperatures is made: e.g. at 60^0C the average bounce height is 43.5cm. This means the GPE after the bounce = mgh = $0.025 \times 10 \times 0.435 = 0.10875$ J; GPE before the bounce = $0.025 \times 10 \times 1 = 0.25$ J; so the efficiency is $0.10875 / 0.25 = 0.435$ or 43.5%. A similar calculation of efficiency at a lower temperature shows a smaller efficiency.
- An alternative numerical treatment would be to realise that GPE = mgh so GPE before the bounce / GPE after the bounce = height before the bounce / height after the bounce. This means that the higher the bounce, the greater the efficiency
- The student's conclusion correctly describes the pattern of the data – the average bounce height increases as the temperature increases so the efficiency does increase with increasing temperature.

Quality of the data

- The repeat readings need to be averaged before any calculations and conclusions are made
- The repeat readings are quite different which suggests there is lots of random error – particularly at the low temperatures.
- One of the readings at 20^0C might be anomalous and would need to be re-checked.

Limitations of the procedure

- The ball's temperature might have decreased between it being in the water bath and being dropped …
- … this would have a greater effect at higher temperatures
- The ball might not have fallen and bounced vertically

 © HarperCollins*Publishers* Limited 2016

- The ball might not be at the same temperature as the water (it doesn't conduct heat very well)
- It is difficult to judge when the ball has reached the maximum height
- There will be error due to not reading the ruler at the right angle (parallax error) as it would be impossible to move your eyes up and down so they are always level with the ball
- The metre ruler might not have been held vertically.

 © HarperCollins*Publishers* Limited 2016

Chapter 8: Global Challenges

Lesson 8.1 Keeping safe on the road

1a 18 m/s

1b 24 m

1c 18 + 24 = 42 m

2 At 22 m/s the thinking distance = 15 m (to the nearest m) and the braking distance is 36 m. Therefore the stopping distance = 15 + 36 = 51 m

3 The braking distance increases as the speed increases at an increasing rate.

4 The braking distance increases at a given speed because there is less friction. Therefore, to keep the same, safe braking distance the car needs to go slower.

5 When the car decelerates rapidly, the brakes heat up. This can damage the brakes which then become worn. Worn out brakes cannot decelerate the car with as much force so the braking distance increases.

6 This increases the time that it takes for the child to change its speed. Therefore, the deceleration is less. Since $F = ma$ then the force on the child becomes less and there is a smaller chance of harm.

7 Estimate the following values:

Mass of car = 1250 kg

Speed of car = 25 m/s

Braking distance = 50m

Kinetic energy of the car before stopping = $\frac{1}{2} mv^2$ = 0.5 × 1250 × 25^2 = 390 625 J

Therefore work done stopping the car = 390 625 J

Force × distance = 390 625

So F = 390 625 / 50 = 7810 N (to three significant figures).

Lesson 8.2 Transmitting electricity

1 If the local power station breaks down or it has to be switched off for maintenance then your house can still receive electrical power from other power stations.

2 They are connected between the power cables and factories or homes.

3a Power input = $V_p \times I_p$ = 3000 × 20 = 60 000 W

Therefore power output = 60 000 W (assuming transformer is 100% efficient)

$V_s \times I_s$ = 60 000

So I_s = 60 000 / 60 000 = 1A

3b $N_s / N_p = V_s / V_p$ = 60 000 / 3000 = 20

So $N_s = 20 \times N_p$ = 20 × 500 = 10 000 turns

Lesson 8.3 Using energy resources

1a e.g. a motor, an electric car, a fan

1b e.g. a lift, an escalator, a drone

2 e.g. coal or wood in a fire, gas in a boiler in a central heating system, solar power in a solar cooker

3 Energy from your food, which originally comes from the Sun.

4 It is easy to install wires above the train tracks that can connect up the train to an electricity supply. It would be very difficult and expensive to install wires that aircraft could use while flying. The aircraft have to carry their energy with them.

5 Any two from: wind, wave, hydroelectric, solar, geothermal, tidal, nuclear.

6 Only some countries are surrounded by water where the waves a large and reliable enough to provide useful amounts of energy.

7 When satellites orbit the Earth they spend part of their orbit in sunlight and the other part in the dark. The solar panels will only provide energy when they are in the sunlight. Therefore, they need to charge up a battery which is able to provide energy to the satellite when it is in the dark.

8 We can plant new trees. As long as the trees produce new wood at the same rate that we use it, then the energy resource is always replenished.

9 Nuclear fuel.

10 Bio-fuels are being replenished at the same rate that they are being used. Fossil fuels were formed over millions of years and we are burning them at a considerably faster rate than they are being created. Therefore, they are running out.

11a **Energy** is the capacity to do work. It is transferred from one energy store to another when an object is doing work. An **energy resource** stores energy in a form where it can easily be transferred to another energy store and thus do useful work.

11b Energy is always conserved because it always gets transferred from one energy store to another energy store – it never just disappears. An energy resource is not conserved since the energy it stores is transferred to other energy stores.

When energy is transferred it usually becomes less useful because some of it is dissipated into the thermal energy stored in the surroundings.

 © HarperCollins*Publishers* Limited 2016

Lesson 8.4 Global energy supplies

1 They are non-renewable resources and so will run out.

2 Coal. The worldwide use of coal dramatically increased between 2000 and 2010 and this is when China's energy use grew rapidly.

3 Although fossil fuels will eventually run out and that there are environmental problems, the world's energy needs to come from somewhere. Renewable energy is unable to meet the demand at present.

4 Environmental considerations: This will release carbon dioxide and sulfur dioxide into the atmosphere which will contribute to global warming and acid rain.

Political considerations: People might not vote to elect the council again if they are angry with the plan.

Ethical considerations: It is not morally right to destroy a beautiful part of the country or harm the health of people in the future.

Social considerations: It might give more people jobs but it might adversely affect the health of lots of people.

Economic considerations: It would bring more income to the area but it might cost a lot of money to set it up.

5 The motor does not burn fuel so the only heat it creates is due to friction. Therefore, less energy is wasted.

6 Efficiency = 80% = 0.8. Input power = output power / efficiency = 3/0.8 = 3.75 kW.

7 The power station needs to provide 3.75kW. Therefore, it needs an input power of 3.75 / 0.4 = 9.375 kW. So the overall efficiency = output power / input power = 3 / 9.375 = 0.32 or 32%. This is less efficient than a diesel engine.

8 Make the car more streamlined; make the car lighter; drive the car with less accelerating and breaking; make sure the moving parts are well lubricated; make sure the tyres are in good condition.

Lesson 8.5 Electricity in the home

1 In a direct potential difference, the current is always pushed in the same direction. In an alternating potential difference, the power supply pushes the current so that it keeps on changing direction.

2 The peaks and the troughs would be about half as big (since the potential difference is about half as much) and there would be less time between them (since the frequency is higher).

3 Plastic is a good insulator of electricity. So this prevents people from receiving an electric shock.

4 It would be very dangerous to wire up a plug the wrong way round. Therefore, you need to be able to identify which wire is which very easily – even if you are colour blind.

5 A battery powered torch only uses a small p.d. and it probably doesn't have a metal case. Therefore there is a much smaller risk of harm if the user receives an electric shock – so it doesn't need an earth wire. A mains lamp, however often has a metal casing and has a high risk of harm if the user receives an electric shock – so it often has an earth wire.

6 The earth wire prevents electric shocks if the metal casing becomes live due to a fault. The earth wire provides a path of low resistance so the current will be very high. This makes sure that the fuse melts and switches off the circuit.

Lesson 8.6 Red-shift

1 Green light has a shorter wavelength than red light.

2 Green light has a longer wavelength than blue light.

3 The light emitted from an object has its wavelengths stretched.

4 It tells us the galaxies are moving away from us.

5 Hubble discovered that the speed that a galaxy is moving away from us is directly proportional to its distance away.

6 This suggests that our Universe is expanding.

7 Observations allow us to test that the Universe behaves in a way that we would expect according to current theories. If the observations do not agree with the theory, then we need to develop the theory or produce an alternative theory that fits in with all of the observations. Therefore, theories develop over time as more observations are made.

8 The Big Bang theory describes the Universe as starting at a very small, very dense and very hot point. It then started to expand at the Big Bang and it has been expanding ever since.

9 It is impossible to tell because all the galaxies would be moving away from us *wherever* we are in the Universe.

Lesson 8.7 The sun and other stars

1 planet, star, Solar System, galaxy

2 Hydrogen, some helium and a small amount of the heavier elements.

 © HarperCollins*Publishers* Limited 2016

3 The gravitational force.

4 Protostars are formed from clouds of dust and gas which are very spread out. The gravity that makes the clouds to contract is very weak.

5 The gravitational force pulls the cloud of gas and dust inwards. This heats up the centre of the cloud and a protostar is formed. As the temperature increases it eventually becomes so hot that nuclear fusion reactions occur at the centre. The pressure produced from the heat emitted from the fusion reactions balances out the inward force of gravity and the star becomes stable.

6 Hydrogen

7 The nuclei repel each other due to their positive charges. Therefore, the need a lot of energy to overcome the repulsion so they become close enough to fuse.

8 Small nuclei (such as hydrogen) join together to form larger nuclei (such as helium).

9 There is a small amount of helium, but this was the helium that was there originally rather than helium formed from fusion reactions.

10 When something is burned, it is reacting with oxygen chemically. This is not happening in the Sun.

Lesson 8.8 Emission and absorption of infrared radiation

1 The surface of the drink. It is the hottest part of the picture.

2 They also have a temperature (above absolute zero).

3 The shiny white side.

4 Shiny copper pipes are poor absorbers of infrared radiation. Painting them black makes them absorb the infrared more effectively so the water is heated more quickly.

5 It would be glowing white hot at a higher intensity than when it was at 7200 ^{0}C. It would also be producing electromagnetic radiation of a much shorter wavelength but this would be invisible to the eyes.

6 The white object would be the same colour as the black object in question 5 only much dimmer because it wouldn't be as effective at emitting radiation.

Lesson 8.9 The solar system

1 Mercury

2 The Sun would look very small and very dim (and just like a typical star looks from Earth) since Pluto is very far away from the Sun.

3 Because if they were drawn to the same scale as the orbits then they would be too small to see.

4 Jupiter

5 Pluto is a dwarf planet.

6 It could have a higher density / higher mass

7 They orbit the Earth or enter the atmosphere and burn up.

8 Pluto is very small; it is a very long distance away from the Earth; it is very dim (because Pluto is very far from the Sun).

Lesson 8.10 Orbits of planets, moons and artificial satellites

1 The force of gravity.

2 The further the distance, the longer it takes to orbit the Sun.

3 The distance the planet has to travel for a complete orbit around the Sun is larger the further away it is from the Sun – so it takes longer to make a complete orbit. Another reason is that the planets further out travel slower as the Sun's gravity is weaker.

4 The force of gravity.

5 a and b

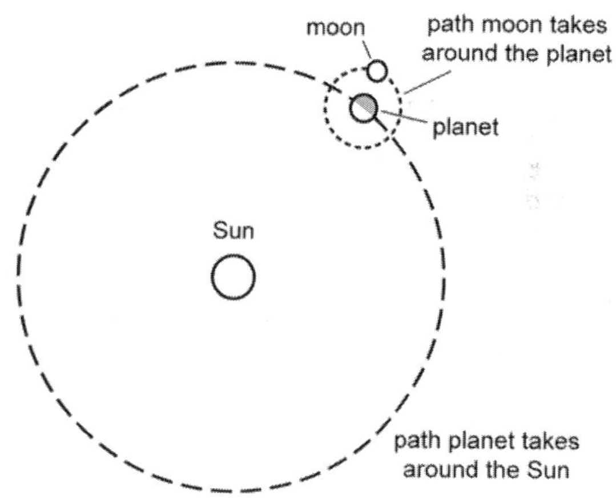

6 Velocity has a magnitude and a direction. Since the direction is changing the velocity is changing.

7 The radius of its orbit and the speed it is going.

8 The faster the satellite is travelling, the closer it is to the surface of the Earth.

Lesson 8.11 Key Concept: Gravity: the force that binds the Universe

1 All of them. Anything that has a mass can exert a force of gravity.

2 Gravity forms galaxies, stars and planets.

 © HarperCollins*Publishers* Limited 2016

3 Our weight is the force of gravity on us, so the stronger the force of gravity the more our weight.

4a Weight = mass × g = 1000 × 9.81 = 9810 N

4b $a = F / m$ = 9810 / 1000 = 9.81 m/s^2

4c Yes – all objects will fall to the ground with an acceleration = g (assuming that there is no air resistance).

5 They found that the mass of the universe was much heavier than the combined mass of all of the objects that they knew about. This suggests that there must be a lot of other matter out there (called dark matter).

Lesson 8.12 Temperature of the Earth

1 The temperature is increasing.

2 White objects are poor absorbers and emitters of radiation. Therefore, the white objects warm up or cool down more slowly than black objects do.

3 It is at a temperature above absolute zero.

4 The mean temperature of the Earth is reached when the amount of infrared radiation it absorbs balances the infrared radiation it emits. Therefore, the mean temperature of the Earth depends on the activity of the Sun and the state of the atmosphere.

5a absorption: how reflective the Earth is – e.g. icy surfaces don't absorb infrared radiation very effectively

reflection: again, icy surfaces reflect infrared radiation more than other surfaces; cloud cover also reflects infrared radiation from the Sun and also from the ground

emission: this depends on the temperature of the Earth and the nature of the gases in the atmosphere which might absorb the radiation on its way back into space.

5b absorption: Humans add gases into the atmosphere which absorb the infrared radiation

reflection: Humans change the surface of the Earth (by building things)

emission: emission is changed due to human activities which affect absorption and reflection

6 Human activities have led to gases being put into the atmosphere which absorb the infrared radiation emitted from the surface of the Earth and prevent them from reaching space.

7 This could decrease the temperature of the Earth because the particles prevent the infrared radiation of the Sun from reaching the surface of the Earth. (This is a larger effect than the greenhouse effect that they might produce.)

8 The temperature of the Earth depends on many factors such as the output of the Sun, the varying weather patterns, oceanic currents etc. Even if there is a correlation between carbon dioxide levels going up and temperature going up, this does not necessarily mean that one of them causes the other. Also the temperatures change all the time over every part of the world and it is difficult to measure the average temperature change to a high degree of accuracy. So the increase in temperature could be a random short term effect.

Lesson 8.13 Measuring wave speeds

1 Time between claps is 23 / 50 = 0.46 s.

Speed = distance / time = 100 m / 0.46 s = 217 m/s.

2 e.g. the wind might be affecting the speed the sound goes; problems with clapping at the same time that you hear an echo; measuring errors in the distance and the time.

3a This is not an accurate method. It would be hard to judge exactly when the wave reached the other end of the swimming pool. There are also likely to be many other waves in the pool which would make the measurement confusing. Also there is only one measurement.

3b You could repeat the experiment several times and take an average. You could also video the wave from above and use software to play the video back in slow motion and to calculate the time it took the wave to travel the 25 m.

4 speed = distance / time = 20 / 10 = 2 m/s.

5a Distance = speed × time = 1500 × 0.1 = 150 m.

The sound has travelled to the fish and back again so the distance between the boat and the fish = 150 / 2 = 75 m.

5b The reflected pulse has come from lots of fish at different depths. So the time it takes for the wave to return becomes spread out.

6a Distance = 220 × 2 = 440 m

6b Time = distance / speed = 440 / 330 = 1.3 s.

7 If the pulse is spread out then the echoes will also be very spread out. This will make it difficult to measure an exact time between the pulse and the echo – so the distance calculation will have more uncertainty.

Lesson 8.14 Seismic waves

1 Gravity

2 The base of the seismometer needs to move around with the rock while the weight remains still.

 © HarperCollins*Publishers* Limited 2016

3a Longitudinal

3b Transverse

4 A P wave travels faster than an S wave.

5 The density of the Earth gets larger the further in you go towards to core and so the speed of the waves increases. As this is a gradual change of density then the speed also gradually changes. The waves gradually refract and so they curve rather than change direction suddenly.

Lesson 8.15 Maths skills: Using scale and standard form

1 A scale of 0.06 mm of A4 paper to 1 million km works:

Place the paper in landscape.

Draw a dot for the Sun on the left hand edge.

Distance from the Sun on the A4 sheet:

2 Distance from Sun to Neptune = 4.5×10^9 km

Distance from Sun to Alpha Centauri = 4.1×10^{12}

Planet	Actual Distance (million km)	Distance on A4 paper (mm)
Mercury	57	3.4
Venus	108	6.5
Earth	150	9.0
Mars	228	14
Jupiter	778	47
Saturn	1429	86
Uranus	2870	172
Neptune	4500	270

km

Diameter of Milky Way galaxy = 1×10^{18} km

3a Distance = speed × time = 3.0×10^8 m/s × (365 × 24 × 60 × 60) s = 9.5×10^{15} m

3b Diameter of Milky Way = 1×10^{18} km = 1×10^{21} m.

Distance in light years = $1 \times 10^{21} / 9.5 \times 10^{15} = 1.1 \times 10^5$ light years (to 2 significant figures).

4 170×10^{-9} m = 1.7×10^{-7} m

5a $1 / 1 \times 10^{-45} = 1 \times 10^{45}$ times bigger.

5b $1 \times 10^{39} / 1 = 1 \times 10^{39}$ times bigger.

5c The proton expands to be 1×10^{45} times bigger which means that the person would expand to be 1×10^{45} m^3. This is $1 \times 10^{45} / 1 \times 10^{39}$ = 1 million times the size of the solar system!

End of Chapter Questions

1 A network of power <u>cables</u> [1 mark] and <u>transformers</u> [1 mark] that connect power stations to consumers.

2 Green and yellow (stripes). [1 mark]

3 The neutral wire and the earth wire are the wrong way round [1 mark]. The cable grip needs to be covering the outer insulation [1 mark].

4 The Sun [1 mark]

5 The force of gravity. [1 mark]

6 c) A galaxy [1 mark]

7 It will become a red giant [1 mark] and then a white dwarf [1 mark]

8 Its speed needs to be faster. [1 mark]

9 Two factors are identified from speed/condition of driver (e.g. tiredness)/condition of road (e.g. icy)/condition of brakes/condition of tyres [1 mark]

Both factors are described:

e.g. the faster the car travels, the further it moves while the driver is reacting; if the driver is tired then they take longer to react and the car moves further; if the road is icy/the brakes are worn/the tyres are worn then the braking force on the car is reduced because there is less friction so the car travels further. [1 mark]

10 Mercury, Venus, Earth, Mars, Jupiter, Saturn, Uranus, Neptune [1 mark]

11 b) Main sequence [1 mark]

12a Jupiter [1 mark] it has a mass that is much bigger than any other planet [1 mark]

12b Mass of the object = W / g = 111 / 3.7 = 30 kg [1 mark]

13 The charge flows from them to the ground. This creates an electric current. [1 mark]

14 If the current becomes <u>too large</u> [1 mark] then the current becomes high enough to <u>melt</u> the wire in the fuse which breaks the circuit and stops the current flowing [1 mark].

15 The wavelength of light becomes longer when the galaxy/star/light source is moving away from us. [1 mark]

16 Any one of: spying, weather forecasting, communications (e.g. TV), gathering scientific data, telescopes, navigation (e.g. GPS/satnav) [1 mark]

17 The star expands [1 mark] and cools down [1 mark].

 © HarperCollins*Publishers* Limited 2016

18 Level 3: A detailed description is made. All the future stages of the stars are identified and linked to the relevant stars with justification. A comparison of the relative timescales is also made, with justification. (5-6 marks)

Level 2: All future stages of the stars are identified and linked to relevant stars with some description of what the stages are. (3-4 marks)

Level 1: Some future stages of the stars are listed and an attempt is made to link them to the correct stars. (1-2 marks)

Indicative content

The future stages

- Proxima Centauri and Tau Ceti will become red giants and then white dwarves
- Zeta Ophioci will become a red supergiant, explode as a supernova ejecting its outer layers, and its core will become a neutron star or a black hole.

Further detail about the stages

- The star enters the red giant/supergiant stage when it runs out of hydrogen at its core
- During the red giant/supergiant stage it starts fusing helium into heavier elements
- The supergiant stage goes through several stages, forming increasingly heavier elements at its core
- The supergiant explodes in a supernova when it starts trying to form elements heavier than iron
- During the white dwarf stage, no fusion happens but the star gradually cools down
- A neutron star is a very small and dense sphere of neutrons
- A black hole is where the gravity is so strong that not even light can escape.

Justification of the stages

- Proxima Centauri and Tau Ceti are less massive than the Sun and are likely to evolve in the same way that the Sun does
- Zeta Ophioci is much more massive than the Sun and so is likely to evolve in the same way as massive stars do.

Relative timescales

- The more massive the star, the quicker it passes through the stages
- Therefore, although Proxima Centauri is likely to pass through the same stages as Tau Ceti, Tau Ceti is likely to evolve to the red giant stage much sooner
- Zeta Ophioci is likely to evolve much quicker still because it is much more massive.

19 Any **two** from: Higher potential difference supply needs a smaller current (for the same power); a current heats the wire and wastes energy; a smaller current means the system is more efficient. [2 marks]

Any **one** from: Transformers are needed to change the pd which makes the system more complicated and prone to faults; a high pd is very dangerous. [1 mark]

20 Work out the power loss in the lines when the potential difference is not transformed:

$P = I^2R = 12^2 \times 50 = 7200$ W [1 mark]

Work out the current in the transmission lines if the p.d. is transformed to 2.25 kV:

Power in = power out

$P = IV$ [1 mark]

So $12 \times 750 = 2250 \times I$ (where I is the current in the transmission lines) [1 mark]

$I = 4$ A [1 mark]

Work out the power loss in the lines when the p.d. is transformed:

$P = I^2R = 4^2 \times 50 = 800$ W [1 mark]

So the power that would be saved = 7200 – 800 = 6400 W [1 mark]

21 Nuclei are fusing together to form heavier nuclei [1 mark]. The pressure from the fusion reactions that tries to expand the star balances out the force of gravity trying to compress the star and so the star remains stable / in equilibrium [1 mark].

22 Level 3: A detailed account of the Big Bang theory is made and the evidence from the data shown is linked to the formulation of the theory. The data is evaluated and its strengths and weaknesses in support of the Big Bang theory are discussed successfully. (5-6 marks)

Level 2: An account of the Big Bang theory is made but it isn't successfully linked to the data shown in the graph. There is some comment about the quality of the data shown in the graph. (3-4 marks)

Level 1: The Big Bang theory is outlined and there is a comment about the quality of the data. However there is no attempt to link the data presented in the graph to the theory. (1-2 marks)

Indicative content

The Big Bang Theory

- The Universe started as a very small size
- It exploded during the Big Bang and has been expanding ever since
- The evidence Hubble used for this is from the red shift of the galaxies

 © HarperCollins*Publishers* Limited 2016

- The red shift shows that the galaxies are moving away from us
- The graph shows that the further away the galaxy the greater the red shift and the faster it is moving
- This suggests that the Universe is expanding
- Which suggests that it was smaller in the past.

The data

- There is only a small number of galaxies plotted out of billions so the data might not represent the overall effect (there is a very small sample size)
- There is a lot of scatter on the graph and some galaxies that are further away are moving more slowly than nearer galaxies
- There is a general trend that suggests that galaxies further away move faster (as the line of best fit has a positive gradient)
- It is very difficult to measure the data and so there is a lot of experimental error in the results
- Lots more data needs to be gathered from other galaxies and readings need to be repeated by other people to see if Hubble is right.

Other relevant detail

- There is other evidence besides the redshift of galaxies that supports the Big Bang theory (such as cosmic microwave background radiation)
- Modern measurements suggest that the Big Bang model needs modifying to account for dark matter and dark energy.

23 Level 3: A detailed account of the absorption, emission and reflection of radiation that is taking place in the diagram. Observations are clearly linked to the diagram and the role of both radiation absorbed from the Sun and radiation emitted from the Earth are explored. The net transfer of energy from the Sun to the Earth due to the role of the greenhouse gases in the atmosphere is explained clearly. (5-6 marks)

Level 2: The absorption, emission and reflection of radiation are discussed but this is not completely clear or detailed and only loosely linked to the diagram. An attempt is made to explain the role of greenhouse gases in global warming and there is some relevant physics present. (3-4 marks)

Level 1: Some ideas about the absorption, emission and reflection of radiation are outlined. (1-2 marks)

Indicative content

Incoming radiation from the Sun

- The Earth is receiving 342 W/m^2 of radiation from the Sun.
- About a third of this (102 W/m^2) is reflected straight back into space.
- About two thirds is (240 W/m^2) is absorbed by the Earth and acts to warm the Earth up.

Outgoing radiation from the Earth

- The Earth re-radiates the energy it has received from the Sun.
- Because the Earth is cooler than the Sun, the wavelength of this radiation is longer than the radiation from the Sun.
- Some radiation transfers straight into space.
- Some radiation is absorbed by the clouds/atmosphere, which re-emits the radiation in all directions.

Global Warming

- Because the atmosphere absorbs some of the Earth's radiation and emits it back down to the surface then not all of it radiates back into space.
- If the intensity of radiation returning to space is less than 240 W/m^2 then the Earth will absorb more energy than it transmits and will warm up.

Greenhouse gas

- Greenhouse gases increase the atmosphere's ability to absorb radiation from the Earth's surface – reducing the amount returning to space.
- Greenhouse gases do not absorb the radiation from the Sun as effectively as they absorb radiation from the Earth (due to the different wavelength) so the radiation the Earth receives from the Sun is still around 342 W/m^2.
- Therefore, the larger the concentration of the greenhouse gas, the smaller the proportion of the energy from the Sun which is radiated back into space.

So a larger concentration of a greenhouse gas will warm the Earth up more.

 © HarperCollins*Publishers* Limited 2016

Lesson number	Lesson title	Lesson objectives	OCR specification reference	Lesson resources (on CD ROM)	Collins Connect resources
Topic 1: Matter					
1.1	Key concept: Developing ideas for the structure of the atom	• Understand how ideas about the structure of the atom have changed. • How evidence is used to test and improve models.	P1.1a	Worksheets 1.1.1, 1.1.2, 1.1.3 and 1.1.4	
1.2	Atomic structure	• Describe the structure of the atom. • Use symbols to represent particles. • Describe ionisation.	P1.1b, c	Worksheets 1.2.1, 1.2.2 and 1.2.3	
1.3	Density	• Use the particle model to explain the different states of matter. • Describe differences in density for different states of matter. • Calculate density.	P1.1d, f	Worksheet 1.3; Practical sheet 1.3; Technician's notes 1.3	
1.4	Key concept: Particle model and changes of state	• Use the particle model to explain states of matter. • Use ideas about energy and bonds to explain changes of state. • Explain the relationship between temperature and energy.	P1.1e	Worksheet 1.4; Practical sheet 1.4; Technician's notes 1.4	
1.5	Practical: To investigate the densities of regular and irregular solid objects and liquids	• Interpret observations and data. • Use spatial models to solve problems. • Plan experiments and devise procedures. • Use an appropriate number of significant figures in measurements and calculations.	Prac P1 (materials)	Worksheet 1.5; Practical sheet 1.5; Technician's notes 1.5	
1.6	Changes of state	• Describe how, when substances change state, mass is conserved. • Describe energy transfer in changes of state. • Explain changes of state in terms of particles.	P1.2a, b	Worksheet 1.6; Practical sheet 1.6; Technician's notes 1.6	
1.7	Internal energy	• Describe the particle model of matter. • Understand what is meant by the internal energy of a system. • Describe the effect of heating on the energy stored within a system.	P1.2c	Worksheet 1.7; Practical sheets 1.7.1, 1.7.2, 1.7.3, 1.7.4, 1.7.5 and 1.7.6; Technician's notes 1.7	

Lesson number	Lesson title	Lesson objectives	OCR specification reference	Lesson resources (on CD ROM)	Collins Connect resources
1.8	Specific heat capacity	• Describe the effect of increasing the temperature of a system in terms of particles. • State the factors that are affected by an increase in temperature of a substance. • Explain specific heat capacity.	P1.2d (part), e	Worksheet 1.8; Practical sheet 1.8; Technician's notes 1.8	
1.9	Specific latent heat	• Explain what is meant by latent heat. • Describe that when a change of state occurs it changes the energy stored but not the temperature. • Perform calculations involving specific latent heat.	P1.2d (part), f	Worksheet 1.9; Practical sheet 1.9; Technician's notes 1.9	
1.10	Maths skills: Drawing and interpreting graphs	• Plot a graph of temperature against time, choosing a suitable scale. • Draw a line of best fit (which may be a curve). • Interpret a graph of temperature against time. • Comment on the specific heat capacity and specific latent heat of a substance.	PM1.2 I, ii	Worksheet 1.10, Practical sheet 1.10, Technician's notes 1.10	
1.11	Practical: Investigating specific heat capacity	• Use theories to develop a hypothesis. • Evaluate a method and suggest improvements. • Perform calculations to support conclusions.	Prac P5	Worksheet 1.11; Practical sheet 1.11; Technician's notes 1.11	
1.12	Particle motion in gases	• Relate the temperature of a gas to the average kinetic energy of the particles. • Explain how a gas has a pressure. • Explain that changing the temperature of a gas held at constant volume changes its pressure.	P1.3 a, b, c	Worksheet 1.12; Practical sheet 1.12; Technician's notes 1.12	
1.13	Increasing the pressure of a gas	• Describe the relationship between the pressure and volume of a gas at constant temperature. • Calculate the change in the pressure or volume of a gas held at constant	P1.3d, e	Worksheet 1.13; Practical sheet 1.13; Technician's notes 1.13	

Lesson number	Lesson title	Lesson objectives	OCR specification reference	Lesson resources (on CD ROM)	Collins Connect resources
		temperature when either the pressure or volume is increased or decreased. • Explain how doing work on a gas can increase its temperature.			
1.14	Pressure in a fluid	• Explain how pressure acts in a fluid. • Calculate pressure at different depths in a liquid. • Explain what causes upthrust.	P1.3h, I, j	Worksheets 1.14.1, 1.14.2 and 1.14.3; Practical sheet 1.14; Technician's notes 1.14	
1.15	Atmospheric pressure	• Show that the atmosphere exerts a pressure. • Explain variations in atmospheric pressure with height. • Describe a simple model of the Earth's atmosphere and atmospheric pressure.	P1.3f, g	Worksheets 1.15.1, 1.15.2 and 1.15.3	
1.6	Maths skills: Handling data	• Recognise the difference between mean, mode and median. • Explain the use of tables and frequency tables. • Explain when to use scatter diagrams, bar charts and histograms.	M2 c, f	Worksheets 1.16.1, 1.16.2 and 1.16.3	
Assessment		End of Chapter test Student Book End of Chapter test Collins Connect			
		Chapter 2: Forces			
2.1	Scalars and vectors	• Define distance, displacement, speed, velocity and acceleration. • Recognise the difference between scalar and vector quantities. • State examples of scalar and vector quantities.	P2.1.2c, d	Worksheets 2.1.1, 2.1.2 and 2.1.3	
2.2	Speed	• Calculate speed using distance travelled divided by time taken. • Calculate speed from a distance–time graph. • Measure the gradient of a distance–time graph at any point.	P2.1a, b, c, e (part), g	Worksheets 2.2.1, 2.2.2 and 2.2.3; Practical sheet 2.2; Technician's notes 2.2	

Lesson number	Lesson title	Lesson objectives	OCR specification reference	Lesson resources (on CD ROM)	Collins Connect resources
2.3	Acceleration	• Describe acceleration. • Calculate acceleration. • Explain motion in a circle.	P2.1h (part)	Worksheets 2.3.1, 2.3.2 and 2.3.3	
2.4	Calculations of motion	• Describe uniform motion. • Use an equation for uniform motion. • Apply this equation to vertical motion.	P2.1h (part)	Worksheet 2.4.1, 2.4.2, 2.4.3, 2.4.4, 2.4.5 and 2.4.6	
2.5	Velocity–time graphs	• Draw velocity–time graphs. • Calculate acceleration using a velocity–time graph. • Calculate displacement using a velocity–time graph.	P2.1f	Worksheets 2.5.1, 2.5.2 and 2.5.3; Practical sheet 2.5; Technician's notes 2.5	
2.6	Maths skills: Making estimates of calculations	• Estimate the results of simple calucations • Round numbers to make an estimate • Calculate order of magnitude	PM2.3i to vi	Worksheets 2.6.1, 2.6.2 and 2.6.3	
2.7	Forces explain how objects interact	• Describe a force • Recognise the difference between contact and non-contact forces	P2.2d, e	Worksheets 2.7.1, 2.7.2 and 2.7.3	
2.8	Forces and motion	• Understand what a force does. • Explain what happens to an object if all the forces acting on it cancel each other out. • Analyse how this applies to everyday situations.	P2.2d, e	Worksheets 2.8.1, 2.8.2 and 2.8.3; Practical sheet 2.8; Technician's notes 2.8	
2.9	Resultant forces	• Calculate the resultant of a number of forces. • Draw free-body diagrams to find resultant forces. • Understand that a force can be resolved into two components acting at right angles to each other.	P2.2e (part), g, h (part)	Worksheets 2.9.1, 2.9.2 and 2.9.3	
2.10	Forces and acceleration	• Explain what happens to the motion of an object when the resultant force is not zero. • Analyse situations in which a non-zero resultant force is acting. • Explain what inertia is.	P2.2i, j	Practical sheets 2.10.1, 2.10.2 and 2.10.3; Technician's notes 2.10	

Lesson number	Lesson title	Lesson objectives	OCR specification reference	Lesson resources (on CD ROM)	Collins Connect resources
2.11	Momentum	• Explain what is meant by momentum. • Apply ideas about rate of change of momentum to safety features in cars. • Use momentum calculations to predict what happens in a collision.	P2.2k	Worksheets 2.11.1, 2.11.2 and 2.11.3	
2.12	Practical: Investigating the acceleration of an object	• Plan an investigation to explore an idea. • Analysing results to identify patterns and draw conclusions. • Compare results with scientific theory.	P2.2i	Practical sheet 2.12; Technician's notes 2.12	
2.13	Newton's third law	• Identify force pairs. • Understand and be able to apply Newton's third law.	P2.2p	Worksheets 2.13.1, 2.13.2 and 2.13.3	
2.14	Work done and energy transfer	• Understand what is meant by work done. • Explain the relationship between work done and force applied. • Identify the transfers between energy stores when work is done against friction.	P2.2m, n	Worksheet 2.14; Practical sheets 2.14.1 and 2.14.2; Technician's notes 2.14	
2.15	Understanding power	• Define power. • Compare the rate of energy transfer by various machines and electrical appliances. • Calculate power.	P2.2o	Worksheet 2.15; Practical sheets 2.15 and 2.15.2; Technician's notes 2.15.1 and 2.15.2	
2.16	Key concept: Forces and acceleration	• Explain what happens to the motion of an object when the resultant force is not zero. • Analyse situations in which a non-zero resultant force is acting. • Explain what inertia is.	KC	Practical sheets 2.16.1, 2.16.2 and 2.16.3; Technician's notes 2.16	
2.17	Forces and energy in springs	• Explain why you need two forces to stretch a spring. • Describe the difference between elastic and inelastic deformation. • Calculate extension, compression and elastic potential energy.	P2.3a, b	Worksheets 2.17.1 and 2.17.2, Practical sheet 2.17, Technician's notes 2.17	
2.18	Practical: Investigate the relationship	• Interpret readings to show patterns and trends.	P2.3c, d, e	Practical sheet 2.18, Practical 2.18, Technician's notes 2.18	

Lesson number	Lesson title	Lesson objectives	OCR specification reference	Lesson resources (on CD ROM)	Collins Connect resources
	between force and the extension of a spring	• Interpret graphs to form conclusions. • Apply the equation for a straight line to the graph.			
2.19	Potential energy	• Consider what happens when a spring is stretched. • Describe what is meant by gravitational potential energy. • Calculate the energy stored by an object raised above ground level.	P2.3f	Worksheet 2.19; Practical sheet 2.19; Technician's notes 2.19	
2.20	Heavy or massive?	• Identify the correct units for mass and weight. • Explain the difference between mass and weight. • Understand how weight is an effect of gravitational fields.	P2.3g, h, l, j	Worksheets 2.20.1, 2.20.2 and 2.20.3, Practical sheet 2.20, Technician's notes 2.20	
2.21	Moments	• Describe the turning effect of a force about a pivot. • Explain and use the principle of moments. • Explain what is meant by the centre of mass of an object.	P2.3k, l	Worksheets 2.21.1, 2.21.2 and 2.21.3; Practical sheet 2.21; Technician's notes 2.21	
2.22	Levers and gears	• Describe how levers and gears can be used to transmit the rotational effect of a force. • Explain how levers and gears transmit forces.	P2.3m	Worksheets 2.22.1, 2.22.2 and 2.22.3	
Assessment		End of chapter test Student Book End of chapter test Collins Connect End of teaching block test Collins Connect			
Topic 3: Electricity					
3.1	Static electricity	• Describe how insulating materials can become charged. • Know that there are two kinds of electric charge. • Explain these observations in terms of electron transfer.	P3.1a, b, c	Worksheet 3.1	
3.2	Electric fields and	• Explain what an electric field is.	P3.1d	Worksheets 3.2.1, 3.2.2 and 3.2.3	

Lesson number	Lesson title	Lesson objectives	OCR specification reference	Lesson resources (on CD ROM)	Collins Connect resources
	currents	• Draw an electric field pattern for a charged sphere. • Use the idea of an electric field to explain electrostatic attraction and sparking. • Recall that an electric current is a rate of flow of electric charge			
3.3	Electric currents and potential difference	• Know circuit symbols. • Recall that current (*I*) depends on resistance (*R*) and potential difference (*V*) • Explain how an electric current passes round a circuit.	P3.1e, f, g	Worksheets 3.3.1, 3.3.2, 3.3.3 and 3.3.4	
3.4	Series and parallel circuits	• Recognise series and parallel circuits. • Describe the changes in the current and potential difference in series and parallel circuits.	P3.2a, i and j (part)	Worksheets 3.4.1, 3.4.2 and 3.4.3	
3.5	Resistance	• Set up a circuit to investigate resistance. • Investigate the changing resistance of a filament lamp. • Compare the properties of a resistor and filament lamp.	P3.2 e, f, g	Practical sheet 3.5; Technician's notes 3.5	
3.6	Practical: Use circuit diagrams to set up and check appropriate circuits for investigating resistance	• Use a circuit to determine resistance. • Gather valid data to use in calculations. • Apply the circuit to determine the resistance of combinations of components.	P3.2h, k (part)	Practical sheet 3.6 Technician's notes 3.6	
3.7	Investigating circuits	• Classify materials as either conducting or insulating. • Use series circuits to test components and make measurements. • Carry out calculations on series circuits.	P3.2j (part)	Worksheets 3.7.1, 3.7.2 and 3.7.3; Practical sheet 3.7; Technician's notes 3.7	
3.8	Control circuits	• Use a thermistor and light-dependent resistor (LDR). • Investigate the properties of thermistors, LDRs and diodes.	P3.2k (part)	Worksheet 3.8; Practical sheet 3.8; Technician's notes 3.8	

Lesson number	Lesson title	Lesson objectives	OCR specification reference	Lesson resources (on CD ROM)	Collins Connect resources
3.9	Power and energy transfers	• Describe the energy transfers in different domestic appliances. • Describe power as a rate of energy transfer. • Calculate the energy transferred.	P3.2l (part)	Worksheets 3.9.1, 3.9.2 and 3.9.3; Practical sheet 3.9; Technician's notes 3.9	
3.10	Calculating power	• Calculate power. • Use power equations to solve problems. • Consider power ratings and changes in stored energy.	P3.2l (part)	Worksheets 3.10.1, 3.10.2 and 3.10.3; Practical sheet 3.10; Technician's notes 3.10	
3.11	Practical: Investigating series and parallel circuits	• Use series circuits to test components and make measurements. • Carry out calculations on series circuits.	PRAC P7	Worksheet 3.11; Practical sheets 3.11.1, 3.11.2 and 3.11.3; Technician's notes 3.11	
3.12	Key concept: What's the difference between potential difference and current?	• Understand and be able to apply the concepts of current and potential difference. • Use these concepts to explain various situations.	P3.2 c, i	Worksheet 3.12, Practical sheets 3.12.1, 3.12.2 and 3.12.3, Technician's notes 3.12	
3.13	Maths skills: Using algebra in electric circuit calculations	• Solve algebraic equations including using appropriate substitutions, numerical values and units. • Change the subject of an equation. • Use the symbols =, <, <<, >>, >, ∝, ~.		Worksheets 3.13.1, 3.13.2 and 3.13.3	
3.14	Maths skills: Using formulae and understanding graphs	• Recognise how algebraic equations define the relationships between variables. • Solve simple algebraic equations by substituting numerical values. • Describe relationships expressed in graphical form.		Worksheets 3.14.1, 3.14.2 and 3.14.3, Practical sheet 3.14, Technician's notes 3.14	
Assessment		End of Chapter test Student Book End of Chapter test Collins Connect			
Topic 4: Magnetism and magnetic fields					
4.1	Magnetism and magnetic forces	• Explain what is meant by the poles of a magnet. • Plot the magnetic field around a bar	P4.1a, b	Worksheet 4.1; Practical sheet 4.1; Technician's notes 4.1; PowerPoint presentation	

Lesson number	Lesson title	Lesson objectives	OCR specification reference	Lesson resources (on CD ROM)	Collins Connect resources
		magnet. • Describe magnetic materials and induced magnetism.			
4.2	Compasses and magnetic fields	• Describe the Earth's magnetic field. • Describe the magnetic effect of a current.	P4.1c, d, e, f	Worksheet 4.2; Practical sheet 4.2; Technician's notes 4.2; PowerPoint presentation	
4.3	Magnetic effects	• Draw the magnetic field around a conducting wire and a solenoid. • Describe the force on a wire in a magnetic field. • Apply the left-hand rule to work out the direction of a magnetic field, a current or a force around a wire.	P4.1g	Worksheets 4.3.1 and 4.3.2; Practical sheet 4.3; Technician's notes 4.3; PowerPoint presentation	
4.4	Calculating the force on a conductor	• Explain the meaning of magnetic flux density, *B*. • Calculate the force on a current-carrying conductor in a magnetic field.	P4.2c	Worksheets 4.4.1 and 4.4.2; Technician's notes 4.4; PowerPoint presentation	
4.5	The generator effect	• Describe how a current is induced in a wire when it moves in a magnetic field. • Identify the factors that affect the size and direction of the induced current or induced potential difference.	P4.2e	Worksheet 4.5; Practical sheet 4.5; Technician's notes 4.5; PowerPoint presentation	
4.6	Electric motors	• List equipment that uses motors. • Describe how motors work. • Describe how to change the speed and direction of rotation of a motor.	P4.2d	Worksheets 4.6.1 and 4.6.2; Practical sheet 4.6; Technician's notes 4.6; PowerPoint presentation	
4.7	Generators and transformers	• Explain how an alternator generates alternating current. • For an alternator, draw and interpret graphs of potential difference generated in the coil against time. • Understand how induced alternating electric currents are used in transformers. • Apply the transformer equation linking the numbers of turns and the potential differences induced.	P4.2g	Worksheets 4.7.1, 4.7.2 and 4.7.3, Practical sheet 4.7, Technician's notes 4.7; PowerPoint 4.7	

Lesson number	Lesson title	Lesson objectives	OCR specification reference	Lesson resources (on CD ROM)	Collins Connect resources
4.8	Loudspeakers and microphones	• Describe how a moving-coil loudspeaker works. • Compare loudspeakers and headphones. • Describe how a moving-coil microphone works.	P4.2j	Worksheets 4.8.1 and 4.8.2; Technician's notes 4.8; PowerPoint presentation 4.8	
4.9	Key concept: The link between electricity and magnetism	• Explore how electricity and magnetism are connected. • Describe how electromagnetic induction occurs. • Describe the principle of the electric motor.	Key concept	Worksheets 4.9.1 and 4.9.2; Practical sheets 4.9.1, 4.9.2, 4.9.3 and 4.13.4; Technician's notes 4.9; PowerPoint presentation	
4.10	Maths skills: Rearranging equations	• Know how to rearrange equations. • Know how to use the transformer equation. • Know how to calculate the force on a conductor.	P4.2h, i	Worksheets 4.10.1 and 4.10.2	
Assessment		End of chapter test Student Book End of chapter test Collins Connect End of teaching block test Collins Connect			
Topic 5: Waves in matter					
5.1	Describing waves	• Describe wave motion. • Define wavelength and frequency. • Apply the relationship between wavelength, frequency and wave velocity.	P5.1a, b, c, d	Worksheets 5.1.1, 5.1.2, 5.1.3, 5.1.4	
5.2	Transverse and longitudinal waves	• Compare the motion of transverse and longitudinal waves. • Explain why water waves are transverse waves. • Explain why sound waves are longitudinal waves.	P5.1e, f, k	Worksheets 5.2.1, 5.2.2 and 5.2.3; PowerPoint presentation	
5.3	Reflection and refraction of waves	• Describe reflection, transmission and absorption of waves. • Construct ray diagrams to illustrate reflection. • Construct ray diagrams to illustrate refraction.	P5.1g	Worksheets 5.3.1, 5.3.2 and 5.3.3; Practical sheets 5.3.1, 5.3.2 and 5.3.3; Technician's notes 5.3.1 and 5.3.2	
5.4	Sound waves	• Describe how we hear sound and state the range of frequencies we can hear.	P5.1h, i	Worksheets 5.4.1, 5.4.2	

Lesson number	Lesson title	Lesson objectives	OCR specification reference	Lesson resources (on CD ROM)	Collins Connect resources
		• Explain that sound travels faster in a denser medium. • Explain about reflection, absorption and transmission of sound.			
5.5	Exploring ultrasound	• Explain what ultrasound is • Describe how ultrasound can be used in industry to investigate or detect hidden or buried objects. • Explain how ultrasound is used in medicine.	P5.2i	Worksheets 5.5.1 and 5.5.2	
5.6	Practical: Measuring the wavelength, frequency and speed of waves in a ripple tank and waves in a solid	• Develop techniques for making observations of waves. • Select suitable apparatus to measure frequency and wavelength. • Use data to answer questions.	P5.1j	Practical sheet 5.6; Technician's notes 5.6	
5.7	Maths skills: Using and rearranging equations	• Select and apply the equations $T = 1/f$ and $v = f\lambda$. • Substitute numerical values into equations using appropriate units. • Change the subject of an equation.	PM5.1i	Worksheets 5.7.1, 5.72 and 5.7.3	
5.8	Key concept: Transferring energy or information by waves	• Understand that all waves have common properties. • Understand how waves can be used to carry information. • Understand various applications of energy transfer by different types of electromagnetic waves.	KC	Worksheet 5.8	
5.9	The electromagnetic spectrum	• Recall the similarities and differences between transverse and longitudinal waves. • Recognise that electromagnetic waves are transverse waves. • Describe the main groupings and wavelength ranges of the electromagnetic spectrum.	P5.2a, b, c, d, e, f	Worksheets 5.9.1, 5.9.2 and 5.9.3	

Lesson number	Lesson title	Lesson objectives	OCR specification reference	Lesson resources (on CD ROM)	Collins Connect resources
5.10	Gamma rays and X-rays	• List the properties of gamma rays and X-rays. • Compare gamma rays and X-rays.	P5.2g (part), h	Worksheets 5.10.1, 5.10.2 and 5.10.3	
5.11	Ultraviolet and infrared radiation	• Describe the properties of ultraviolet and infrared radiation. • Describe some uses and hazards of ultraviolet radiation. • Describe some uses of infrared radiation.	P5.2g (part), h (part)	Worksheet 5.11; Practical sheet 5.11; Technician's notes 5.11	
5.12	Microwaves	• List some properties of microwaves. • Describe how microwaves are used for communications.	P5.2g (part), i (part)	Worksheet 5.12	
5.13	Radio and microwave communication	• Describe how radio waves are used for television and radio communications. • Describe how microwaves are used in satellite communications. • Describe the reflection and refraction of radio waves.	P5.2g (part), j	Worksheets 5.13.1, 5.13.2 and 5.13.3	
5.14	Reflection, refraction and wave fronts	• Explain reflection and refraction and how these may vary with wavelength. • Construct ray diagrams to illustrate refraction. • Use wave front diagrams to explain refraction in terms of the difference in velocity of the waves in different substances.	P5.3a (part), b	Worksheets 5.14.1, 5.14.2 and 5.14.3; Practical sheet 5.14; Technician's notes 5.14	
5.15	Practical: Investigate how the amount of infrared radiation absorbed or radiated by a surface depends on the nature of that surface	• Explain reasons for the equipment used to carry out an investigation. • Explain the rationale for carrying out an investigation. • Apply ideas from an investigation to a range of practical contexts.	P5.3a (part)	Practical 5.15, Technician's notes 5.15	
5.16	Practical: Investigate the reflection of light by	• Make and record observations of how light is reflected and transmitted at different surfaces.	P5.3c (part)	Worksheets 5.16.1 and 5.16.2; Practical sheets 5.16.1, 5.16.2, 5.16.3	

Lesson number	Lesson title	Lesson objectives	OCR specification reference	Lesson resources (on CD ROM)	Collins Connect resources
	different types of surface and the refraction of light by different substances	• Measure angles and discuss the method, apparatus and uncertainty in measurements. • Draw conclusions from experimental results.		and 5.16.4; Technician's notes 5.16	
5.17	Lenses	• Understand what a lens does. • Draw ray diagrams to show the formation of images by lenses. • Describe the difference between a real and a virtual image.	P5.3c (part)	Worksheets 5.17.1, 5.17.2 and 5.17.3; Technician's notes 5.17	
5.18	Images and magnification	• Draw ray diagrams to show the formation of real and virtual images by lenses. • Calculate the magnification of an image.	P5.3c (part)	Worksheets 5.18.1, 5.18.2 and 5.18.3	
5.19	Colour	• Describe what happens when light of different wavelengths lands on an object. • Explain what determines the colour of an opaque object. • Explain the effect of coloured filters.	P5.3e	Worksheets 5.19.1, 5.19.2 and 5.19.3	
Assessment		End of chapter test Student Book End of chapter test Collins Connect			
		Chapter 6: Radioactivity			
6.1	Atomic structure	• Describe the structure of the atom. • Use symbols to represent particles. • Describe ionisation.	P6.1a, b, c, h	Worksheets 6.1.1, 6.1.2 and 6.1.3	
6.2	Radioactive decay	• Describe radioactive decay. • Describe the types of nuclear radiation. • Understand the processes of alpha decay and beta decay.	P6.1d, e	Worksheets 6.2.1, 6.2.2 and 6.2.3	
6.3	Nuclear equations	• Understand nuclear equations. • Write balanced nuclear equations.	P6.1f, g	Worksheets 6.3.1, 6.3.2 and 6.3.3	
6.4	Radioactive half-life	• Explain what is meant by radioactive half-life. • Calculate half-life.	P6.1j	Worksheets 6.4.1, 6.4.2 and 6.4.3; Practical sheet 6.4; Technician's notes 6.4	

Lesson number	Lesson title	Lesson objectives	OCR specification reference	Lesson resources (on CD ROM)	Collins Connect resources
		• Choose the best radioisotope for a task.			
6.5	Background radiation	• Recall sources of background radiation. • Describe how different types of radiation have different ionising power.	P6.1l	Worksheets 6.5.1, 6.5.2 and 6.5.3	
6.6	Maths skills: Using ratios and proportional reasoning	• Draw a curve of best fit to calculate radioactive half-life. • Calculate the net decline	M1c, M3c		
6.7	Hazards and uses of radiation	• Describe radioactive contamination. • Give examples of how radioactive tracers can be used.	P6.2a (part), b, c (part)	Worksheets 6.7.1, 6.7.2 and 6.7.3	
6.8	Irradiation	• Explain what is meant by irradiation. • Understand the distinction between contamination and irradiation. • Appreciate the importance of communication between scientists.	P6.2a (part)	Worksheets 6.8.1, 6.8.2 and 6.8.3	
6.9	Uses of radiation in medicine	• Compare gamma rays and X-rays. • Describe some uses of nuclear radiation for medical diagnosis and therapy.	P6.2c	Worksheets 6.9.1, 6.9.2, 6.9.3 and 6.9.4	
6.10	Using nuclear radiation	• Explore the risks and benefits of using nuclear radiation. • Describe how internal organs can be explored. • Understand how nuclear radiation can control or destroy unwanted tissue.	P6.2c	Worksheets 6.10.1, 6.10.2 and 6.10.3	
6.11	Nuclear fission	• Describe nuclear fission. • Explain how a chain reaction occurs. • Explain how fission is used.	P6.2d	Worksheets 6.11.1, 6.11.2 and 6.11.3	
6.12	Nuclear fusion	• Explain nuclear fusion. • Describe the conditions needed for fusion. • Describe how nuclear fusion may be an attractive energy source.	P6.2e	Worksheet 6.12	
Assessment		End of chapter test Student Book End of chapter test Collins Connect End of teaching block test Collins Connect			
		Topic 7: Energy			

Lesson number	Lesson title	Lesson objectives	OCR specification reference	Lesson resources (on CD ROM)	Collins Connect resources
7.1	Investigating kinetic energy	• Describe how the kinetic energy store of an object changes as its speed changes • Calculate kinetic energy. • Consider how energy is transferred.	P7.1b (part), e (part)	Worksheets 7.1.1, 7.1.2 and 7.1.3; Practical sheets 7.1.1 and 7.1.2; Technician's notes 7.1	
7.2	Work done and energy transfer	• Understand what is meant by work done. • Explain the relationship between work done and force applied. • Identify the transfers between energy stores when work is done against friction.	P7.1b (part), c (part)	Worksheets 7.2.1, 7.2.2 and 7.2.3 Practical sheets 7.1.1 and 7.1.2; Technician's notes 7.4	
7.3	Specific heat capacity	• Understand how things heat up. • Find out about heating water. • Find out about specific heat capacity.	P7.1c (part)	Worksheets 7.3.1, 7.3.2 and 7.3.3; Practical sheet 7.3; Technician's notes 7.3	
7.4	Electrical energy and power	• Recall that energy cannot be created or destroyed, only transferred. • Describe the energy transfers in different domestic appliances. • Describe power as a rate of energy transfer.	P7.2a, b, c	Worksheets 7.4.1, 7.4.2 and 7.4.3, Practical sheet 7.4, Technician's notes 7.4	
7.5	Dissipation of energy	• Explain ways of reducing unwanted energy transfer. • Describe what affects the rate of cooling of a building. • Understand that energy is dissipated.	P7.2a, g	Worksheets 7.5.1, 7.5.2 and 7.5.3, Practical sheets 7.5.1, and 7.5.2, Technician's notes 7.5	
7.6	Energy efficiency	• Explain what is meant by energy efficiency. • Calculate the efficiency of energy transfers. • Find out about conservation of energy.	P7.1a	Worksheets 7.6.1 and 7.6.2, one of Practical sheets 7.6.1–7.6.4	
7.7	Practical: Investigating ways of reducing the unwanted energy transfers in a system	• Use scientific ideas to make predictions • Analyse data to identify trends. • Evaluating an experimental procedure.	P7.2f	Worksheet 7.7; Practical sheet 7.7; Technician's notes 7.7	
7.8	Key concept: Energy transfer	• Understand why energy is a key concept in science.	KC	Worksheet 7.8, Practical sheets 7.8.1 and 7.8.2, Technician's notes 7.8.1	

Lesson number	Lesson title	Lesson objectives	OCR specification reference	Lesson resources (on CD ROM)	Collins Connect resources
		• Use ideas about stores and transfers to explain what energy does. • Understand why accounting for energy transfers is a useful idea.		and 7.8.2	
7.9	Maths skills: Calculations using significant figures	• Substitute numerical values into equations and use appropriate units.		Worksheets 7.9.1, 7.9.2 and 7.9.3, Technician's notes 7.9	
Assessment		End of chapter test Student Book End of chapter test Collins Connect			
		Topic 8: Global challenges			
8.1	Keeping safe on the road	• Explain the factors that affect stopping distance. • Explain the dangers caused by large deceleration. • Estimate the forces involved in the deceleration of a road vehicle. • Apply the idea of rate of change of momentum to explain safety features.	P8.1d, e, f, g, h, i	Worksheets 8.1.1, 8.1.2 and 8.1.3	
8.2	Transmitting electricity	• Describe how electricity is transmitted using the National Grid. • Explain why electrical power is transmitted at high potential differences. • Understand the role of transformers.	P8.2c, d, e	Worksheet 8.2	
8.3	Using energy resources	• Describe the main energy resources available for use on Earth. • Distinguish between renewable and non-renewable resources. • Explain the ways in which the energy resources are used.	P8.2b	Worksheet 8.3, Practical sheet 8.3, Technician's notes 8.3	
8.4	Global energy supplies	• Analyse global trends in energy use. • Understand what the issues are when using energy resources.	P8.2b	Worksheets 8.4.1, 8.4.2 and 8.4.3, Practical sheet 8.4, Technician's notes 8.4	
8.5	Electricity in the home	• Recall that the domestic supply in the UK is	P8.2g, h, i, j	Worksheets 8.5.1, 8.5.2 and 8.5.3	

Lesson number	Lesson title	Lesson objectives	OCR specification reference	Lesson resources (on CD ROM)	Collins Connect resources
		a.c. at 50 Hz and about 230 V. • Describe the main features of live, neutral and earth wires.			
8.6	Red-shift	• Describe red-shift. • Describe evidence for the expanding Universe.	P8.3a, b	Worksheet 8.6, Practical sheet 8.6, Technician's notes 8.6	
8.7	The Sun and other stars	• Describe how the Sun and other stars formed. • Describe the nuclear fusion reactions in the Sun.	P8.7c	Worksheet 8.7, Practical sheet 8.7, Technician's notes 8.7	
8.8	Emission and absorption of infrared radiation	• Realise that all bodies emit and absorb infrared radiation. • Compare emission and absorption of radiation from different surfaces. • Define a perfect black body. • Explain that the intensity and distribution of wavelengths of any emission depend on the temperature of the body.	P8.3d	Worksheets 8.8.1, 8.8.2 and 8.8.3, Practical sheet 8.8, Technician notes 8.8	
8.9	The Solar System	• Describe the orbits of planets and moons in the Solar System. • Distinguish between planets, dwarf planets and moons.	P8.3e (part)	Worksheet 8.9, Practical sheets 8.9.1, 8.9.2 and 8.9.3, Technician's notes 8.9	
8.10	Orbits of planets, moons and artificial satellites	• Compare the orbital motion of moons, artificial satellites and planets in the Solar System. • Describe what keeps bodies in orbit around planets and stars. • Explain how for circular orbits an object can have a changing velocity but unchanged speed. • Explain why bodies must move at a particular speed to stay in orbit at a particular distance.	P8.3e (part), f, g	Worksheet 8.10, Practical sheets 8.10.1 and 8.10.2, Technician's notes 8.10.1 and 8.10.2	
8.11	Gravity: the force that binds the	• Understand that gravity provides the force that keeps planets and satellites in orbits.	KC	Worksheet 8.11, Practical sheets 8.11.1 and 8.11.2, Technician's notes	

Lesson number	Lesson title	Lesson objectives	OCR specification reference	Lesson resources (on CD ROM)	Collins Connect resources
	Universe	• Understand that gravity is necessary at the start of a star's life cycle and to maintain equilibrium in a star. • Describe how the weight of an object depends on the gravitational field strength. • Recognise that there is still much about the universe that is not understood, e.g dark mass and dark energy.		8.11.1 and 8.11.2	
8.12	Temperature of the Earth	• Describe how the atmosphere absorbs radiation in a way that varies with wavelength. • List the factors affecting the temperature of the Earth. • Explain how the temperature of an object is related to the radiation absorbed and radiation emitted.	P8.3h	Worksheets 8.12.1, 8.12.2, 8.12.3 and 8.12.4	
8.13	Measuring wave speeds	• Explain how the speed of sound in air can be measured. • Explain how the speed of water ripples can be measured. • Describe the use of echo sounding.	P8.3i (part)	Worksheets 8.13.1, 8.13.2 and 8.13.3, Practical sheet 8.13, Technician's notes 8.13	
8.14	Seismic waves	• Describe how earthquakes are detected. • Describe the properties of P waves and S waves. • Explain how the properties of seismic waves allow us to investigate the inside of the Earth.	P8.3i (part)	Worksheets 8.14.1, 8.14.2 and 8.14.3	
8.15	Maths skills: Using scale and standard	• Understand the scale of objects in the Universe. • Use standard form.		Worksheet 8.15	
Assessments		End of chapter test Student Book End of chapter test Collins Connect End of teaching block test Collins Connect End of course test Collins Connect			

Using Collins OCR Gateway GCSE (9-1) Physics Student Book and Teacher Pack to teach OCR GCSE (9-1) Combined Science

The following table gives an overview of the Student Book pages and Lesson Plans that can be used to teach Combined Science.

Student Book and Lesson Plan Reference	**Chapter**	**Notes on how to use the resources for Combined Science**
Chapter 1	**Matter**	
Student Book only	Chapter Introduction	Combined Science students do not need to know the content in the following boxes: Pressure in a fluid, How are pressure and volume connected?
1.1	Key concept: Developing ideas for the structure of the atom	Combined Science students need to know and understand all content.
1.2	Model of the atom	Combined Science students do not at this point need to understand the detail of the material concerning mass number, atomic number, isotopes and electrons and energy levels. These topics are developed further in Chapter 6 Radioactivity content.
1.3	Density	Combined Science students need to know and understand all content.
1.4	Key concept: Particle model and changes of state	Combined Science students need to know and understand all content.
1.5	Practical: To investigate the densities of regular and irregular solid objects and liquids	Combined Science students need to know and understand all content.
1.6	Changes of state	Combined Science students need to know and understand all content.
1.7	Internal energy	Combined Science students need to know and understand all content.
1.8	Specific heat capacity	Combined Science students need to know and understand all content.
1.9	Specific latent heat	Combined Science students need to know and understand all content.
1.10	Maths skills: Drawing and interpreting graphs	Combined Science students need to know and understand all content.
1.11	Practical: Investigating specific heat capacity	Combined Science students need to know and understand all content.
1.12	Particle motion in gases	Combined Science students do not need to know the content in the section 'Compressing or expanding gases', although questions 7 and 8 can still be answered based on the material in the earlier sections.

1.13	Increasing the pressure of a gas	Combined Science students do not need to know any of the content on this page.
1.14	Pressure in a fluid	Combined Science students do not need to know any of the content on this page.
1.15	Atmospheric pressure	Combined Science students do not need to know any of the content on this page.
1.16	Maths skills: Handling data	Combined Science students need to know and understand all content.
	Check your progress/Worked example	Combined Science students do not need to know the content in the following Check your progress boxes: [green] Recall that gases can be compressed or expanded by pressure changes [blue] Use the equation pV = constant to calculate the pressure or volume of a gas at constant temperature [pink] Use the particle model to explain that increasing the volume of a gas, at constant temperature, can lead to a decrease in pressure [green] Describe how a fluid exerts a pressure on a surface [green] Describe how pressure varies with depth in a fluid [blue] Calculate pressure at any depth in a fluid and explain what causes atmospheric pressure [pink] Explain how a partially (or totally) submerged object experiences upthrust and why atmospheric pressure decreases with height [pink] Describe the factors which influence floating and sinking
	End of chapter questions	Combined Science students do not need to answer the following questions: 12c, 18, 19, 20

Chapter 2	**Forces**	
Student Book only	Chapter Introduction	Combined Science students need to know and understand all content.
2.1	Scalars and vectors	Combined Science students need to know and understand all content.
2.2	Speed	Combined Science students need to know and understand all content.
2.3	Acceleration	Combined Science students need to know and understand all content.
2.4	Calculations of motion	Combined Science students need to know and understand all content.
2.5	Velocity–time graphs	Combined Science students need to know and understand all content.
2.6	Maths skills: Making estimates of calculations	Combined Science students need to know and understand all content.
2.7	Forces explain how objects interact	Combined Science students need to know and understand all content.
2.8	Forces and motion	Combined Science students need to know and understand all content.
2.9	Resultant forces	Combined Science students need to know and understand all content.
2.10	Forces and acceleration	Combined Science students need to know and understand all content.
2.11	Momentum	Combined Science students need to know and understand all content.
2.12	Practical: Investigating the acceleration of an object	Combined Science students need to know and understand all content.
2.13	Newton's third law	Combined Science students need to know and understand all content.
2.14	Work done and energy transfer	Combined Science students need to know and understand all content.
2.15	Understanding power	Combined Science students need to know and understand all content.
2.16	Key concept: Forces and acceleration	Combined Science students need to know and understand all content.
2.17	Forces and energy in springs	Combined Science students need to know and understand all content.
2.18	Practical: Investigate the relationship between force and the extension of a spring	Combined Science students need to know and understand all content.
2.19	Potential energy	Combined Science students need to know and understand all content.
2.20	Heavy or massive?	Combined Science students need to know and understand all content.

2.21	Moments	Combined Science students do not need to know any of the content on this page.
2.22	Levers and gears	Combined Science students do not need to know any of the content on this page.
	Check your progress/Worked example	Combined Science students do not need to know the content in the following Check your progress boxes: [green] Explain that a moment is the turning effect of a force [blue] Calculate the size and direction of a moment [blue] Explain how gears and levers transmit the rotational effect of a force [pink] Apply the idea of moments to contexts such as the balancing of a seesaw
	End of chapter questions	Combined Science students do not need to answer questions 9, 12 and 16

Chapter 3	**Electricity**	
Student Book only	Chapter Introduction	Combined Science students need to know and understand all content.
3.1	Static electricity	Combined Science students need to know and understand all content.
3.2	Electric fields and currents	Combined Science students do not need to know and understand the content on electric fields.
3.3	Electric circuits and potential difference	Combined Science students need to know and understand all content.
3.4	Series and parallel circuits	Combined Science students need to know and understand all content.
3.5	Resistance	Combined Science students need to know and understand all content.
3.6	Practical: Use circuit diagrams to set up and check appropriate circuits to investigate the factors affecting the resistance	Combined Science students need to know and understand all content.
3.7	Investigating circuits	Combined Science students need to know and understand all content.
3.8	Control circuits	Combined Science students need to know and understand all content.
3.9	Power and energy transfers	Combined Science students need to know and understand all content.
3.10	Calculating power	Combined Science students need to know and understand all content.
3.11	Practical: Investigating series and parallel circuits	Combined Science students need to know and understand all content.
3.12	Key concept: What's the difference between potential difference and current?	Combined Science students need to know and understand all content.
3.13	Maths skills: Using algebra in electric circuit calculations	Combined Science students need to know and understand all content.
3.14	Maths skills: Using formulae and understanding graphs	Combined Science students need to know and understand all content
	Check your progress/Worked example	Combined Science students do not need to know the content in the following Check your progress boxes: [pink] explain static electricity in terms of electric fields
	End of chapter questions	Combined Science students do not need to answer question 6

Chapter 4	**Magnetism and magnetic fields**	
Student Book only	Chapter Introduction	Combined Science students do not need to know the content in the following boxes: • How can a magnetic field be used to produce an electric current? • What is a transformer? • How are electromagnets used? Note that use of transformers to transmit electrical power at high potential differences is covered in Chapter 8 spread 8.2, not in chapter 4.
4.1	Magnetism and magnetic forces	Combined Science students need to know and understand all content.
4.2	Compasses and magnetic fields	Combined Science students need to know and understand all content.
4.3	Magnetic effects	Combined Science students need to know and understand all content.
4.4	Calculating the force on a conductor	Combined Science students need to know and understand all content.
4.5	The generator effect	Combined Science students do not need to know any of the content on this page.
4.6	Electric motors	Combined Science students need to know and understand all content.
4.7	Generators and transformers	Combined Science students do not need to know any of the content on this page.
4.8	Loudspeakers and microphones	Combined Science students do not need to know any of the content on this page.
4.9	Key concept: The link between electricity and magnetism	Combined Science students do not need to know the content in the last section: Electromagnetic induction
4.10	Maths skills: Rearranging equations	Combined Science students do not need to know the content in the last section: The Transformer equation.
	Check your progress/Worked example	Combined Science students do not need to answer worked example 3. Combined Science students do not need to know the content in the following Check your progress boxes: [pink] Explain how electromagnets are used in devices [green] State what the generator effect is [blue] Recognise that the size of the induced potential difference (and so induced current) can be increased by increasing the speed of movement or by increasing the strength of the magnetic field [pink] Explain that if the direction of motion of the conductor or the polarity of the magnetic field is reversed, the direction of the induced potential difference and any induced current is reversed

		[green] State that a dynamo generates direct current and an alternator generates alternating current [blue] Describe that when a coil is rotated in a magnetic field an alternating current is induced in the coil [blue] Draw graphs of potential difference generated in the coil against time [pink] Explain how the generator effect is used in a dynamo to generate d.c. with the use of a commutator; and is used in an alternator to generate a.c. [pink] Describe how to draw and interpret graphs of potential difference generated in the coil against time [green] State what a basic transformer consists of and the difference between step-up transformer and a step-down transformer [blue] Explain how the potential differences across the two coils depend on the number of turns on each coil and how the potential difference is induced pink] Use and apply the expression $V_p/V_s = n_p/n_s$
	End of chapter questions	Combined Science students do not need to answer the following questions: 7, 9, 12, 15, 16, 17, 18, 19

Chapter 5	**Waves in matter**	
Student Book only	Chapter Introduction	Combined Science students do not need to know about: Visible light box, bullet point: • Different colours are absorbed or reflected by different surfaces Waves can travel through different solids box, bullet point: • Ultrasound is defined as sound with a pitch too high for humans to hear Lenses in the eye and the camera box, bullet points: • A convex lens refracts light to form an image. • Your eye contains a convex lens which focuses images on your retina. • A camera can focus an image on a light-sensitive material such as photographic film. Are there any waves beyond the visible spectrum? box, bullet point: • The colour of an object depends on which wavelengths it reflects most strongly How do waves allow us to detect structures we cannot see? box, all bullet points How do lenses work? box, all bullet points
5.1	Describing waves	Combined Science students need to know and understand all content.
5.2	Transverse and longitudinal waves	Combined Science students do not need to know the content of the last section 'Change in speed of sound waves', although question 9 can be answered using the other material on these pages.
5.3	Reflection and refraction of waves	Combined Science students do not need to know how to construct ray diagrams or the calculation of angles of reflection and refraction, although these ideas are helpful in gaining a better understanding of reflection, refraction, absorption and transmission.
5.4	Sound waves	Combined Science students do not need to know any of the content on this page.
5.5	Exploring ultrasound	Combined Science students do not need to know any of the content on this page.
5.6	Practical: Measuring the wavelength, frequency and speed of waves in a ripple tank and waves in a solid	Combined Science students need to know and understand all content.
5.7	Maths skills: Using and rearranging equations	Combined Science students do not need to know the content of the last section: Changes in velocity, frequency and wavelength, although question 7 is appropriate.
5.8	Key concept: Transferring energy or information by waves	Combined Science students need to know and understand all content.
5.9	The electromagnetic spectrum	Combined Science students need to know and understand all content.

5.10	Gamma rays and X-rays	Combined Science students need to know and understand all content, but do not need to be able to answer questions 1, 2, 8 and 9 from this spread
5.11	Ultraviolet and infrared radiation	Combined Science students need to know and understand all content.
5.12	Microwaves	Combined Science students do not need to know the content in the last section: Radar.
5.13	Radio and microwave communication	Combined Science students need to know and understand all content.
5.14	Reflection, refraction and wave fronts	Combined Science students do not need to construct ray diagrams to illustrate refraction, so do not need to answer question 5.
5.15	Practical: Investigate how the amount of infrared radiation absorbed or radiated by a surface depends on the nature of that surface	Combined Science students do not need to know any of the content on this page.
5.16	Practical: Investigate the reflection of light by different types of surface and the refraction of light by different substances	Combined Science students do not need to know any of the content on this page.
5.17	Lenses	Combined Science students do not need to know any of the content on this page.
5.18	Images and magnification	Combined Science students do not need to know any of the content on this page.
5.19	Colour	Combined Science students do not need to know any of the content on this page.
	Check your progress/Worked example	Combined Science students do not need to answer worked examples 3 and 4. Combined Science students do not need to know the content in the following Check your progress boxes: [pink] Explain how to calculate the depth of water using echo sounding [green] Draw a labelled ray diagram to illustrate reflection of a wave at a boundary [blue] Construct ray diagrams to illustrate refraction at a boundary [green] Describe the range of normal human hearing [green] Define the term ultrasound [blue] Describe how ultrasound waves can be used for medical and industrial imaging [pink] Explain how P and S waves can be used to deduce information about the structure of the Earth [green] State that each colour in the visible spectrum has its own narrow band of wavelength [blue] Describe that colour filters absorb certain wavelengths and transmit other wavelengths

		[pink] Explain that the colour of an opaque object depends on which wavelengths are more strongly reflected [green] State that in a convex lens parallel rays of light are brought to a focus at the principal focus [blue] Use ray diagrams to determine the nature of the image formed by a lens [pink] Use ray diagrams to determine the position and magnification of images
	End of chapter questions	Combined Science students do not need to answer the following questions: 2, 8, 12, 14, 15, 16, 17, 18, 19, 20

Chapter 6	**Radioactivity**	
Student Book only	Chapter Introduction	Combined Science students do not need to know the following: Nuclear fission occurs when a nucleus splits into the nuclei of two different atoms, emitting nuclear radiation and sometimes other particles; Nuclear fusion occurs when two small atomic nuclei join to form a larger nucleus; The nuclear radiation emitted from nuclei has many uses in medicine and elsewhere.
6.1	Atomic structure	Combined Science students need to know and understand all content.
6.2	Radioactive decay	Combined Science students need to know and understand all content.
6.3	Nuclear equations	Combined Science students need to know and understand all content.
6.4	Radioactive half-life	Combined Science students need to know and understand all content.
6.5	Background radiation	Combined Science students need to know and understand all content.
6.6	Maths skills: Using ratios and proportional reasoning	Combined Science students need to know and understand all content.
6.7	Hazards and uses of radiation	Combined Science students do not need to know the content in the last section: Using medical tracers, however there are relevant and useful points and questions here for 'good practice' in making measurements. In any case, Combined Science students do not need to answer questions 5 and 6.
6.8	Irradiation	Combined Science students need to know and understand all content.
6.9	Uses of radiation in medicine	Combined Science students do not need to know any of the content on this page.
6.10	Using nuclear radiation	Combined Science students do not need to know any of the content on this page.
6.11	Nuclear fission	Combined Science students do not need to know any of the content on this page.
6.12	Nuclear fusion	Combined Science students do not need to know any of the content on this page.
	Check your progress/Worked example	Combined Science students do not need to know the content in the following Check your progress boxes: blue] List some uses of radioisotopes in medicine [pink] Describe how specific radioisotopes are used in medicine [green] Define nuclear fission [blue] Describe nuclear fission using a nuclear equation [pink] Describe, using a nuclear equation, how an uncontrolled nuclear fission chain reaction can result in a huge release of energy [green] Define nuclear fusion

		[blue] Describe nuclear fusion using a nuclear equation [pink] Explain the potential benefits of nuclear fusion
	End of chapter questions	Combined Science students do not need to answer the following questions: 5, 6, 9, 11, 19

Chapter 7	**Energy**	
Student Book only	Chapter Introduction	Combined Science students need to know and understand all content.
7.1	Investigating kinetic energy	Combined Science students need to know and understand all content.
7.2	Work done and energy transfer	Combined Science students need to know and understand all content.
7.3	Specific heat capacity	Combined Science students need to know and understand all content.
7.4	Electrical energy and power	Combined Science students need to know and understand all content.
7.5	Dissipation of energy	Combined Science students need to know and understand all content.
7.6	Energy efficiency	Combined Science students need to know and understand all content.
7.7	Practical: Investigating ways of reducing the unwanted energy transfers in a system	Combined Science students need to know and understand all content.
7.8	Key concept: Energy transfer	Combined Science students need to know and understand all content.
7.9	Maths skills: Calculations using significant figures	Combined Science students need to know and understand all content.
	Check your progress/Worked example	Combined Science students need to know and understand all content.
	End of chapter questions	Combined Science students need to be able to answer all the questions:

Chapter 8	**Global challenges**	
Student Book only	Chapter Introduction	Combined Science students do not need to know about the boxes: Our Sun is a star The force of gravity is important to the Earth What can we learn about stars? What is the role of gravity in space physics?
8.1	Keeping safe on the road	Combined Science students do not need to know and understand the content on estimating the forces involved in the Higher tier box at the end of the last section, although they do need to be able to explain the dangers caused by large decelerations.
8.2	Transmitting electricity	Combined Science students do not need to know and understand the content on transformer calculations in the Higher tier box at the end of the last section.
8.3	Using energy resources	Combined Science students need to know and understand all content.
8.4	Global energy supplies	Combined Science students need to know and understand all content.
8.5	Electricity in the home	Combined Science students need to know and understand all content.
8.6	Red-shift	Combined Science students do not need to know any of the content on this page.
8.7	The Sun and other stars	Combined Science students do not need to know any of the content on this page.
8.8	Emission and absorption of infrared radiation	Combined Science students do not need to know any of the content on this page.
8.9	The Solar System	Combined Science students do not need to know any of the content on this page.
8.10	Orbits of planets, moons and artificial satellites	Combined Science students do not need to know any of the content on this page.
8.11	Key concept: Gravity: the force that binds the Universe	Combined Science students do not need to know any of the content on this page.
8.12	Temperature of the Earth	Combined Science students do not need to know any of the content on this page.
8.13	Measuring wave speeds	Combined Science students do not need to know the content in the last section: Echo sounding.
8.14	Seismic waves	Combined Science students do not need to know any of the content on this page.
8.15	Maths skills: Using scale and standard form	Combined Science students do not need to know any of this content, although the mathematical skills are all appropriate to the Combined Science course.
	Check your progress/Worked example	Combined Science students do not need to answer any of these worked examples.

		Combined Science students do not need to know the content in the following Check your progress boxes: [green] State that the hotter the body the more radiation it emits in a given time [blue] Explain that a perfect black body absorbs all the radiation incident on it, and does not reflect or transmit any radiation [pink] Explain how the temperature of a body is related to the balance between incoming radiation absorbed and radiation emitted [green] Describe how planets and dwarf planets are arranged in the Solar System [blue] Describe how moons are arranged in the Solar System in relation to planets [pink] Describe the relationship between a solar system and a galaxy [green] Recall what stars are formed from [blue] Explain the role of gravity in star formation [pink] Explain the balance of forces in a stable star [green] Recall that stars go through a life cycle [blue] Describe the life cycle of stars similar in size to our Sun and much larger star [pink] Explain the role of fusion in the life cycle of a star [green] Recall that elements heavier than iron are produced in a supernova [blue] Explain how heavy elements become distributed throughout the Universe [pink] Explain the role of fusion in the formation of elements [green] Compare the orbital motions of planets, moons and artificial satellites [blue] Explain the role of gravity in enabling objects to describe circular orbits [pink] Explain the relationship between speed and radius in stable orbits [green] Recall that an observed increase in wavelength of light results in red-shift [blue] Recall that more distant galaxies are travelling away faster [pink] Explain what this evidence suggests about the Universe [green] Recall that the Big Bang suggests the Universe started from a small dense region [blue] Explain how red-shift provides evidence for the Big Bang [pink] Explain how scientists use evidence such as red-shift to support the Big Bang theory
	End of chapter questions	Combined Science students do not need to be able to answer questions 4, 5, 6, 7, 8, 10, 11, 12, 15, 16, 17, 18, 21, 22, 23

Acknowledgements

The Publishers gratefully acknowledge the permissions to reproduce copyright material in this book. Every effort has been made to contact the holders of copyright material, but if any have been inadvertently overlooked, the publisher will be pleased to make necessary arrangements at the first opportunity.

Worksheets

Worksheet 2.7.1, 2.7.2, 2.7.3 EpicStock Media/Shutterstock, Photo Works/Shutterstock, Bipedbones/Shutterstock, Rustam Shanov, sportpoint/Shutterstock, YMZK-Photo/Shutterstock; Worksheet 2.3.1, 2.3.2, 2.3.3 Fingerhut/Shutterstock, Stuart G Porter/Shutterstock, sportpoint/Shutterstock

PowerPoints

51041_OCR_phys_ch4_pres4.2 © sciencephotos / Alamy Stock Photo; 51041_OCR_phys_ch4_pres4.6 © rurality / Alamy Stock Photo; 51041_OCR_ch4_pres4.8 Gwen Shockey/SCIENCE PHOTO LIBRARY; All other images © Shutterstock

 © HarperCollins*Publishers* Limited 2016